10-28-96

Large-Scale Neuronal Theories of the Brain

Large-Scale Neuronal Theories of the Brain

edited by Christof Koch and Joel L. Davis

A Bradford Book
The MIT Press
Cambridge, Massachusetts
London, England

Second printing, 1995

© 1994 Massachusetts Institute of Technology

This book was set in Palatino by TechBooks and was printed and bound in the United States of America.

Library of Congress Cataloging-in-Publication Data

Large-scale neuronal theories of the brain/edited by Christof Koch and
    Joel L. Davis.
        p. cm. -- (Computational neuroscience)
    "A Bradford book."
    Includes bibliographical references and index.
    ISBN 0-262-11183-7
    1. Brain. 2. Neurons. 3. Cerebral cortex. 4. Cognitive
neuroscience. 5. Neural circuitry. I. Koch, Christof. II. Davis,
Joel L., 1942-   . III. Series.
QP376.L33   1994   93-29736
612.8'2--dc20   CIP

Contents

Series Foreword

Computational neuroscience is an approach to understanding the information content of neural signals by modeling the nervous system at many different structural scales, including the biophysical, the circuit, and the systems levels. Computer simulations of neurons and neural networks are complementary to traditional techniques in neuroscience. This book series welcomes contributions that link theoretical studies with experimental approaches to understanding information processing in the nervous system. Areas and topics of particular interest include biophysical mechanisms for computation in neurons, computer simulations of neural circuits, models of learning, representations of sensory information in neural networks, systems models of sensory-motor integration, and computational analysis of problems in biological sensing, motor control, and perception.

Terrence J. Sejnowski
Tomaso A. Poggio

Introduction

This book originated at a small and informal workshop held in December of 1992 in Idyllwild, a relatively secluded resort village situated amid forests in the San Jacinto Mountains above Palm Springs in Southern California. Eighteen colleagues from a broad range of disciplines, including biophysics, electrophysiology, neuroanatomy, psychophysics, clinical studies, mathematics and computer vision, discussed "Large Scale Models of the Brain," that is, theories and models that cover a broad range of phenomena, including early and late vision, various memory systems, selective attention, and the neuronal code underlying figure-ground segregation and awareness (for a brief summary of this meeting, see Stevens 1993). The bias in the selection of the speakers toward reseachers in the area of visual perception reflects both the academic background of one of the organizers as well as the (relative) more mature status of vision compared with other modalities. This should not be surprising given the emphasis we humans place on "seeing" for orienting ourselves, as well as the intense scrutiny visual processes have received due to their obvious usefullness in military, industrial, and robotic applications.

What distinguishes this volume from the myriad of edited books on brains, neural networks, and consciousness that currently flood the market is the ambitious—some would say overly ambitious—attempt at constructing theories from the *bottom-up*, that is firmly based on nerve cells, their firing properties, and their anatomical connections. Such theorizing stands in marked contrast to earlier attempts by psychologists—going back to Sigmund Freud—and by theorists in the artificial intelligence community—such as David Marr—to understand the brain from a primarily psychological or computational point of view, that is from the *top-down*.

In the top-down approach, modules derived from psychological or mathematical constraints are imposed onto the brain, without knowing or caring to what extent the nervous system actually implements such structures. A case in point is the distinction between the various forms of short-lasting memories such as iconic, working, and short-term memory. It may well be that each part of the brain has some ability to change its response as a function of its previous history, without requiring the existence of a small number of discrete memory modules as postulated by cognitive science.

Another misconception may be the division of the structure-from-motion module into two separate ones, one for solving the correspondence problem and one for deriving three-dimensional structure from the moving two-dimensional image.

While cognitive, psychophysical, and computational considerations are obviously crucial for understanding the brain—how else would we even know about focal attention or the mathematical problems associated with computing optical flow—they are by themselves not sufficiently powerful enough to uniquely derive, for instance, the specific algorithms underlying short-range motion perception. To achieve this, we need to know about direction-selective cells in cortex, their distribution along the V1-MT-MST pathway, and the representation of velocity at the single cell level. Thus, in the long run, memory, perception, and awareness can be solved only by explanations at the neuronal level, explanations that can be tested using the tools of electrophysiology and imaging, in combination with psychophysical and theoretical studies. To the chagrin of many a theorist, however, this emphasis on neuronally based models does rule out a number of seductive, but from the point of view of the brain irrelevant, topics such as Schrödinger's cat, quantum gravity or whether or not the brain is a Turing machine.

Est ubi gloria nunc Babylonia? After all has been said and done, what has remained of earlier brain theories? In the years since the end of the Second World War, many interdisciplinary meetings dedicated to the experimental and theoretical study of the brain have occurred. Three prominent ones were the MIT Endicott House Symposium on the "Principles of Sensory Communications" in 1959 (Rosenblith 1961), the Neurosciences Research Program work session on "Theoretical Approaches in Neurobiology" in Boston in 1978 (Reichardt and Poggio 1981), and the recent Dahlem Workshop on "Exploring Brain Functions: Models in Neuroscience" that took place in a Berlin without a wall (Poggio and Glaser 1993). Yet almost all of the theories and models proposed and discussed in at least the first two volumes have fallen out of favor and have ceased to be part of the current scientific debate!

In fact, with the exception of the Hodgkin and Huxley model of action potential generation and propagation (Hodgkin and Huxley 1952), as well as the correlation model of motion perception in beetles and flies (Hassenstein and Reichardt 1956), no theory or model of brain function has survived its birth by more than a decade (most of these models, in fact, die in infancy!). Yet for all their ephemeral nature, models profoundly affect the way we think about the brain. For instance, the idea of a *Hebbian* synapse whose strength increases during a conjunction of pre- and postsynaptic activity determines and shapes the LTP field. In visual perception, such theoretical notions as multiple spatial scales, the correspondence problem, epipolar lines, and the aperture problem attest to the legacy of theories of computational vision. Given the pre-Copernican state of brain sciences, this is the best we can hope for: that the models and theories presented in this volume will shape and influence the way we think about the brain, the mind, and the interactions among the two in the years to come.

We wish to gratefully acknowledge the Office of Naval Research, which has the vision and foresight to fund such interdisciplinary meetings as our Idyllwild Workshop, and Candace Hochenedel for "sweating it out at the keyboard" and converting all chapters into the appropriate dialect of LaTeX. Danke schön.

Christof Koch
California Institute of Technology
Pasadena

Joel Davis
Office of Naval Research
Arlington

1 What Is the Computational Goal of the Neocortex?

Horace Barlow

INTRODUCTION

The human species originated very recently and has been changing very rapidly. Since the neocortex is the main structure that enlarged in primates and now makes us (for our body size) the biggest brained of all animals, its selective advantage is probably responsible for this extraordinarily rapid evolution. Figure 1.1 attempts to give a perspective on all this by displaying the history of our species on a cosmic time scale, and it shows both that our status has been changing at a breathtaking rate over the past 10,000 years, and that there is now a serious threat of overpopulation of the earth by humans. Does this mean that the neocortex has done its job too well? And if it has, is there any alternative to further trust in its supposed product—rational action planned by rational thought—to avert the overpopulation threat? How the neocortex evolved so rapidly and what it does are important problems.

This chapter starts by emphasizing the inadequacy of the historical account of the evolution of the human neocortex, and the insufficiency of the neurophysiological account of it as providing a processed representation of the current sensory input. Next a role for it is suggested that combines and reconciles the neurophysiologal view with that of comparative anatomists, who have told us that it acquires and stores knowledge of the world. At first these views appear to be quite different, but the hypothesis that the neocortical representation is specialized to facilitate the identification and learning of new associations amalgamates them. The middle part of the chapter sets out the requirements for such a specialized representation, and it is shown that a working model or cognitive map of the world is entailed in its production. This map or model would be used automatically in representing sensory information, but the knowledge that the code embodies might also be accessible by a different route for imagery and recall. I think the hypothesis provides a new and illuminating way of looking at the key role of perception in mediating between sensation and learning. The last part of the chapter outlines collaborative work, still incomplete, prompted by the hypothesis and done with A. R. Gardner-Medwin and D. J. Tol-

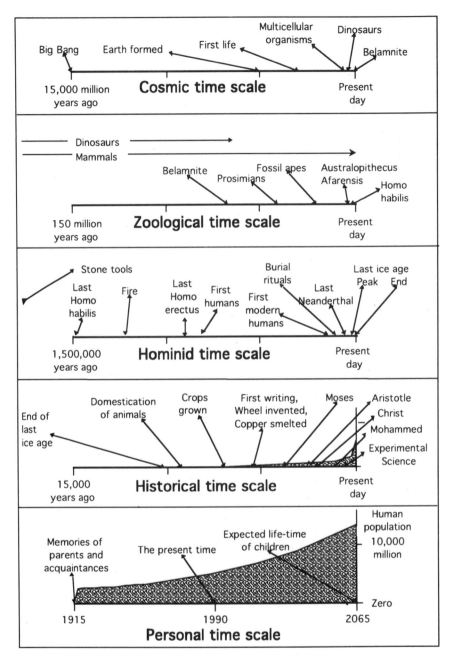

Figure 1.1 The top line shows the prominent events from the big bang to the present day on a linear time scale. The next scale shows the prominent events in the last 1/100th of this enlarged 100 times, and so on for the next two scales. The final scale also enlarges the final 1/100th, but places 1990 at the center. The shaded area under the curve gives total human poulation estimated from Thatcher (1983). It has currently reached a density of 25 per square mile, averaged over the whole surface of the earth, oceans, arctic wastes, and all. Human culture is very recent and has been accompanied by an explosive growth in human population.

hurst, respectively. The two questions were (1) How easy is it to identify the association of reward or punishment with the logical conjunction of two or more active representational elements? This is the "Yellow Volkswagen" problem posed by Harris (1980); Gardner-Medwin has shown that this can be done with reasonable efficiency in the case of frequently occurring conjunctions in sparse representations, but rare conjunctions in dense distributed representations will be masked by noise resulting from accidental associations with the separate constituents of the conjunction. (2) What features should be directly represented by single elements in order to promote the efficient identification of associations? It is argued that one should choose as primitives conjunctions of active elements that actually occur often, but would be expected to occur only infrequently by chance. Tolhurst has done measurements on natural images confirming that edges, which the brain certainly does use as representational elements, are aptly described as such "suspicious coincidences."

Inadequacy of the Historical View of Cortical Evolution

A historical explanation is usually given for our large brains and intellectual domination of the world. Our earliest mammalian ancestors, it is said, were ground-living creatures with smell as their dominant sense, but when they colonized the trees smell became less useful, whereas sight, sound, and muscular dexterity became more important. Smell formerly dominated the forebrain, and when it lost its importance this freed the protocortex for other purposes, so the small regions previously devoted to vision, hearing, touch, and muscular movement rapidly expanded and thus formed the primitive neocortex. This organization enabled our ancestors to expand into new ecological niches, and the improved associative power of the new organ gave us the intellectual advantages, including versatility, insight, and adaptability, that have enabled us to dominate the world. The outline of this view dates back at least to Elliot Smith (1924), but many details have been added (Allman 1987; Jerison 1991).

This crude sketch does not do justice to several nice aspects of this story, but it is basically unsatisfactory because the neocortex appears to have led the evolution of mammals, primates, and man, and not to have followed passively as a result of a series of historical accidents. What selective advantage could the forebrain, or future neocortex, provide that other brain regions could not? What is meant by improved associative power, and why should an organ formerly dominated by smell have it? These are the interesting questions, and the history of man's evolution is not the right place to look for the answers.

The supposed origin of neocortex in a region specializing in olfaction is interesting, but that is a difficult fact to interpret and would not make a good starting point. Instead we look at the account of neocortex that neurophysiology has given us.

Inadequacy of the Neurophysiological Account of Neocortex

As many have recognized, the view of cortical function derived from neurophysiology is unsatisfactory. Something like 60% of the monkey cortex seems to be directly connected with vision (Van Essen and Maunsell 1980), but so far no one has really tried to understand how it does anything but *represent* the current visual scene. The same is true in other modalities—it is the function of representing the current input that has received attention. But we do not have a homunculus to look at these representations: our cortex and associated structures form the representation, look at it, analyze it, store results about it, use it, and continuously add to it. Animals learn almost everything they know through their senses, and academic knowledge apart, the same is true for us; but the means of acquisition, storage, and utilization of this knowledge have been little thought about or studied, and I think it is time to accept that the neocortex must do more for us than merely represent the current scene.

A Hypothesis about the Computational Goal of Neocortex

The outstanding question is: How does neocortex give the great selective advantage that must lie behind our rapid evolution? Herrick (1926) said that the cerebral cortex provided the "filing cabinets of the central executive," and he also called it the "organ of correlation." Jerison (1991) summarizes its role as "knowing about the world." The importance for higher mental function of forming working models and cognitive maps of the world was pointed out by Craik (1943) and Tolman (1948), and—although heretical at that time—these ideas from psychology fit the view from comparative anatomy very well and are now widely accepted. Tolman was thinking primarily of representing the geographic layout of the world, and Craik's working models imitated the dynamics of interactions in the material world, but as Humphrey (1976) pointed out, the interactions between people are the most complex and important things we have to understand, and the cortex is therefore likely to be much concerned with this aspect.

Thus the hypothesis is that the cerebral cortex confers skill in deriving useful knowledge about the material and social world from the uncertain evidence of our senses, it stores this knowledge, and gives access to it when required. This extremely complex and difficult task specifies a definite computational goal for neocortex, providing a useful framework for thinking about its structure, organization, and function. First consider the problem of acquiring such knowledge.

Information and Knowledge

We understand the problem of acquiring knowledge of the world better now than in Herrick's day. It is not a matter of simply recording or videotaping the succession of messages from the outside world that our senses provide, but is a much more analytic process. For present purposes it is convenient to distinguish two aspects of the stream of sensory data, *information* and *knowledge*. *Information* is unpredictable, both from previous parts of the stream of data and from other parts of the current stream. As Shannon told us (Shannon and Weaver 1949), all this genuine information can in principle be encoded on to a channel of much lower capacity than that which is required for the physical data transduced by the sense organs. The *structure* and *regularity* in the stream of data are *redundancy* in terms of information theory, but this part constitutes the *knowledge* that the neocortex must continuously acquire and use. Both parts, information and knowledge, are important to the brain: it must recognize the structure and regularity both to distinguish what is new information and to make useful interpretations and predictions about the world. Finding the structure and regularity is the analytic part of dealing with the succession of sensory impressions that the brain receives, and this is the part that the neocortex performs better than other brain structures according to the current hypothesis: it gives meaning to the stream of sensory data.

The Salience of Structure

Look for a moment at the top left part of figure 1.2. It consists of a random array of dots. Compare it with the top right, where each dot has been paired at a position symmetric about the center line. The random parts of these two figures are identical, but this is not obvious. It is the symmetric structure on the right that leaps to the eye, while the structureless array means nothing to us—unless we look at it long enough and start to impose structure on it, such as faces or other imaginary forms. In the lower two figures the structure resulting from other pairing rules stands out equally clearly, and it is pretty obvious what these pairing rules are. However, it is not at all obvious that a pairing rule is solely responsible for the structure seen; it is hard to believe that the vivid streaks and swirls result just from pairs of dots, with no longer concatenations, but that is the case (for the first description of these figures see Glass 1969).

One can detect structures of this sort when they are overlaid by a huge number of completely randomly placed dots (Maloney et al. 1987), so the suggestion is that our perceptual system grabs simple examples of world-knowledge of this sort and uses them to construct its representation of the world. It is plausible to suppose that mirror symmetry and translational symmetry are so abundant in our sensory diet that an animal is *certain* to encounter them; hence mechanisms for detecting these forms of structure will always prove useful, and their universal provision by ontogenetic mechanisms has selective advantage.

But much of the knowledge we acquire is not like this at all; it consists of arbitrary forms whose regularity or structure results simply from the fact that they occur often, or are repeatedly associated with reward and

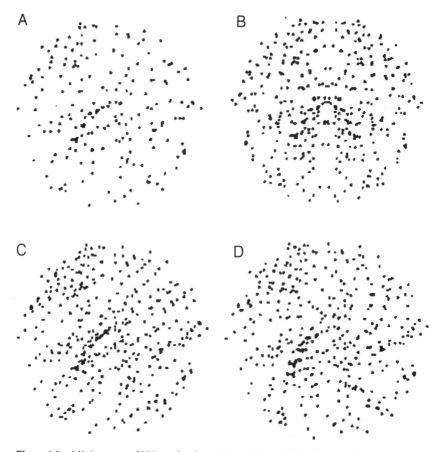

Figure 1.2 (*A*) An array of 200 randomly positioned dots. (*B*) Each dot in the array of *A* has been paired at a position mirror-symmetric about the vertical midline. (*C*) Each dot has been paired at a position up and to the left of the original position. (*D*) Each dot has been paired at a position displaced radially and tangentially from the center. It is the structure that leaps to the eye, though this is technically a form of *redundancy*; the random positions of the dots contain much more *information*, but the eye gives them less prominence.

gratification. Each individual system has to discover these for itself, and we spend our lives finding, storing, and using knowledge of these regularities in our sensory diet. They range from the often-repeated experience of our parent's smell, voice, and appearance, through the geographic details of our environment and the acoustical specificities of our language, to the customs, myths, and true knowledge of our culture. Much of this process of acquisition is fostered by teaching, but each individual brain has to do a lot of discovering for itself.

So far we have been considering the goal of the computations the cortex performs on the current sensory input, arguing that it prepares a representation suitable for discovering associative structure, and that this process entails storing world knowledge. But each cortex not only has its own experience and history, but also an evolutionary history. Evolution results from natural selection acting on variants produced genetically, so perhaps the pattern of variants produced by the cortex has enabled it to excel in the evolutionary acquisition of world knowledge.

Evolutionary Learning and Neocortex

Most people will accept the fact that there is such a thing as inherited knowledge of the world. Many of the most striking examples are found in insects—for example, the yucca moth could not fertilize the yucca plant and use its ovaries as incubators for its own eggs without such knowledge, nor could the ichneumon select a particular species of caterpillar to lay its eggs in. But it occurs in mammals too—the specialized skills of a retriever are quite different from those of a sheepdog or a greyhound—and no one doubts that these skills have a large inherited component.

Now a characteristic cannot play an important role in the evolution of a species unless it is controlled genetically and subject to genetic variability. Therefore the view that neocortex is responsible for our rapid evolution implies that its function must be controlled genetically, for otherwise it could not have brought us to the position we are in. The full hypothesis must therefore be that the neocortex gives us useful knowledge of the world in two ways: not only does it discover the structure of its world by experience during its lifetime, but it also has mechanisms, adapted through the process of genetic selection, that confer skills for doing this. These mechanisms and skills are sometimes highly specialized and amount to inherited knowledge of the world. On this view both extreme schools of thought about the origin of our mental powers are correct: the neocortex acquires knowledge of the world by nature as well as by nurture, but these methods work toward the same end rather than being the mutually exclusive alternatives that we tend to think. For this reason they can be considered together when trying to define the computational goal of the neocortex.

REPRESENTATIONS DESIGNED FOR KNOWLEDGE ACQUISITION

What we know of the neurophysiology of neocortex does not at first suggest that it is concerned with acquiring, storing, and utilizing knowledge of the world. Instead, it seems to form *representations* of the current scene in the sensory areas, and perhaps the motor area could be thought of as forming a *representation* of current motor actions. But this representational function does not necessarily conflict with the hypothesis about acquisition of knowledge. Different types of representation are suitable for different purposes, and the cortical representation may be one that is specially adapted to facilitate the learning of new associations. It turns out that storage and utilization of knowledge about the world is necessary to form such a representation, so the comparative anatomists' view that neocortex provides the filing cabinets of the central executive could be nicely reconciled with the neurophysiological facts about representation.

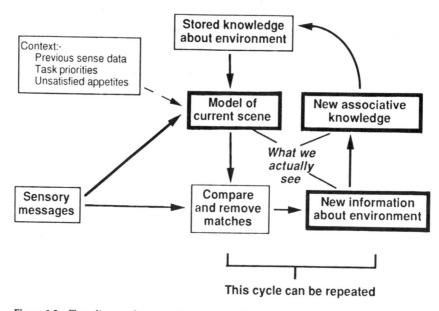

Figure 1.3 Flow diagram for perception suggested by the hypothesis that the cerebral cortex creates a representation of the current sensory scene that facilitates the identification of new associations. To separate new *information* from *knowledge* (i.e., redundancy) there must already be a store of the known structure and regularities found in sensory inputs, and this must be used to form a model that accounts for as much as possible of the current sensory input; what this accounts for is then removed from the input representation. Though we think we experience sensory messages directly, what we see corresponds better to the contents of the heavily outlined boxes.

Figure 1.3 illustrates a flow diagram for perception according to this hypothesis. The sensory messages are combined with a store of knowledge of the world to find the best model of the current sensory scene. This model is then compared with the sensory messages being received, and those parts that match are removed. The residue represents the part of the current sensory input that is unaccounted for by preexisting knowledge. Ideally this would correspond to new information, together of course with noise of random origin. We can be aware of this residue, but the subjectively salient and objectively useful parts of the sensory flow consist mainly of items that have been accounted for (i.e., items that have been successfully modeled), and also new regularities or structure in the parts that the current model does not account for. This flow diagram has features related to the "matching response" of MacKay (1955), the thalamic "active blackboard" of Mumford (1991, 1992), and the adjusting feedback of Daugman (1988) and Pece (1992).

Changing the Code Stores Associative Structure

The suggested operation can be thought of in a different way, as a recoding to reduce redundancy. The presence of one type of associative structure

Figure 1.4 The image on the left was "whitened" by making the power spectrum of the spatial Fourier transform level, thus producing the image on the right which has a much narrowed autocorrelation function. The types of statistical structure that occur at the borders of objects survive whitening and can be more easily analyzed and detected in the absence of the autocorrelations that whitening removes. (From Tolhurst and Barlow 1993)

in a body of data makes it more difficult to detect another type, so to detect this new type it is desirable to recode the messages to eliminate the first type. It is certainly very often the case that removing a known type of associative structure makes it easier to identify a new type, and figure 1.4 provides an illustration. The left part is a normal image and thus has an autocorrelation function that extends over a large fraction of the whole image. The counterpart of this is the great excess of low spatial frequencies in the power spectrum of the Fourier transform, and these can be removed by applying an inverse spatial filter to make the power spectrum level. This process of "whitening" eliminates the correlations, estimated over the whole image, between pairs of points with any fixed separation, and the result is shown in the right image. It is clear that the higher order structures, whatever they are, that correspond to borders and edges survive and can be more easily examined in this image.

In outline then, the idea is that associative structure one already knows about should be removed from the data stream to make it easier to detect new associative structure. Knowledge of the old associations should be used to change the code and thus modify the representation so that these old associations are no longer present. This is the idea of recoding to reduce redundancy (Barlow 1959; Watanabe 1960) or, if you like a simpler analogy, it is like calculating the regression that corresponds to an already recognized correlation to make it easier to find further relationships in the residuals. Of course the modifications will not generally be as simple as subtracting out an expected regression, but a set of modifications aimed at accounting for and reducing the known structure in a set of images would constitute stored knowledge about those images.

To an outside observer, a system performing these operations would look like one that constructed Craik's working models (Craik 1943) and Tolman's cognitive maps of the environment (Tolman 1948), for it would show evidence of finding and using the associative structure that underlies such models and maps. Obviously this store of knowledge has many other potential uses, particularly in the processes of imagination and recall where we experiment and play with what we know. Possibly it could be made accessible in the absence of sensory input by lowering neural thresholds in the box marked "stored knowledge about the environment" in figure 1.3, but this possibility cannot be pursued here. Using this knowledge to discount the expected in the representation of the current scene would have enormous selective advantage by improving learning and the acquisition of new knowledge, though it certainly seems wasteful not to use it for imagination and recall as well.

Anything that improves the appropriateness and speed of learning must have immense competitive advantage, and the main point about this proposal is that it would explain the enormous selective advantage of the neocortex. Such an advantage, together with appropriate genetic variability, could in turn account for its rapid evolution and the subsequent growth of our species to its dominant position in the world.

Although these notions do not obviously follow from the neurophysiological facts, I think suggestive evidence in support can be found from the changes in neural connectivity that occur in the sensitive period early in the life of cats and monkeys (Hubel and Wiesel 1970; Movshon and Van Sluyters 1981), and in the known phenomena of pattern-selective adaptation discussed elsewhere (Barlow 1990, 1991). There are aspects of the evolution and neurophysiology of the cortex that we certainly do not yet understand properly, and the new hypothesis can give us a fresh viewpoint if we examine what it requires in more detail.

Acquiring Knowledge from Representations of Features

Acquiring knowledge means finding out about the regularities and patterns in the sensory input. It's a vast task to determine the associational structure of the continuous stream of sensory messages that we receive, and table 1.1 lists some of the requirements, starting with the point above about the desirability of removing evidence for the associations you already know about. The next items have been discussed before (Barlow 1991) but will be summarized below.

Suppose that the representation of the current scene consists of reports of features, of which there can be a wide variety. For instance, one of them might be a point in the image having a luminance value above the mean for the neighborhood of that point, and this would correspond approximately to the feature that causes the firing of an on-center ganglion cell in the retina.

Table 1.1 What would make it easier to identify new associations?

Remove evidence of the associations you already know about	To facilitate detecting new ones
Make available the probabilities of the features currently present	To determine chance expectations
Choose features that occur independently of each other in the normal environment	To determine chance expectations of combinations of them
Choose "suspicious coincidences" as features	To reduce redundancy and ensure appropriate generalization

Or it might be the occurrence of a visual pattern resembling a monkey's face, which would correspond to the occurrence of the trigger feature of a so-called face cell in inferotemporal cortex. Thus almost any representation one can imagine can be described as reporting the occurrence of features.

There must be many levels in the actual representational system in the brain, and more complex features are presumably represented at higher levels. The first item in table 1.1 suggests that recoding to take account of identified regularities in sensory messages will be an important step in progressing to higher levels in the perceptual system. But for present purposes let us consider a single level and examine what is needed to identify a new association. We hope that the repetition of this one operation may lead to a system that identifies the complex associations that we undoubtedly use all the time.

The Need for Prior Probabilities

To identify new associations a representation must do more than just report the occurrence of features: it must also signal the unexpectedness of the features reported, or at least make this information immediately accessible. This might be done by adjusting the threshold for a unit so that, averaged over a long period, it fires once in a particular period; when it fires, it then signals an event that has a probability of occurring once in that period. Alternatively, the number of impulses in the volley signaling an event might be an inverse function of its probability such as $-\log p$. Either of these would appear as forms of habituation, which is of course often observed in sensory systems.

The reason prior probabilities are needed is obvious: to show that two features are associated one must show that they occur together at a rate different from that expected by chance, and to calculate this expected rate one needs to know the expected rates of the constituent individual features. Of course one also needs to know how often the features occur together, but one can justifiably regard this as a requirement of the associative mechanism itself, while it seems more natural to suppose that the representation is responsible for storage and access to the rates of the individual features.

Two points may need clarifying. First we are assuming that the probability of occurrence of a single feature can be estimated from its rate of occurrence over some period in the recent past. This would not always be justified, but in some cases it will be and to let the argument proceed let us assume it is. Second, the features we shall be dealing with will usually have prior probabilities well below half; this means that the predicted rate of occurrence of *joint* features will be low, and their expected number may be close to zero. Under these circumstances it becomes difficult to establish a *negative* association, and one must therefore look for joint features that occur *more often* than expected by chance. That is why they were called suspicious coincidences or clichés (Phillips et al. 1984; Barlow 1985), but the basic property is their nonaccidental nature.

The Need for Independence

Knowing and using the unexpectedness of features seem unavoidable for efficient associative learning, but there is another highly desirable property when detecting new associations, namely statistical independence, in the environment to which the system is adapted, of the features represented. Even in a simple case, such as finding a new association between a special occurrence such as reinforcement and an individual element of the representation, one would start by assuming they were independent to estimate the expected number of coincidences. The alternative would be to take account of the known correlations, but this would become difficult when detecting associations between arbitrary pairs of elements, and virtually impossible if one wished to find an association with some logical function of a group of elements. In that case one either has to know the associational structure within that group, or else one must again assume independence, and if one is going to do the latter it is important to make sure that the events represented are in fact as nearly independent as possible. While it is plausible for a representation to store the rate of occurrence of its individual elements, one cannot suppose that it stores the associational structure of arbitrary groups of such elements.

Are we actually able to detect new associations with logical functions of representational elements? For simple functions, surely we can, and so can most animals. We learn to stop at red traffic lights and not at green ones, for example. In this case one might suppose that there are different representational elements for red and green lights, but it would be a great restriction on the utility of a representation if this was always necessary before separate associations could be formed. Harris (1980) brought this out very nicely when discussing contingent adaptation, for he noted that almost any contingency that had ever been tested seemed to produce adaptive effects. How could this be, he said, if contingent adaptation requires neurons specifically sensitive to each contingency? We might have neurons signaling *yellowness*, and perhaps *Volkswagens*, but surely we cannot have neurons reserved for signaling *yellow Volkswagens*! This problem will be considered again, but the advantage of distributed, as opposed to grandmother-cell, representations results from their supposed ability to utilize the vast number of combinations of active elements, and this advantage would vanish if, as a result of their prior probabilities being unavailable or grossly misleading, one could not form associations with these combinations efficiently.

Following the idea that one should discount the expected, a possible course of action would be to devise a code in which the elements occur as nearly as possible independently, and some ways of doing this have been suggested elsewhere (Barlow 1959, 1989; Barlow and Földiák 1989; Hentschel and Barlow 1991), together with evidence that something of the sort may be happening (Barlow 1990). As already pointed out, the codes that are required to obtain independence embody knowledge about the associational structure of the environment, and an outside observer watching behavior based on this modified representation should suspect that some kind of cognitive map or working model of the environment had been constructed.

Forms of Representation

A distributed representation is one in which the features that can be utilized effectively for further processing are represented by combinations of activity of the elements, rather than directly by the activity of neurons or elements specifically and selectively sensitive to each of these features. The 7-bit ASCII code provides a familiar example of a distributed representation, and because one must perform a logical manipulation on the representational elements before one can decide if the represented feature has occurred, we say that they represent the features implicitly or indirectly rather than directly. Tony Gardner-Medwin and I have been exploring a limitation on the use of implicit representations for learning (Gardner-Medwin and Barlow 1992, 1994).

The limitation arises as follows. Consider classical conditioning, where an initially neutral sensory feature is "reinforced" by being presented repeatedly in conjunction with a pleasant reward such as some food in the mouth. When the animal has identified the association, it uses the initially neutral feature to predict the reward. Now to determine whether there is a genuine association one must form a 2×2 contingency table for the feature and the reward, counting the numbers in each box of this table. If the feature is directly represented there is no great conceptual difficulty in obtaining all these numbers: assuming that knowledge of reinforcement is available everywhere, then local mechanisms at the element can estimate how often the feature and reinforcement occur by themselves, how often they occur together, and how often nothing occurs. A calculation equivalent to a chi-squared test can then be done on these numbers to decide if the association is genuine. In the case of implicitly represented features this is not so straightforward.

The difficulty is that there is no point in the system where all the information is available to estimate the necessary numbers. One can imagine the reinforcement signal being available at all the elements that carry the information telling one that the feature has occurred, but one of these elements by itself is not enough to determine whether that feature occurred or not: one must evaluate the logical function using all the elements before one knows this. One can postulate an element that does this logical decoding, but such an element would directly represent the feature and it would no longer be only implicitly represented. What can be done?

Even though accurate counts of the required numbers are not available at any one spot, a relatively simple mechanism could collect together information for an innaccurate estimate. In the logical representation of the feature there will be some elements that are positively correlated with the presence of the feature, and others that are negatively correlated. The appropriately weighted sum of the activities of these elements will give an indication of the presence of the feature, and the average over time of this measure can be used to estimate how often it occurs. We know from the limitations of perceptrons (Minsky and Papert 1969) that there are many logical functions that this method will be incapable of detecting correctly, but we also know there are many cases where it works satisfactorily. An approximation of this sort actually seems to be involved in most learning in artificial neural networks. We have been trying to determine for what types of representation such learning will be reasonably fast and efficient, and under what conditions it is bound to be slow and unreliable.

Statistical Efficiency as a Measure of Performance

To assess the merit of one representation against another we need a measure of associational performance, and we have used the statistical efficiency defined by R. A. Fisher (1925) for this purpose. To make any statistical decision up to a required standard of reliability a sample of a certain minimum size is necessary, but if the method is inefficient a larger sample will be needed to obtain the same standard of reliability. Fisher's efficiency is simply the ratio of these two sample sizes. In our case the sample size is given by the maximum number of occurrences or coincidences that could occur in the time for which the counts were made, so if the method is inefficient it will take longer to determine that an association is present, or more mistakes will be made if a decision is made in the same time. It is pretty clear that speedy and reliable learning about new causative factors in the environment will have high survival value, and the statistical efficiency attainable in a particular type of representation gives a very direct measure of how useful it would be for enabling an animal to detect new associations and acquire new knowledge of the world.

Explicit Representation

So far we have referred to "directly represented features," where there are selectively sensitive representatational elements that respond when and only when the feature is present, and "implicitly represented features," for which there are no such selectively sensitive elements but the appropriate logical analysis of the pattern of activity in the whole representation nonetheless allows one to decide if the feature is present. We now introduce an intermediate type, explicitly represented features, for which we have been able to solve the problem of determining the statistical efficiency for detecting associations.

An explicitly represented feature is one whose presence can be determined by a simple logical operation performed on a subset of the elements in the representation, rather than on the whole representation. So far the simple logical operation we have analyzed is the presence of a particular pattern in the elements of the specified subset, since this seems both the simplest and perhaps the most interesting case. It turns out that inactive elements carry very little information if the representation is reasonably sparse (i.e., the average proportion of elements active at any one time is low, say less than 10%), so one need consider only the active elements. Each of these directly represents a different feature, so the occurrence of the pattern corresponds to the conjunction or joint occurrence of certain specific features. To return to Harris's example, if one element at a particular point in the visual field directly represented Volkswagens, and another element at that position directly represented Yellowness, Yellow Volkswagens would not be directly represented, but they would be explicitly represented by the joint occurrence of the above two elements. The question we think we have answered is "Under what conditions can a representation in which there are Yellowness (Y) units and Volkswagen (V) units, but no Yellow Volkswagen (YV) units, nonetheless be used to detect efficiently an association with Yellow Volkswagens?" The answer is, however, more general than this, for it applies to multiple conjunctions and patterns in subsets with more than a pair of active elements.

The outline of the analysis is as follows. To determine if an association is present between a feature and reinforcement (R) one does a chi-squared test on a 2×2 contingency table in which the feature (Y, V, or YV) is one of the variables and reinforcement (R) is the other. Because of the sampling errors in the numbers in such a table the result will be variable, and this variability determines how large a sample is required before one can confidently assert that an association is present.

If there are no YV units one must look at the 2×2 tables for Y vs. R and V vs. R, and combine the results to assess whether YV is associated with R. Now for each of the Y vs. R and V vs. R tables there is a perturbing factor: Yellowness can occur with reinforcement even if there is no Yellow Volkswagen present, and likewise for nonyellow Volkswagens. These intrusive extra occurrences will not bias the result if one knows the unexpectedness

What Is the Computational Goal of the Neocortex?

of Y and V, but they will add to the variability of the two subtables, and even when optimally combined the decision about the association of reinforcement with Yellow Volkswagens cannot be made as reliably as if they were represented directly.

Sparse Coding Helps

How serious is this factor? Note first that the problem arises when the same representational element is active in more than one of the features that may be reinforced. In the current example, the Y unit is active for all yellow things, not just yellow Volkswagens, and similarly for the V unit. The extent of this overlap depends on the *sparseness* of the representation, which is defined by the average fraction of the elements that are active. If it is very sparse, then only a small proportion of the units will be active for any input, and there will be little overlap. Indeed, if it is sparse enough there will be only a single unit active for each input, and each will therefore be directly represented. On the other hand if it is dense, a given unit will be active for a high proportion of inputs, and the overlap problem will be serious.

When there is overlap, what matters is the number of these intrusive extra occurrences relative to the number of genuine occurrences of Yellow Volkswagens, and this in turn depends upon the probability of the joint event (YV) relative to the single features (Y and V). If Yellowness and Volkswagens are both common, then the reinforcement of rare Yellow Volkswagens would be masked by the quite frequent chance reinforcement of other yellow things and other colored Volkswagens.

One can show that the efficiency for detecting a feature X depends to a good approximation upon the value of a parameter Γ_x that is equal to $\alpha_x p_x Z / \langle \alpha \rangle$, where α_x is the fraction of representative elements in the subset active for the representation of the feature X, $\langle \alpha \rangle$ the average fraction active for all inputs, p_x is the probability of the feature X, and Z is the number of representative elements in the subset. Figure 1.5 shows how efficiency varies with the value of this parameter.

Improbable Features Need Denser Representation

Note first that efficiency increases with Γ, and, as expected, Γ increases with the number of neurons in the subset and the average sparseness (low $\langle \alpha \rangle$). It also increases both with the probability p_x of the feature X, and with the activity ratio α_x for the feature X; in fact these two are reciprocally related, so a very rare feature X can still form associations efficiently if it causes an unusually large proportion of the units in a subset of the representation to be active. Clearly there is scope here for genetic factors to improve selectively the performance of a learning network: factors of biological importance should cause many units in a learning network to become active. Another way of putting this is to say that many of the

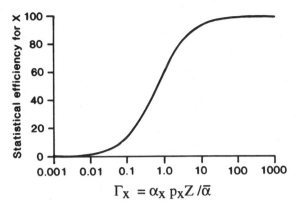

$$\Gamma_X = \alpha_X \, p_X Z / \bar{\alpha}$$

Figure 1.5 The curve shows how statistical efficiency for detecting associations with a feature X varies with the value of a parameter defined as follows: $\Gamma = \alpha_X p_X Z / \langle \alpha \rangle$, where α_X, $\langle \alpha \rangle$ are the activity ratio for feature X and the average activity ratio, p_X is the probability of X, and Z is the number of neurons in the subset under consideration. For instance, one could identify an association with any one of the 45 possible pairs of active neurons in a subset of 10 with an efficiency of 50% provided that the neurons were active independently, the pair caused two neurons to be active, the probability of the pair occurring was 0.1, and the average fraction active was 0.2. (From Gardner-Medwin and Barlow 1994)

directly represented features should correspond to features possessed by biologically important objects; then, when one of these objects appears, it will cause a high level of activity.

Next look at the actual efficiencies attainable for various values of Γ. Although one needs Γ to be 10 or 100 for efficiencies in the 90% to 100% range, useful efficiencies of about 50% are obtained with $\Gamma = \sim 1$; this is the order of magnitude of the efficiency of human subjects detecting bilateral symmetry in dot patterns such as those shown in figure 1.2 (Barlow and Reeves 1979). Consider a subset of 10 elements in a network; if one could specify 10 mutually exclusive features, the elements of the subset could each handle one of them and associations with them could be formed with 100% efficiency. Now suppose that the features of interest do not all cause firing of only single elements among the ten. If a particular feature X does cause firing of just one element ($\alpha_x = 0.1$) but this element is also active in conjunction with other elements when other features occur, then if $p_x = \langle \alpha \rangle = 0.1$ we will have $\Gamma = 1$ and the efficiency for detecting associations will have dropped to around 50%. This reduction occurs because intrusive or accidental reinforcement occurs in conjunction with the activity of any given element, but this is a small price to pay for the increased versatility resulting from the possibility of using and learning associations of combinations of the features, as illustrated below.

Suppose the feature X is represented by the conjunction of two elements ($\alpha_x = 0.2$). If again $p_x = 0.1$ but we suppose $\langle \alpha \rangle$ is now 0.2, the same as the new value of α_x, then Γ still has the value 1, corresponding to the same efficiency, ca. 50%. There are 45 such conjunctions of pairs of elements among 10 elements, so a much wider range of features can be used to form associations efficiently, and there is not an enormous loss of efficiency compared with the direct representation of features on single elements. Notice that the above applies to features represented by pairs of active units, but a particular merit of such a system is that it can form associations with patterns containing three or more active elements. Even if such multiple conjunctions of directly represented features are rare, provided that they cause activity in a high proportion of elements, they will be learned with reasonable efficiency.

What this shows is that it is possible to learn about explicit conjunctions of any number of elements in known subsets of a representation, provided that the representation is sparse, provided that these conjunctions do not occur too infrequently and activate a substantial proportion of elements when they do, and provided that the representative elements can be considered, a priori, to occur independently. How to achieve this, and read out the results in a useful way, cannot be gone into here.

The analysis we have done so far is only a beginning. What can be done using the union rather than conjunction of representational elements in a subset? What can be done with threshold logic functions on the activity of members of a subset? We do not know the answer to these questions, but one point does seem evident.

We have already seen that the features that are directly represented should (1) occur as closely as possible independently of each other in the environment to which the representative system is adapted; (2) occur sufficiently frequently so that the representation is neither too dense nor too sparse. These requirements might not be too difficult to meet if one could postulate an indefinite number of directly represented features, but such an indefinitely large representation would have none of the capacity to generalize sensibly that is needed in a representation to be used for learning. This introduces another requirement for the selection of directly represented features: they must each represent as much as possible of the incoming stream of data from the environment and must occur frequently so that generalization occurs usefully. Some tests of this prediction on digitized images of natural scenes will now be described.

SELECTION OF DIRECTLY REPRESENTED FEATURES

It is generally agreed that the neurons of the primary visual cortex respond selectively to the borders and edges of objects in the visual image. There is argument about whether they should be regarded as edge-detectors, Gabor filters, or wavelet functions, but there is no disagreement that they do in fact respond to the oriented patterns of light that occur at the borders of objects. If the arguments (Barlow 1985) about the importance of nonchance associations are correct, then measurements of the distribution of light at the borders of objects should show that edges qualify as "suspicious coincidences." We set out to test whether this was so, and the main result confirms that it is (Barlow and Tolhurst 1992).

We took a selection of digitized images and removed the correlations between pairs of points, averaged over the whole image, by the "whitening" process described before and shown in figure 1.4; this leaves behind the image structures we are now interested in that occur at the borders of objects. The distribution of pixel values in such whitened images gives us the basis for the chance expectation of combinations of pixel values, and what the hypothesis says is that at the borders of objects we shall find combinations of pixel values that occur more frequently than this chance expectation.

Perhaps it is already obvious by inspection of the whitened figure that this is the case, for you would not expect to find by chance the rows of high or low values you can see in figure 1.4. To confirm this we measured the distribution of the sum of nine pixel values selected at random from all over the whitened image to provide the chance distribution, and from nine adjacent spots in a row to show what actually occurs. Figure 1.6 shows the result: the distributions are strikingly different. For the sum of nine randomly selected pixels the range is from about 980 to 1320 on the horizontal scale, but as you can see values outside this range are very common for the sum of nine pixels in a row.

Do these extreme values occur at the borders? Yes they do, as shown in figure 1.7, which marks the positions in the image where these extremes occur. As you see they occur at edges.

Is this a consistent feature of the sum of pixels in a line? To answer this we looked at a varied selection of 15 images, and estimated the *kurtosis excess* for the sums in a line compared with randomly selected pixels and sums over square regions. This measure (Weatherburn 1961) is based on the fourth moment and values greater than 0 can be crudely taken to indicate that the distribution has an excess of extreme values compared with a gaussian. As shown in table 1.2, the kurtosis excess is much greater for the line sum than for the other distributions, though it has to be said that we do not understand why patches, and even single pixels, also show kurtosis excess. The large excess for lines, combined with the fact that the extreme values occur at the borders, vindicates the hypothesis that the features we use to represent an image are suspicious coincidences—at least in the case of the orientationally selective units of V1.

SUMMARY AND CONCLUSIONS

It was suggested initially that we dominate the world because we know more about it than other animals, and that it is the neocortex that is responsible for this. How to acquire and store knowledge of the world is a vast problem, but although we have only scratched the surface we may be beginning to discover how the neocortex could, as the combined result of genetic selection and individual experience, provide us with a representation of the current scene that automatically stores, gives access to, and adds to such knowledge.

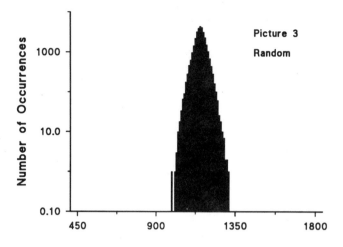

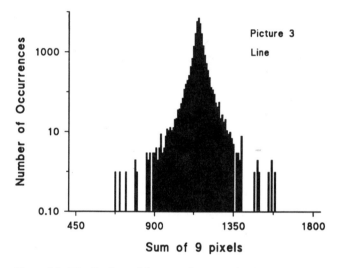

Figure 1.6 Distributions of the sum of nine randomly selected pixels (*top*), and nine pixels in a line at four orientations (*bottom*) from the right hand (whitened) image of figure 1.4. The lower distribution has an excess of extreme values—that is, values unexpected on the basis of the distribution of individual pixel values in the whitened image. (Tolhurst and Barlow 1994)

1. The first principle, suggested diagrammatically in figure 1.3, is that neocortex removes associative structure that has already been identified through past experience. This is analogous to discounting the mean luminance in light adaptation, or removing a known regression when trying to make sense of residual deviations. Identified structure would be stored, and when recognized in the current scene it would form part of a matching model; the unmatched residue would contain new information about that scene. Stored knowledge of the associative structure of the world would be used continuously and automatically in this way, but there could be other methods of accessing it for purposes of imagery and recall.

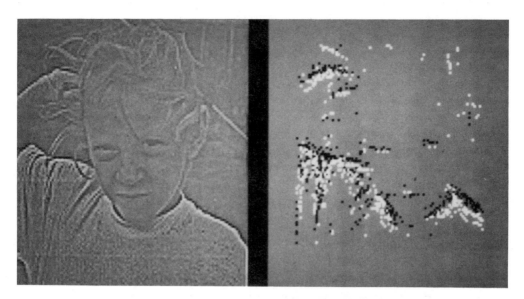

Figure 1.7 The left panel is the whitened image from figure 1.4. In the right panel white dots are placed at the positions of the upper extreme values of the distributions of the sum of pixels in a line shown in figure 1.6, and dark dots at lower extreme values. The extreme values obviously occur at the borders of objects in the image. Hence the combinations of pixel values that occur at edges are ones that would not be expected by chance.

Table 1.2 Kurtosis excess for 15 images

	Mean ± SE (n = 15)
Sum of pixels over line 9 × 1	15.82 ± 3.66
Sum of 9 random pixels	0.989 ± 0.76
Sum over square 3 × 3	7.30 ± 1.68
Single pixels	8.81 ± 1.70

2. To prove that an association between two features exists you need to know their individual frequencies of occurrence, because you must estimate the chance frequency of joint occurrence to show that the actual frequency is significantly greater.

3. To detect associations with combinations of features, the features should be chosen so that they occur as nearly as possible independently of each other in the environment to which the sytem is adapted, for otherwise the expected frequency of occurrence of a combination is hard to determine.

4. In distributed representations, detecting associations with conjunctions of features is difficult because accidental associations with the constituents of the conjunctions mask associations with the conjunctions themselves. This problem tends to make the identification of associations with conjunc-

tions inefficient in dense distributed representations, and such representations are therefore unlikely to be useful.

5. On the other hand in sparse distributed representations associations can be identified reasonably efficienctly (say 50%) with features represented by the conjunction of directly represented features, provided that these conjunctions are not too sparsely represented and occur with a frequency not too far below that of the directly represented features.

6. To generate a reasonably economical representation of the current scene the directly represented features should be suspicious coincidences—combinations of signals from lower levels that occur frequently but would be rarely expected by chance. The representational elements at higher levels should be matched to the biological importance and statistical structure of occurrences at lower levels.

7. This notion that the representational elements used by the brain correspond to suspicious coincidences, or combinations of simpler events that occur more often than expected by chance, has received some support from the statistical analysis of edges in images.

8. The features that are directly represented at any level in the hierarchy will have a strong effect on the performance of a representational network, including the way that generalization occurs. The selection of these features is likely to be one way that genetic factors exert their influence. In addition, ontogenetic control of the connections between levels probably determines the way that information of different types is segregated and brought together according to nontopographic principles. In these two ways, and possibly others, genetic factors must influence how the cortex handles sensory information, and they can be regarded as an inherited store of world knowledge; the genetic variability that has enabled such a store to be formed may be at least as important as the ability of the cortex to acquire world knowledge by its own direct experience.

2   A Critique of Pure Vision[1]

Patricia S. Churchland, V. S. Ramachandran, and Terrence J. Sejnowski

INTRODUCTION

Any domain of scientific research has its sustaining orthodoxy. That is, research on a problem, whether in astronomy, physics, or biology, is conducted against a backdrop of broadly shared assumptions. It is these assumptions that guide inquiry and provide the canon of what is reasonable—of what "makes sense." And it is these shared assumptions that constitute a framework for the interpretation of research results. Research on the problem of how we see is likewise sustained by broadly shared assumptions, where the current orthodoxy embraces the very general idea that the business of the visual system is to create a detailed replica of the visual world, and that it accomplishes its business via hierarchical organization and by operating essentially independently of other sensory modalities as well as independently of previous learning, goals, motor planning, and motor execution.

We shall begin by briefly presenting, in its most *extreme* version, the conventional wisdom. For convenience, we shall refer to this wisdom as the Theory of Pure Vision. We then outline an alternative approach, which, having lurked on the scientific fringes as a theoretical possibility, is now acquiring robust experimental infrastructure (see, e.g., Adrian 1935; Sperry 1952; Bartlett 1958; Spark and Jay 1986; Arbib 1989). Our characterization of this alternative, to wit, *interactive vision*, is avowedly sketchy and inadequate. Part of the inadequacy is owed to the nonexistence of an appropriate vocabulary to express what might be involved in interactive vision. Having posted that caveat, we suggest that systems ostensibly "extrinsic" to literally seeing the world, such as the motor system and other sensory systems, do in fact play a significant role in what is literally seen. The idea of "pure vision" is a fiction, we suggest, that obscures some of the most important computational strategies used by the brain. Unlike some idealizations, such as "frictionless plane" or "perfect elasticity" that can be useful in achieving a core explanation, "pure vision" is a notion that impedes progress, rather like the notion of "absolute downness" or "indivisible atom." Taken individually, our criticisms of "pure vision" are neither new nor convincing; taken collectively in a computational context, they make a rather forceful case.

These criticisms notwithstanding, the Theory of Pure Vision together with the Doctrine of the Receptive Field have been enormously fruitful in fostering research on functional issues. They have enabled many programs of neurobiological research to flourish, and they have been crucial in getting us to where we are. Our questions, however, are not about past utility, but about future progress. Has research in vision now reached a stage where the orthodoxy no longer works to promote groundbreaking discovery? Does the orthodoxy impede really fresh discovery by cleaving to outdated assumptions? What would a different paradigm look like? This chapter is an exploration of these questions.

PURE VISION: A CARICATURE

This brief caricature occupies one corner of an hypothesis-space concerning the computational organization and dynamics of mammalian vision. The core tenets are logically independent of one another, although they are often believed as a batch. Most vision researchers would wish to amend and qualify one or another of the core tenets, especially in view of anatomical descriptions of backprojections between higher and lower visual areas. Nevertheless, the general picture, plus or minus a bit, appears to be rather widely accepted—at least as being correct in its essentials and needing at most a bit of fine tuning. The approach outlined by the late David Marr (1982) resembles the caricature rather closely, and as Marr has been a fountainhead for computer vision research, conforming to the three tenets has been starting point for many computer vision projects.[2]

1. *The Visual World.* What we see at any given moment is in general a fully elaborated representation of a visual scene. The goal of vision is to create a detailed model of the world in front of the eyes in the brain. Thus Tsotsos (1987) says, "The goal of an image-understanding system is to transform two-dimensional data into a description of the three-dimensional spatio-temporal world" (p. 389). In their review paper, Aloimonos and Rosenfeld (1991) note this characterization with approval, adding, "Regarding the central goal of vision as scene recovery makes sense. If we are able to create, using vision, an accurate representation of the three-dimensional world and its properties, then using this information we can perform any visual task" (p. 1250).

2. *Hierarchical Processing.* Signal elaboration proceeds from the various retinal stages, to the LGN, and thence to higher and higher cortical processing stages. At successive stages, the basic processing achievement consists in the extraction of increasingly specific features and eventually the integration of various highly specified features, until the visual system has a fully elaborated representation that corresponds to the visual scene that initially caused the retinal response. Pattern recognition occurs at that stage. Visual leaning occurs at later rather than earlier stages.

3. *Dependency Relations.* Higher levels in the processing hierarchy depend on lower levels, but not, in general, vice versa. Some problems are

Churchland, Ramachandran, and Sejnowski

early (low level) problems; for example, early vision involves determining what is an edge, what correspondences between right and left images are suitable for stereo, what principle curvatures are implied by shading profiles, and where there is movement (Yuille and Ullman 1990). Early vision does not require or depend on a solution to the problems of segmentation or pattern recognition or gestalt.[3]

Note finally that the caricature, and, most especially, the "visual world" assumption of the caricature, gets compelling endorsement from common sense. From the vantage point of how things seem to be, there is no denying that at any given moment we seem to see the detailed array of whatever visible features of the world are in front of our eyes. Apparently, the world is there to be seen, and our brains do represent, in essentially all its glory, what is there to be seen. Within neuroscience, a great deal of physiological, lesion, and anatomical data are reasonably interpretable as evidence for some kind of hierarchical organization (Van Essen and Anderson 1990). Hierarchical processing, moreover, surely seems an eminently sensible engineering strategy—a strategy so obvious as hardly to merit ponderous reflection. Thus, despite our modification of all tenets of the caricature, we readily acknowledge their prima facie reasonableness and their appeal to common sense.

INTERACTIVE VISION: A PROSPECTUS

What is vision for? Is a perfect internal recreation of the three-dimensional world really necessary? Biological and computational answers to these questions lead to a conception of vision quite different from pure vision. Interactive vision, as outlined here, includes vision with other sensory systems as partners in helping to guide actions.

1. *Evolution of Perceptual Systems.* Vision, like other sensory functions, has its evolutionary rationale rooted in improved motor control. Although organisms can of course see when motionless or paralyzed, the visual system of the brain has the organization, computational profile, and architecture it has in order to facilitate the organism's thriving at the four Fs: feeding fleeing, fighting, and reproduction. By contrast, a pure visionary would say that the visual system creates a fully elaborated model of the world in the brain, and that the visual system can be studied and modeled without worrying too much about the nonvisual influences on vision.

2. *Visual Semiworlds.* What we see at any given moment is a partially elaborated representation of the visual scene; only immediately relevant information is explicitly represented. The eyes saccade every 200 or 300 msec, scanning an area. How much of the visual field, and within that, how much of the foveated area, is represented in detail depends on many factors, including the animal's interests (food, a mate, novelty, etc.), its long- and short-term goals, whether the stimulus is refoveated, whether the stimulus is simple or complex, familiar or unfamiliar, expected or unexpected, and so on. Although unattended objects may be represented in some min-

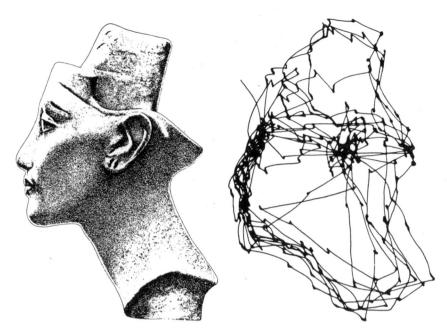

Figure 2.1 The scan path of saccadic eye movements made by a subject viewing the picture. (Reprinted with permission from Yarbus 1967.)

imal fashion (sufficient to guide attentional shifts and eye movements, for example) they are not literally seen in the sense of "visually experienced."

3. *Interactive Vision and Predictive Visual Learning*. Interactive vision is exploratory and predictive. Visual learning allows an animal to predict what will happen in the future; behavior, such as eye movements, aids in updating and upgrading the predictive representations. Correlations between the modalities also improve predictive representations, especially in the murk and ambiguity of real-world conditions. Seeing an uncommon stimulus at dusk such as a skunk in the bushes takes more time than seeing a common animal such as a dog in full light and in full, canonical view. The recognition can be faster and more accurate if the animal can make exploratory movements, particularly of its perceptual apparatus, such as whiskers, ears, and eyes. There is some sort of integration across time as the eyes travel and retravel a scan path (figure 2.1), foveating again and again the significant and salient features. One result of this integration is the strong but false introspective impression that at any given moment one sees, crisply and with good definition, the whole scene in front of one. Repeated exposure to a scene segment is connected to greater elaboration of the signals as revealed by more and more specific pattern recognition [(e.g., (1) an animal, (2) a bear, (3) a grizzly bear with cubs, (4) the mother bear has not yet seen us].

4. *Motor System and Visual System*. A pure visionary typically assumes that the connection to the motor system is made only after the scene is fully elaborated. His idea is that the decision centers make a decision

about what to do on the basis of the best and most complete representation of the external world. An interactive visionary, by contrast, will suggest that motor assembling begins on the basis of preliminary and minimal analysis. Some motor decisions, such as eye movements, head movements, and keeping the rest of the body motionless, are often made on the basis of minimal analysis precisely in order to achieve an upgraded and more fully elaborated visuomotor representation. Keeping the body motionless is not doing nothing, and may be essential to getting a good view of shy prey. A very simple reflex behavior (e.g., nociceptive reflex) may be effected using rather minimal analysis, but planning a complex motor act, such as stalking a prey, may require much more. In particular, complex acts may require an antecedent "inventorying" of sensorimotor predictions: what will happen if I do a, b, and g; how should I move if the X does p, and so forth.

In computer science, pioneering work exploring the computational resources of a system whose limb and sensor movements affect the processing of visual inputs is well underway, principally in research by R. Bajcsy (1988), Dana Ballard (Ballard 1991; Ballard et al. 1992; Ballard and Whitehead, 1991; Whitehead and Ballard, 1991, Randall Beer (1990) and Rodney Brooks (1989). Other modelers have also been alerted to potential computational economies, and a more integrative approach to computer vision is the focus of a collection of papers, *Active Vision* (1993), edited by Andrew Blake and Alan Yuille.

5. *Not a Good-Old-Fashioned Hierarchy Recognition.* The recognition (including predictive, what-next recognition) in the real-world case depends on richly recurrent networks, some of which involve recognition of visuomotor patterns, such as, roughly, "this critter will make a bad smell if I chase it," "that looks like a rock but it sounds like a rattlesnake, which might bite me." Consequently, the degree to which sensory processing can usefully be described as hierarchical is moot. Rich recurrence, especially with continuing multicortical area input to the thalamus and to motor structures, appears to challenge the conventional conception of a chiefly unidirectional, low-to-high processing hierarchy. Of course, temporally distinct stages between the time photons strike the retina and the time the behavior begins do exist. There are, as well, stages in the sense of different synaptic distances from the sensory periphery and the motor periphery. Our aim is not, therefore, to gainsay stages per se, but only to challenge the more theoretically emburdened notion of a strict *hierarchy*. No obvious replacement term for "hierarchy" suggests itself, and a new set of concepts adequate to describing interactive systems is needed. (Approaching the same issues, but from the perspective of neuropsychology, Antonio Damasio also explores related ideas [see Damasio 1989 b,d]).

6. *Memory and Vision.* Rich recurrence in network processing also means that stored information from earlier learning plays a role in what the animal literally sees. A previous encounter with a porcupine makes a difference to how a dog sees the object on the next encounter. A neuroscientist and a rancher do not see the same thing in figure 2.2. The neuroscientist cannot

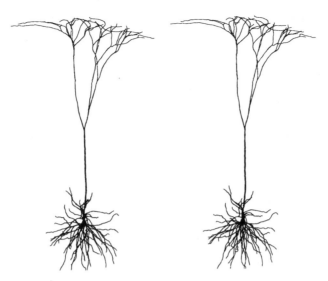

Figure 2.2 Stereo pair of a reconstructed layer five pyramidal neuron from cat visual cortex (courtsey of Rodney Douglas). The apical dendrite extends through the upper layers of the cortex and has an extensive arborization in layer 1. This neuron can be fused by placing a sheet of cardboard between the two images and between your two eyes. Look "through" the figure to diverge your eyes sufficiently to bring the two images into register. The basal dendrites, which receive a majority of the synapses onto the cell, fill a ball in three-dimensional space. Apical dendritic tufts form clusters.

help but see it as a neuron; the rancher wonders if it might be a kind of insect. A sheep rancher looking over his flock recognizes patterns, such as a ewe with lambing troubles, to which the neuroscientist is utterly blind. The latency for fusing a Julesz random-dot stereogram is much shorter with practice, even on the second try. Some learning probably takes place even in very early stages.

7. *Pragmatics of Research.* In studying nervous systems, it seems reasonable to try to isolate and understand component systems before trying to see how the component system integrates with other brain functions. Nevertheless, if the visual system is intimately and multifariously integrated with other functions, including motor control, approaching vision from the perspective of sensorimotor representation and computations may be strategically unavoidable. Like the study of "pure blood" or "pure digestion," the study of "pure vision" may take us only so far.

Our perspective is rooted in neuroscience (see also Jeannerod and Decety 1990). We shall mainly focus on three broad questions: (1) Is there empirical plausibility—chiefly, neurobiological and psychological plausibility—to the interactive perception approach? (2) What clues are available from the nervous system to tell us how to develop the interactive framework beyond its nascent stages? and (3) What computational advantage would such an interactive approach have over traditional computational approaches? Under this aegis, we shall raise issues concerning possible reinterpretation of existing neurobiological data, and concerning the implications for the problem of learning in nervous systems. Emerging from this exploration is a general direction for thinking about interactive vision.

IS PERCEPTION INTERACTIVE?

Visual Psychophysics

In the following subsections, we briefly discuss various psychophysical experiments that incline us to favor the interactive framework. In general, these experiments tend to show that whatever stages of processing are really involved in vision, the idea of a largely straightforward hierarchy from "early processes" (detection of lines, shape from shading, stereo) to "later processes" (pattern recognition) is at odds with the data (see also Ramachandran 1986; Nakayama and Shimojo 1992; Zijang and Nakayama 1992).

Are There Global Influences on Local Computation? Subjective Motion Experiments
Seeing a moving object requires that the visual system solve the problem of determining which features of the earlier presentation go with which features of the later presentation (also known as the Correspondence Problem). In his work in computer vision, Ullman (1979) proposed a solution to this problem that avoids global constraints and relies only on local information. His algorithm solves the problem by trying out all possible matches and through successive iterations it finds the set of matches that yields the minimum total distance. A computer given certain correspondence tasks and running Ullman's algorithm will perform the task. His results show that the problem can be solved locally, and insofar it is an important demonstration of possibility. To understand how biological visual systems really solve the problem, we need to discover experimentally whether global factors play a role in the system's perceptions. In the examples discussed in this section, "global" refers to broad regions of the visual field as opposed to "local," meaning very small regions such as the receptive fields of cells in the parafoveal region of V1 ($\sim 1°$) or V4 ($\sim 5°$).

1. *Bistable Quartets.* The displays shown in figure 2.3 are produced on a television screen in fast alternation—the first array of dots (A: coded as filled), then the second array of dots (B: coded as open), then A then B, as in a moving picture. The brain matches the two dots in A with dots in B, and subjects see the dots moving from A position to B position. Subjects see either horizontal movement or vertical movement; they do not see diagonal movement. The display is designed to be ambiguous, in that for any given A dot, there is both a horizontal B dot and also a vertical B dot, to which it could correspond. Although the probability is 0.5 of seeing any given A–B pair oscillating in a given direction, in fact observers always see the set of dots moving as a group—they all move vertically or all move horizontally (Ramachandran and Anstis, 1983). Normal observers do not see a mixture of some horizontal and some vertical movements. This phenomenon is an instance of the more general class of effects known as motion capture, and

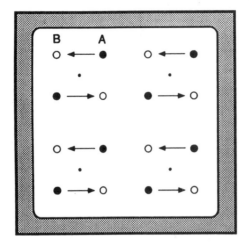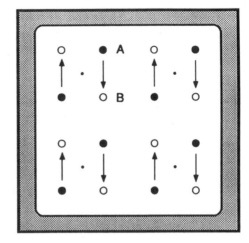

Figure 2.3 Bistable quartets. This figure shows that when the first array of dots (represented by filled circles, and indicated by A in the top left quartet) alternate with the second array of dots (represented by open circles, indicated by B in the top left quartet). Subjects see either all vertical or all horizontal oscillations. Normal observers do not see a mixture of some horizontal and some vertical movements, nor do they see diagonal movement. (Based on Ramachandran and Anstis 1983)

it strongly suggests that global considerations are relevant to the brain's strategy for dealing with the correspondence problem. Otherwise, one would expect to see, at least some of the time, a mix of horizontal and vertical movements.

2. *Behind the Occluder* (figure 2.4). Suppose both the A frame and the B frame contain a shaded square on the righthand side. Now, if all dots in the A group blink off and only the uppermost and lowermost dots of the B1 group blink on, subjects see all A dots, move to the B1 location, including the middle A dot, which is seen to move behind the "virtual" occluder. (It works just as well if the occluder occupies upper or lower positions.) If, however, A contains only one dot in the middle position on the left plus the occluding square on its right, when that single dot merely blinks off, subjects do not see the dot move behind the occluder. They see a square on the right and a blinking dot on the left. Because motion behind the occluder is seen in the context of surrounding subjective motion but not in the context of the single dot, this betokens the relevance of surrounding subjective motion to subjective motion of a single spot. Again, this suggests that the global properties of the scene are important in determining whether subjects see a moving dot or a stationary blinking light (Ramachandran and Anstis 1986).

3. *Cross-Modal Interactions.* Suppose the display consist of a single blinking dot and a shaded square (behind which the moving dot could "hide"). As before, A and B are alternately presented—first A (dot plus occluder), then B (occluder only), then A, then B. As noted above, the subject sees no motion (figure 2.4 III). Now, however, change conditions by adding an auditory stimulus presented by earphones. More exactly, the change is this:

Churchland, Ramachandran, and Sejnowski

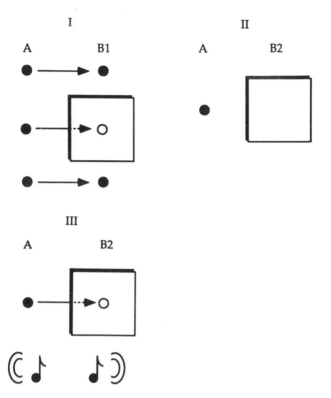

Figure 2.4 This figure shows the stimuli used to elicit the phenomenon of illusory motion behind an occluder. When the occluder is present, the subjects perceive all the dots move to the right, including the middle left dot, which is seen to move to the right and behind the square. In the absence of the occluder, the middle dot appears to move to the upper right. When the display is changed so that only the middle dot remains while upper and lower dots are removed, the middle dot is seen to merely blink off and on, but not to move behind the occluder. When, however, a tone is presented in the left ear simultaneously with the dot coming on, and in the right ear simultaneously with the dot going off, subjects do see the single dot move behind the occluder. (Based on Ramachandran and Anstis 1986)

Simultaneous with the blinking on of the light, a tone is sounded in the left ear; simultaneous with the blinking off, a tone is sounded in the right ear. With the addition of the auditory stimulus, subjects do indeed see the single dot move to the right behind the occluder. In effect, the sound "pulls" the dot in the direction in which the sound moves (Ramachandran, Intriligator, and Cavanaugh, unpublished observations). In this experiment, the cross-modal influence on what is seen is especially convincing evidence for some form of interactive vision as opposed to a pure, straight through, noninteractive hierarchy. (A weak subjective motion effect can be achieved when the blinking of the light is accompanied by somatosensory left–right vibration stimulation to the hands. Other variations on this condition could be tried.)

It comes as no surprise that visual and auditory information is integrated at some stage in neural processing. After all, we see dogs barking and drummers drumming. What is surprising in these results is that the

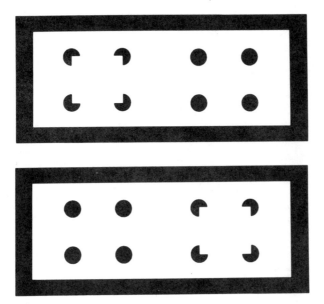

Figure 2.5 Two frames in an apparent motion display. The four Pacmen give rise to the perception of an occluding square that moves from the left circles to the right circles.

auditory stimulus has an effect on a process (motion correspondence) that pure vision orthodoxy considers "early." In this context it is appropriate to mention also influence in the other direction—of vision on hearing. Seeing the speaker's lips move has a significant effect on auditory perception and has been especially well documented in the McGurk effect.

4. *Motion Correspondence and the Role of Image Segmentation.* Figure 2.5 shows two frames of a movie in which the first frame has four Pacmen on the left, and the second has four Pacmen on the right. In the movie, the frames are alternated, and the disks are in perfect registration from one frame to the next. What observers report seeing is a foreground opaque square shifting left and right, occluding and revealing the four black disks in the background. Subjects never report seeing pacmen opening and closing their mouths; they never report seeing illusory squares flashing off and on. Moreover, when a template of this movie was then projected on a regular grid of dots, the dots inside the subjective square appeared to move with the illusory surface even though they were physically stationary (figure 2.6). "Outside" dots did not move (Ramachandran 1985).

These experiments imply that the human visual system does not always solve the correspondence problem independently of the segmentation problem (the problem of what features are parts belonging to the same thing), though pure visionaries tend to expect that solving segmentation is a late process that kicks in after the correspondence problem is solved. Subjects' overwhelming preference for the "occluding square" interpretation over the "yapping Pacmen" interpretation indicates that the solution to the

Churchland, Ramachandran, and Sejnowski

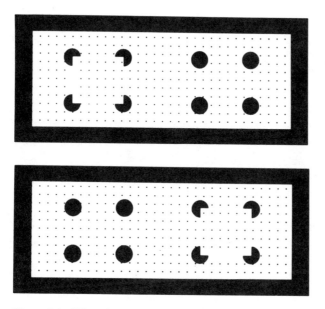

Figure 2.6 When dots are added to the background of figure 2.5, those dots internal to the occluding square appear to move with it when it occludes the right side circles. The background dots, however, appear stationary. (Based on Ramachandran 1985)

segmentation problem itself involves large-scale effects dominating over local constraints. If seeing motion in this experiment depended on solving the correspondence problem at the local level, then presumably yapping Pacmen would be seen. The experiment indicates that what are matched between frames are the larger scale and salient features; the smaller scale features are pulled along with the global decision.

Are the foregoing examples really significant? A poo-pooing strategy may downplay the effects as minor departures ("biology will be biology"). To be sure, a theory can always accommodate any given "anomaly" by making some corrective adjustment or other. Nevertheless, as anomalies accumulate, what passed as corrective adjustments may come to be deplored as ad hoc theory-savers. A phenomenon is an anomaly only relative to a background theory, and if the history of science teaches us anything, it is that one theory's anomaly is another theory's prototypical case. Thus "retrograde motion" of the planets was an anomaly for geocentric cosmologists but a typical instance for Galileo; the perihelion advance of Mercury was an anomaly for Newtonian physics, but a typical instance for Einsteinian physics. Any single anomaly on its own may not be enough to switch investment to a new theoretical framework. The cumulative effect of an assortment of anomalies, however, is another matter.

Can Semantic Categorization Affect Shape-from-Shading? Helmholtz observed that a hollow mask presented from the "inside" (the concave view, with the nose extending away from the observer) about 2 m from the observer is invariably perceived as a convex mask with features pro-truding (nose coming toward the observer). In more recent experiments, Ramachandran (1988) found that the concave mask continues to be seen as convex even when is it is illuminated from below, a condition that often suffices to reverse a perception of convexity to one of concavity. This remains true even when the subject is informed about the direction of illumination of the mask. Perceptual persistence of the convex mask as a concave mask shows a strong top-down effect on an allegedly early visual task, namely determining shape from shading.

Does this perceptual reversal of the hollow mask result from a generic assumption that many objects of interest (nuts, rocks, berries, fists, breasts) are usually convex or that faces in particular are typically convex? That is, does the categorization of the image as a face override the shading cues such that the reversal is a very strong effect? To address this question, Ramachandran, Gregory, and Maddock (unpublished observations) presented subjects with two masks: one is right side up and the other is upside down. Upside-down faces are often poorly analyzed with respect to features, and an upside-down mask may not be seen as having facial features at all. In any case, upright faces are what we normally encounter. In the experiment, subjects walk slowly backward away from the pair of stimuli, starting at 0.5 m, moving to 5.0 m. At a close distance of about 0.5 m subjects correctly see both inverted masks as inverted (concave). At about 1 m, subjects usually see the upright mask as convex; the upside-down mask, however, is still seen as concave until viewing distance is about 1.5–2.0 m, whereupon subjects tend to see it too as convex. The stimuli are identical save for orientation, yet one is seen as concave and the other as convex. Hence this experiment convincingly illustrates that an allegedly "later" process (face categorization) has an effect on an allegedly "earlier" process (the shading predicts thus and such curvatures) (figure 2.7).

Can Subjective Contours Affect Stereoscopic Depth Perception? Stereo vision has been cited (Poggio et al. 1985) as an early vision task, one that is accomplished by an autonomous module prior to solving segmentation and classification. That we can fuse Julesz random dot stereograms to see figures in depth is evidence for the idea that matching for stereo can be accomplished with matching of local features only, independently of global properties devolving from segmentation or categorization decisions. While the Julesz stereogram is indeed a stunning phenomenon, the correspondence problem it presents is entirely atypical of the correspondence problem in the real world. The logical point here should be spelled out: "Not always dependent on a" does not imply "Not standardly dependent on a," let alone, "Never dependent on a." Hence the question remains whether in typical real world conditions, stereo vision might in fact make use of top-down, global information. To determine whether under some conditions the segmentation data might be used in solving the correspondence problem, Ramachandran (1986) designed stereo pairs

Figure 2.7 The hollow mask, photographed from its concave orientation (as though you are about to put it on). In (*A*) the light comes from above; in (*B*) light comes from both sides.

where the feature that must be matched to see stereoptic depth is some high-level property. The choice was subjective contours, allegedly the result of "later" processing (figure 2.8).

In the monocular viewing condition, illusory contours can be seen in any of the four displays (above). The top pair can be stereoscopically fused so that one sees a striped square standing well in front of a background consisting of black circles on a striped mat. The bottom pair can also be stereoptically fused. Here one sees four holes in the striped foreground mat, and through the holes, well behind the striped mat, one sees a partially occluded striped square on a black background. These are especially surprising results, because the stripes of the perceived foreground and the perceived background are at zero disparity. The only disparity that exists on which the brain can base a stereo depth perception comes from the subjective contour.

According to pure vision orthodoxy, perceiving subjective contours is a "later" effect requiring global integration, in contrast to finding stereo correspondences for depth, which is considered an "earlier" effect. This result, however, appears to be an example of "later" influencing—in fact enabling—"earlier." It should also be emphasized that the emergence of qualitatively different percepts (lined square in front of disks versus lined square behind portholes) cannot be accounted for by any existing stereo algorithms that standardly predict a reversal in sign of perceived depth only

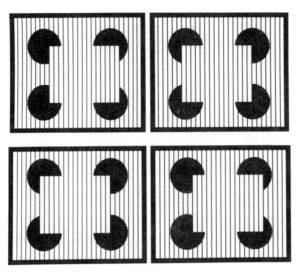

Figure 2.8 By fusing the upper stereo pairs, one sees a striped square standing well fore of a background consisting of black circles on a striped mat. By fusing the bottom pair, one sees four holes in the striped foreground mat, and through the holes, a partially occluded striped square on a black background. (This assumes fusion by divergence. The opposite order is available to those who fuse by convergence.) In both cases, the stripes are at zero disparity. (Based on Ramachandran 1986)

if the disparities are reversed. At the risk of repetition, we note again that in figure 2.8 (top and bottom), the lines are at zero disparity (Ramachandran 1986; Nakayama and Shimojo 1992).

Can Shape Recognition Affect Figure-Ground Relationships? Figure-ground identification is generally thought to precede shape recognition, but recent experiments using the Rubin vase/faces stimulus demonstrate that shape recognition can contribute to the identification of figure-ground (Peterson and Gibson, 1991).

Does the discovery of cells in V1 and V2 that respond to subjective contours (see below, p. 45) mean that detecting subjective contours is an early achievement after all? Not necessarily. The known physiological facts are consistent both with the "early effects" possibility as well as with a "later effect backsignaled" possibility. Further neurobiological and modeling experiments will help answer which possibility is realized in the nervous system.

Visual Attention

An hypothesis of interactive vision claims that the brain probably does not create and maintain a visual world representation that corresponds detail-by-detail to the visual world itself. For one thing, it need not, since the world itself is highly stable and conveniently "out there" to be sampled and resampled. On any given fixation, the brain can well make do with a partially elaborated representation of the world (O'Regan 1992; Ballard 1991; Dennett 1992). As O'Regan (1992) puts it, "the visual environment functions as a sort of outside memory store."

For another thing, as some data presented below suggest, the brain probably does not create and maintain a picture-perfect world representation. We conjecture that the undeniable feeling of having whole scene visual representation is the result mainly of (1) repeated visual visits to stimuli in the scene, (2) short-term semantic memory on the order of a few seconds that maintains the general sense of what is going on without creating and maintaining the point-by-point detail, (3) the brain's "objectification" of sensory perception such that a signal processed in cortex is represented as being about an object in space, i.e., feeling a burn on the hand, seeing a skunk in the grass, hearing a train approaching from the north, etc., and (4) the predictive dimension of pattern recognition, i.e., recognizing something as a burning log involves recognizing that it will burn my hand if I touch it, that smokey smells are produced, that water will quench the fire, that sand will smother it, that meat tastes better when browned on it, that the fire will go out after a while, and so on and on.

Evidence supporting the "partial-representation per glimpse" or "semi-world" hypothesis derives from research using on-line computer control to change what is visible on a computer display as a function of the subject's eye movements. When major display changes are made during saccades, those changes are rarely noticed, even when they involve bold alterations of color of whole objects, or when the changes consist in removal, shifting about, or addition of objects such as cars, hats, trees, and people (McConkie 1990). The exception is when the subject is explicitly paying attention to a certain feature, watching for a change.

Many careful studies using text-reading tasks elegantly support the "partial-representation per glimpse" hypothesis. These studies use a "moving window paradigm" in which subjects read a line of text that contains a window of normal text surrounded fore and aft by "junk" text. As readers move their eyes along the line, the window moves with the eyes (McConkie and Rayner 1975; Rayner et al. 1980; O'Regan 1990) (figure 2.9). The strategy is to discover the spatial extent of the zone from which useful information is extracted on a given fixation by varying the size of the window and testing using reading rate and comprehension measures. This zone is called the "perceptual" or "attentional" span. If at a given window width reading rate or comphrehension declines from a reader's baseline, it is presumed that surrounding junk text has affected reading, and hence that reader's attentional span is wider than the size of the window. By finding the smallest width at which reading is unaffected, a reader's attentional span can be quite precisely calibrated.

In typical subjects, reading text the size you are now reading, the attentional span is about 17–18 characters in width, and it is asymmetric about the point of fixation, with about 2–3 characters to the left of fixation and about 15 characters to the right. On the other hand, should you be reading

This sentence shows the nature of the perceptual span.
                        *

|_____|

xxxxxxxxxxx shows the nature xxxxxxxxxxxxxxxxxxxx
                        *

Maximum Perceptual Span

2-3 character spaces left (beginning of current word).

15 character spaces right (2 words beyond current word).

Figure 2.9   The attentional ("perceptual") span is defined as that zone from which useful information can be extracted on a given fixation. Fixation point is indicated by an asterisk. This displays the width of the attentional span and the asymmetry of the span (Courtesy John Henderson)

Hebrew instead of English, and hence traveling from page right to page left, the attention span will be about 2–3 characters to the right and 15 to the left (Pollatsek et al 1981), or reading Japanese, in which case it is asymmetric in the vertical dimension (Osaka and Oda 1991). This means that subjects read as well when junk text surround the 17–18 character span as when the whole line is visible, but read less well if the window is narrowed to 14 or 12 characters. At 17–18 character window width, the surrounding junk text is simply never noticed. Interestingly, it remains entirely unnoticed even when the reading subject is told that the moving window paradigm is running (McConkie 1979; O'Regan 1990; Henderson 1992).

Further experiments using this paradigm indicate that a shift in visual attention precedes saccadic eye movement to a particular location, presumably guiding it to a location that low-level analysis deems the next pretty good landing spot (Henderson et al. 1989). Henderson (1993) proposes that visual attention binds; inter alia, it binds the visual stimulus to a spatial location to enable a visuo-motor representation that guides the next motor response. When the fovea has landed, some features are seen.

Experiments along very different lines suggest that the information capacity of attention per glimpse is too small to contain a richly detailed whole-scene icon. Verghese and Pelli (1992) report results concerning the amount of information an observer's attention can handle. Based on their results, they conclude that the capacity of the attention mechanism is limited to about 44 ± 15 bits per glimpse. Preattentive mechanisms (studied by Treisman and by Julesz) presumably operate first, and operate in parallel. Verghese and Pelli calculated that the preattentive information capacity is much greater—about 2106 bits. The attentional mechanism, in contrast to the preattentive mechanism, they believe to be low capacity. (Verghese and Pelli define a preattentive task as "one in which the probability of detecting the target is independent of the number of distracter elements" and an attentive task as one in which "the probability of detecting the target is

Churchland, Ramachandran, and Sejnowski

inversely proportional to the number of elements in the display" [p. 983].)

Verghese and Pelli ran two subjects on a number of attention tasks of varying difficulty, and compared results across tasks. In a paradigm they call "finding the dead fly," subjects are required to detect the single stationary spot among moving spots. The complementary task of finding the live fly—the moving object among stationary objects—is a preattentive task in which the target "pops out." They note that their calculation of 44 ± 15 bits is consistent with Sperling's (1960) estimate of 40 bits for the iconic store. In Sperling's technique, an array of letters was flashed to the observer. He found that subjects could report only part of the display, roughly 9 letters (= 41 bits).

There are important dependencies between visual attention, visual perception, and iconic memory. To a first approximation: (1) if you are not visually attending to a then you do not see a (have a visual experience of a), and (2) if you are not attending to a and you do not have a visual experience of a, then you do not have iconic memory for a. Given the limited capacity of visual attention, these assumptions imply that the informational capacity of visual perception (in the rough and ready sense of "literally seeing") is approximately as small (see also Crick and Koch 1990b)

Nevertheless, some motor behavior—and goal-directed eye movement in particular—apparently does not require conscious perception of the item to which the movement is directed, but does require some attentional scanning and some parafoveal signals that presumably provide coarse, easy to extract visual cues. During reading a saccade often "lands" the fovea near the third letter of the word (close enough to the "optimal" viewing position of the word), and small correction saccades are made when this is not satisfactory. This implies that the eyes are aiming at a target, and hence that at least crude visual processing has guided the saccade (McConkie et al. 1988; Rayner et al. 1983; Kapoula 1984).

In concluding this section, we emphatically note that what we have discussed here is only a small part of the story since, as Schall (1991) points out, orienting to a stimulus often involves more than eye movements. It often also involves head and whole body movements.

Considerations from Neuroanatomy

The received wisdom concerning visual processing envisages information flows from stage to stage in the hierarchy until it reaches the highest stage, at which point the brain has a fully elaborated world model, ready for motor consideration. In this section, we shall draw attention to, though not fully discuss, some connectivity that is consistent with a loose, interactive hierarchy but casts doubt on the notion of a strict hierarchy. We do of course acknowledge that so far these data provide only suggestive signs that the interactive framework is preferable. (For related ideas based on back-projection data in the context of neuropsychological data, see Damasio 1989b and Van Hoesen 1993.)

Backprojections (Corticocortical) Typically in monkeys, forward axon projections (from regions closer in synaptic distance to the sensory periphery to regions more synaptically distant; e.g., V2 to V4) are equivalent to or outnumbered by projections back (Rockland and Pandya 1979; Rockland and Virga 1989; Van Essen and Maunsell 1983; Van Essen and Anderson 1990). The reciprocity of many of these projections (P to Q and Q to P) has been documented in many areas, including connections back to the LGN (figure 2.10). It has begun to emerge that some backprojections, however, are not merely reciprocating feedforward connections, but appear to be widely distributed, including distribution to some areas from which they do not receive projections. Thus Rockland reports (1992a,b) injection data showing that some axons from area TE do indeed project reciprocally to V4, but sparser projections were also seen to V2 (mostly layer 1, but some in 2 and 5) and V1 (layer 1). These TE axons originated mainly in layers 6 and 3a (figure 2.11).

Diffuse Ascending Systems In addition to the inputs that pass through the thalamus to the cortex, there are a number of afferent systems that arise in small nuclei located in the brainstem and basal forebrain. These systems include the locus coeruleus, whose noradrenergic axons course widely throughout the cortical mantle, the serotonergic raphe nuclei, the ventral tegmental area, which sends dopamine projections to the frontal cortex, and cholinergic inputs emanating from various nuclei, including the nucleus basalis of Meynert. These systems are important for arousal, for they control the transition from sleep to wakefulness. They also provide the cortex with information about the reward value (dopamine) and salience (noradrenaline) of sensory stimuli. Another cortical input arises from the amygdala, which conveys information about the affective value of sensory stimuli to the cortex, primarily to the upper layers. Possible computational utility for these diffuse ascending system will be presented later.

Corticothalamic Connections Sensory inputs from the specific modalities project from the thalamus to the middle layers (mainly layer 4) of the cortex. Reciprocal connections from each cortical area, mainly originating in deep layers, project back to the thalamus. In visual cortex of the cat it is

Figure 2.10 (*Top*) Schematic diagram of some of the cortical visual areas and their connections in the macaque monkey. Solid lines indicate projections involving all portions of the visual field representation in an area; dotted lines indicate projections limited to the representation of the peripheral field. Heavy arrowheads indicate forward projections; light arrowheads indicate backward projections. (Reprinted with permission from Desimone and Ungerleider 1989) (*Bottom*) Laminar patterns of cortical connectivity used for making "forward" and "backward" assignments. Three characteristic patterns of termination are indicated in the central column. These include preferential termination in layer 4 (the F pattern), a columnar (C) pattern involving approximately equal density of termination in all layers, and a multilaminar (M) pattern that preferentially avoids layer 4. There are also characteristic patterns for cells of origin in different pathways. Filled ovals, cells bodies; angles, axon terminals. (Reprinted with permission from Felleman and Van Essen 1991)

Churchland, Ramachandran, and Sejnowski

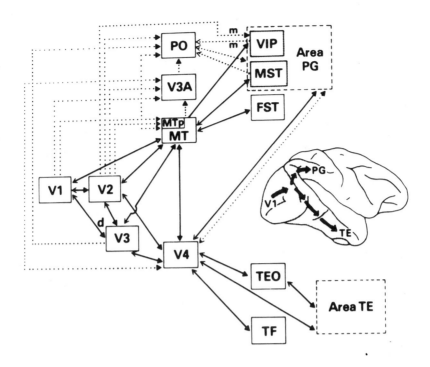

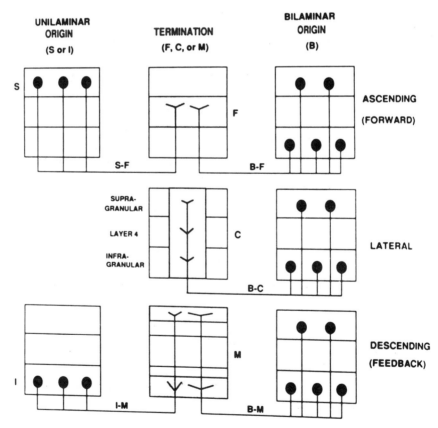

UNILAMINAR ORIGIN (S or I)

TERMINATION (F, C, or M)

BILAMINAR ORIGIN (B)

ASCENDING (FORWARD)

LATERAL

DESCENDING (FEEDBACK)

SUPRA-GRANULAR

LAYER 4

INFRA-GRANULAR

known that the V1 projections back to the LGN of the thalamus outnumber thalamocortical projections by about 10:1.

Corticofugal projections have collaterals in the reticular nucleus of the thalamus. The reticular nucleus of the thalamus is a sheet of inhibitory neurons, reminiscent of the skin of a peach. Both corticothalamic axons as well as thalamocortical projection neurons have excitatory connections on these inhibitory neurons whose output is primarily back to the thalamus. The precise function of the reticular nucleus remains to be discovered, but it does have a central role in organizing sleep rhythms, such as spindling and delta waves in deep sleep (Steriade et al. 1993b).

Connections from Visual Cortical Areas to Motor Structures Twenty-five cortical areas (cat) project to the superior colliculus (SC) (Harting et al. 1992). These include areas 17, 18, 19, 20a, 20b, 21a, and 21b. Harting et al. (1992) found that the corticotectal projection areas 17 and 18 terminate exclusively in the superficial layers, while the remaining 23 areas terminate more promiscuously (figure 2.12). The SC has an important role in directing saccadic eye movements, and, in animals with orientable ears, ear movements.

Nearly every area of mammalian cortex has some projections to the striatum, with some topological preservation. Although the functions of the striatum are not well understood, the correlation between striatal lesions and severe motor impairments is well known, and it is likely that the striatum has an important role in integrating sequences of movements. Lesion studies also indicate that some parts of the striatum are relevant to producing voluntary eye movements, as opposed to sensory-driven or reflex eye movements. It appears that the striatum can veto some reflexive responses via an inhibitory effect on motor structures, whereas voluntary movements are facilitated by disinhibitory striatal output to motor structures.

What is frustrating about this assembly of data, as with neuroanatomy generally, is that we do not really know what it all means. The number of neurons and connections is bewildering, and the significance of projections to one place or another, of distinct cell populations, and so on, is typically puzzling. (See Young 1992 for a useful startegy for clarifying the significance.) Neuroanatomy is, nonetheless, the observational hard-

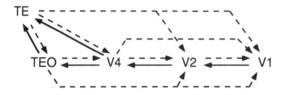

Figure 2.11 Schematic diagram of the feedforward connections (solid lines) and backprojections (broken lines) in the monkey. What is especially striking is that fibers from visual cortical areas TE (inferior temporal cortex) and TEO (posterior to TE and anterior to V4) project all the way back to V2 and V1. (Based on Rockland et al. 1992.)

Churchland, Ramachandran, and Sejnowski

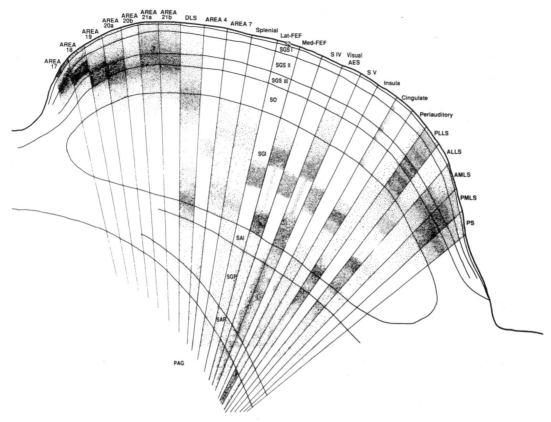

Figure 2.12 Summary diagram in the sagittal plane of the superior colliculus (SC) showing the laminar and sublaminar distribution of axons from cortical areas to the SC in the cat, as labeled above each sector. (Reprinted with permission from Harting et al. 1992)

pan for neuroscience, and the data can be provocative even when they are not self-explanatory. The prevalence and systematic character of feedback loops are particularly provocative, at least because such loops signify that the system is dynamic—that it has time-dependent properties. Output loops back to affect new inputs, and it is possible for a higher areas to affect inputs of lower areas. The time delays will matter enormously in determining what capacities the system display.

The second point is that all cortical visual areas, from the lowest to the highest, have numerous projections to lower brain centers, including motor-relevant areas such as the striatum, superior colliculus, and cerebellum. The anatomy is consistent with the idea that motor assembly can begin even before sensory signals reach the highest levels. Especially for skilled actions performed in a familiar context, such as reading aloud, shooting a basket, and hunting prey, this seems reasonable. Are the only movements at issue here eye movements? Probably not. Distinguishing gaze-related movements from extra-gaze movements is anything but straightforward, for the eyes are in the head, and the head is attached to the rest of the body. Foveating an object, for example, may well involve movement of the eyes, head, and neck—and on occasion, the entire body. Watching Michael Jordan play basketball or a group of ravens steal a caribou corpse from a wolf tends to underscore the integrated, whole-body character of visuomotor coordination.

Considerations from Neurophysiology

In keeping with the foregoing section, this section is suggestive rather than definitive. It is also a bit of a fact salad, since at this stage the evidence does not fit together into a tight story of how interactive vision works. Such unity as does exist is the result of to the data's constituting evidence for various interactions between so-called "higher" and "lower" stages of the visual system, and between the visual and nonvisual systems. (see also Goldman-Rakic 1988; Van Hoesen 1993.)

Connections from Motor Structures to Visual Cortex   Belying the assumption that the representation of the visual scene is innocent of nonvisual information, certain physiological data show interactive effects even at very early stages of visual cortex. For example, the spontaneous activity of V1 neurons is suppressed according to the onset time of saccades. The suppression begins about 20–30 msec after the saccade is initiated, and lasts about 200 msec (Duffy and Burchfield 1975). The suppression can be accomplished only by using oculomotor signals, perhaps efference copy, and hence this effect supports the interactive hypothesis. Neurons sensitive to eye position have been found in the LGN (Lal and Friedlander 1989), visual cortical area V1 (Trotter et al. 1992; Weyand and Malpeli 1989), and V3 (Galleti and Battaglini 1989). Given the existence and causal efficacy of various nonvisual V1 signals, Pouget et al. (1993) hypothesized that visual features are encoded in egocentric (spatiotopic) coordinates at early stages of visual processing, and that eye-position information is used in computing where in egocentric space the stimulus is located. Their network model demonstrates the feasibility of such a computation when the network takes as input both retinal and eye-position signals.

Consider also that a few V1 cells and a higher percentage of V2 cells show an enhanced response to a target to which a saccade is about to be made (Wurtz and Mohler 1976). Again, these data indicate some influence of motor system signals, specifically motor planning signals, on cells in early visual processing. As further evidence, note that some neurons in V3A show variable response as a function of the angle of gaze; response was enhanced when gaze was directed to the contralateral hemifield (Galletti and Battaglini 1989).

Inferior Parietal Cortex and Eye Position   Caudal inferior parietal cortex (IPL) has two major subdivisions: LIP and 7a (figure 2.13). LIP is directly connected to the superior colliculus, the frontal eye fields. Area 7a has a different connectivity: mainly polymodal cortex, limbic, and some prestriate.

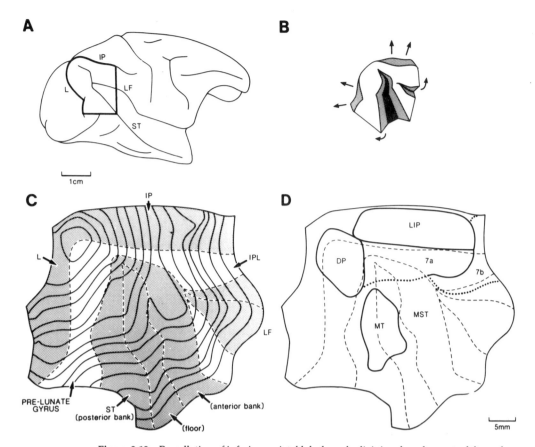

Figure 2.13 Parcellation of inferior parietal lobule and adjoining dorsal aspect of the prelunate gyrus. The cortical areas are represented on flattened reconstructions of the cortex. (*A*) Lateral view of the monkey hemisphere. The darker line indicates the area to be flattened. (*B*) The same cortex isolated from the rest of the brain. The stippled areas are cortex buried in sulci, and the blackened area is the floor of the superior temporal sulcus. The arrows indicate movement of local cortical regions resulting from mechanical flattening. (*C*) The completely flattened representation of the same area. The stippled areas represent cortical regions buried in sulci and the contourlike lines are tracings of layer IV taken from frontal sections through this area. (*D*) Locations of several of the cortical fields. The dotted lines indicate borders of cortical fields that are not precisely determinable. IPL, inferior parietal lobule; IPS, intraparietal sulcus; LIP, lateral intraparietal region; STS, superior temporal sulcus. (Reprinted with permission from Andersen 1987).

LIP responses are correlated with execution of saccadic eye movements; area 7a cells respond to a stimulus at a certain retinal location, but modulated by the position of the eye in the head (Zipser and Andersen 1988). Hardy and Lynch (1992) report that both LIP and area 7a receive the majority of their thalamic inputs from distinct patches in the medial pulvinar nucleus (figure 2.14).

Illusory Contours and Figure Ground In 1984 von der Heydt et al. reported that neurons in visual area V2 of the macaque will respond to il-

case 1

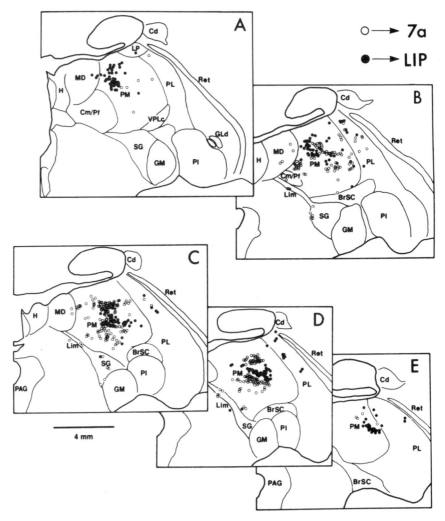

Figure 2.14 Diagram of sections of the thalamus showing the distribution of retrogradely labeled neurons within the thalamus resulting from inferior parietal lobule (IPL) injections in parietal areas 7a (open circles) and lateral intraparietal region (LIP) (filled circles). A single injection (1.2 μ) of the fluorescent dye DY (diamidino-dihydrochloride yellow) was made in 7a and a single injection (0.5 μ) of fast blue in LIP. Each individual symbol denotes a singled labeled neuron. The densest labeling is in medial pulvinar nucleus of the thalamus (PM), with LIP and 7a showing a distinct projection pattern. BrSC, brachium of the superior colliculus; Cd, caudate nucleus; Cm/Pf, centromedian and parafascicular nuclei; GM, medial geniculate nucleus, pars parvicellularis; H, habenula; Lim, nucleus limitans; MD, mediodorsal nucleus; PAG, periaqueductal gray; PL, lateral pulvinar nucleus; Ret, reticular nucleus. (Reprinted with permission from Hardy and Lynch 1992)

lusory contours. More recently, Grosof et al. (1992) report that some orientation selective cells in V1 respond to a class of illusory contours. As noted earlier, these data are consistent with the possibility that low-level response depends on higher level operations whose results are backprojected to lower levels. This is lent plausibility by the facts that detection of

Churchland, Ramachandran, and Sejnowski

illusory contours will depend on previous operations involving interpolation across a span of the visual field.

Neurons in area MT respond selectively to direction of motion but not to wavelength. Nonetheless, color can have a major effect on how these cells respond by virtue of how the visual stimulus is segmented (Dobkins and Albright 1993).

Cross-Modal Interactions   The responses of cells in V4 to a visual stimulus can be modified by somatosensory stimuli (Maunsell et al. 1991). Fuster (1990) has shown similar task-dependent modifications for cells in somatosensory cortex, area S1.

Dynamic Mapping in Exotropia

In this section we discuss an ophthalmic phenomenon observed in human subjects. Conventionally, this is a truly surprising phenomenon, and it seems to demonstrate that processing as early as V1 can be influenced by top-down factors. The phenomenon has not been well studied, to say the least, and much more investigation is required. Nevertheless, we mention it here partly because it is intriguing, but mainly because if the description below is accurate, then we must rethink the Pure Vision's conventional assumptions about the Receptive Field.

Exotropia is a form of squint in which both eyes are used when fixated on small objects close by (e.g., 12 in from the nose) but when looking at distant objects, the *squinting* eye deviates outward by as much as 45° to 60°. Curiously, the patient does not experience double vision—the deviating eye's image is usually assumed to be *suppressed*. It is not clear, however, at what stage of visual processing the suppression occurs.

Ophthalmologists have claimed that, contrary to expectations, in a small subset of these patients, *fusion* occurs not only during inspection of near objects, but also when the squinting eye deviates. This phenomenon, called *anomalous retinal correspondence* or ARC, has been reported frequently in the literature of ophthalmology and orthoptics. The accuracy of the reports, and hence the existence of ARC, has not always been taken seriously, since ARC implies a rather breathtaking lability of receptive fields. Clinicians and physiologists raised in the Hubel–Wiesel tradition usually take it as basic background fact that (1) binocular connections are largely established in area 17 in early infancy and that (2) binocular *fusion* is based exclusively on anatomical correspondence of inputs in area 17. For instance, if a squint is surgically induced in a kitten or an infant monkey, area 17 displays a complete loss of binocular cells (and two populations of monocular cells) but the maps of the two eyes never change. No apparent compensation such as *anomalous correspondence* has been observed in area 17 and this has given rise to the conviction that it is highly improbable that an ARC phenomenon truly exists.

On the possibility that there might be more to the ARC reports, Ramachandran, Cobb, and Valente (unpublished) recently studied two patients who had intermittent exotropia. These patients appeared to fuse images both during near vision and during far vision—when the left eye deviated outward—a condition called *intermittent exotropia with anomalous correspondence*. To determine whether the patients do indeed have two (or more) separate binocular *maps* of the world, Ramachandran, Cobb, and Valenti devised a procedure that queried the alignment of the subject's afterimages where the afterimage for the right eye was generated independently of the afterimage for the left eye. Here is the procedure: (1) The subject (with squint) was asked to shut one eye and to fixate on the bottom of a vertical slit-shaped window mounted on a flashgun. A flash was delivered to generate a vivid monocular afterimage of the slit. The subject was then asked to shut this eye and view the top of the slit with the other eye (and a second flash was delivered). (2) The subject opened both eyes and viewed a dark screen, which provided a uniform background for the two afterimages.

The results were as follows: (1) The subject (with squint) reported seeing afterimages of the two slits that were perfectly lined up with each other, so long as the subject was deliberately verging within about arm's length. (2) On the other hand, if the subject relaxed vergence and looked at a distant wall (such that the left eye deviated), the upper slit (from the anomalous eye) vividly appeared to move continuously outward so that the two slits became misaligned by several degrees. Then this experiment was repeated on two normal control subjects and it was found that no misalignment of the slits occurred for any ordinary vergence or conjugate eye movements. Nor could misalignments of the slits be produced by passively displacing one eyeball in the normal individuals to mimic the exotropia. It appears that eye position signals from the deviating eye selectively influence the egocentric localization of points for that eye alone.

In the next experiment, a light point was flashed for 150 msec either to the right eye alone or the left eye alone; the subject's task was merely to point to the location of the light point. Subjects became quite skilled at deviating their anomalous eye by between about $1°$ and $40°$, and the afterimage alignment technique could be used to calibrate the deviation. Tests were made with deviations between $1°$ and $15°$. It was found that regardless of the degree of deviation of the anomalous eye, and regardless of which eye was stimulated, subjects made only marginally more errors than normal subjects in locating the light point. Is the remapping sufficiently fine-grained to support stereopsis? Testing for accuracy of stereoptic judgments using ordinary stereograms under conditions of anomalous eye deviations between $1°$ and $12°$, Ramachandran, Cobb, and Valenti found that disparities as small as 20 min of arc could be perceived correctly even though the anomalous eye deviated by as much as $12°$. Even when the half-images of the two eyes were exciting noncorresponding retinal points separated by $12°$, very small retinal disparities could be detected.

Ramachandran and his colleagues have dubbed this phenomenon *dynamic anomalous correspondence*. Their results suggest that something in the ARC reports is genuine, with a number of implications.

First, binocular correspondence can change continuously in *real time* in a single individual depending on the degree of exotropia. Hence, binocular correspondence (and *fusion*) cannot be based exclusively on the anatomical convergence of inputs in area 17. The relative displacement observed between the two afterimages also implies that the *local sign* of retinal points (and therefore binocular correspondence) must be continuously updated as the eye deviates outward.

Second, since the two slits would always be *lined up* as far as area 17 is concerned, the observed misalignment implies that feedback (or feedforward) signals from the deviating eye must somehow be extracted separately for each eye and must then influence the egocentric location of points selectively for that eye alone. This is a somewhat surprising result, for it implies that time *remapping* of egocentric space must be done very early—before the *eye of origin* label is lost—i.e., before the cells become completely binocular. Since most cells anterior to area 18 (e.g., MT or V4) are symmetrically binocular we may conclude that the correction must involve interaction between reafference signals and the output of cells as early as 17 or 18.

Nothing in the psychophysical results suggests what the mechanism might be by which these interactions occur. Whatever the ultimate explanation, however, the results do imply that even as simple a perceptual process as the localization of an object in X/Y coordinates is not strictly and absolutely a *bottom-up* process. Even the output of early visual elements—in this case the monocular cells of area 17—can be strongly modulated by back projections from eye movement command centers.

If indeed a complete remapping of perceptual space can occur selectively for one eye's image simply in the interest of preserving binocular correspondence, this is a rather remarkable phenomenon. It would be interesting to see if this remapping process can be achieved by algorithms of the type proposed by Zipser and Andersen (1988) for parietal neurons or by *shifter-circuits* of the kind proposed by Anderson and Van Essen (1987; see also chapter 13).

COMPUTATIONAL ADVANTAGES OF INTERACTIVE VISION

So far we have discussed various empirical data that lend some credibility to an interactive-vision approach. But the further question is this: Does it make sense computationally for a nervous system to have an interactive style rather than a hierarchical, modular, modality-pure, and motorically unadulterated organization? In this section, we briefly note four reasons, based on the computational capacities of neural net models, why evolution might have selected the interactive modus operandi in nervous systems. As more computer models in the interactive style are developed and explored, additional factors, for or against, may emerge. The results from neural net models also suggest experiments that could be run on real nervous systems to reveal whether they are in fact computationally interactive.

Figure-Ground Segmentation and Recognition Are More Efficiently Achieved in Tandem Than Strictly Sequentially Segmentation is a difficult task, especially when there are many objects in a scene partially occluding one another. The problem is essentially that global information is needed to make decisions at the local level concerning what goes with what. At lower levels of processing such as V1, however, the receptive fields are relatively small and it is not possible locally to decide which pieces of the image belong together. If lower levels can use information that is available at higher levels, such as representation of whole objects, then feedback connections could be used to help tune lower levels of processing. This may sound like a chicken-and-egg proposal, for how can you recognize an object before you segment it from its background? Just as the right answer to the problem "where does the egg come from" is "an earlier kind of chicken," so here the the answer is "use partial segmentation to help recognize, and use partial recognition to help segment." Indeed, interactive segmentation–recognition may enable solutions that would otherwise be unreachable in short times by pure bottom-up processing.

It is worth considering the performance of machine reading of numerals. The best of the "pure vision" configured machines can read numerals on credit card forms only about 60% of the time. They do this well only because the sales slip "exactifies" the data: numerals must be written in blue boxes. This serves to separate the numerals, guarantee an exact location, and narrowly limit the size. Carver Mead (in conversation) has pointed out that the problem of efficient machine reading of zip codes is essentially unsolved, because the preprocessing regimentations for numeral entry on sales slips do not exist in the mail world. Here the machine readers have to face the localization problem (where are the numerals and in what order?) and the segmentation problem (what does a squiggle belong to?) as well as the recognition problem (is it a 0 or a 6?).

Conventional machines typically serialize the problem, addressing first the segmentation problem and then, after that is accomplished, addressing the recognition problem. Should the machine missolve or fail to solve the first, the second is doomed. In the absence of strict standardization of location, font, size, relation to other numerals, relation of zip code to other lines, and so forth, classical machines regularly fumble the segmentation problem. Unlike engineers working with the strictly serial problem-design, Carver Mead and Federico Faggin (in conversation) have found that if networks can address segmentation and recognition in parallel, they well outperform their serial competitors.

The processing of visual motion is another example of how segregation may proceed in parallel with visual integration. Consider the problem of trying to track a bird flying through branches of a tree; at any moment only parts of the bird are visible through the occluding foliage, which may itself be moving. The problem is to identify fleeting parts of the bird that may be combined to estimate the average velocity of the bird and to keep this information separate from information about the tree. This is a global problem in that no small patch of the visual field contains enough information to unambiguously solve the segregation problem. However, area MT of the primate visual cortex has neurons that seem to have "solved" this segregation problem. A recent model of area MT that includes two parallel streams, one that selects regions of the visual field that contain reliable motion information, and another that integrates information from that region, exhibits properties similar to those observed in area MT neurons (Nowlan and Sejnowski 1993). This model demonstrates that segmentation and integration can to some extent be performed in parallel at early stages of visual processing.

It would not be surprising if evolution found the interactive strategy good for brains. So long as the segmentation problem is partially solved, a good answer can be dumped out of the visual "pipeline" very quickly. When, however, the task is more difficult, iterations and feedback may be essential to drumming up an adequate solution. To speed up processing in the difficult cases—which will be the rule, not the exception, in real-world vision—the system may avail itself of learning. If, after frequent encounters, the brain learns that certain patterns typically go together, thereafter the number of iterations needed to find an adequate solution is reduced (Sejnowski 1986). Humans probably "overlearn" letter and word patterns, and hence seasoned readers are faster and more accurate than novice readers. Even when text is degraded or partially occluded, a good reader may hardly stumble.

Movement (of Eye, Head, Body) Makes Many Visual Computations Simpler A number of reasons support this point. First, the smooth pursuit system for tracking slowly moving objects supports image stability on the retina, simplifying the tasks of analyzing and recognizing. Second, head movement during eye fixation yields cues useful in the task of separating figure from ground and distinguishing one object from another. Motion parallax (the relative displacement of objects caused by change in observer position) is perhaps the most powerful cue to the relative depth of objects (closer objects have greater relative motion than more distant objects), and it continues to be critical for relative depth judgments even beyond about 10 m from the observer, where stereopsis fades out. Head bobbing is common behavior in animals, and a visual system that integrates across several glimpses to estimate depth has computational savings over one that tries to calculate depth from a single snapshot.

Another important cue is optical flow (figure 2.15). When an animal is running, flying, or swimming, for example, the speed of an image moving radially on the retina is related to the distance of the object from the observer.[4] This information allows the system to figure out how fast it is closing in on a chased object, as well as how fast a chasing object is closing

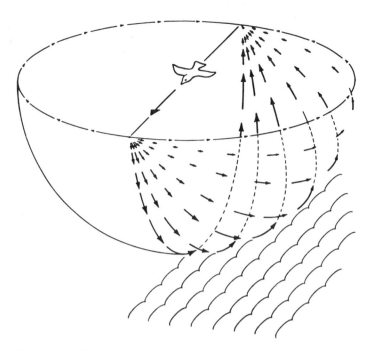

Figure 2.15 Optical flow represented by a vector field around a flying bird provides information about self-movement through the environment. (From Gibson 1966)

in. Notice that any of these movements (eyeball, head, and whole body) on its own means computational economies. In combination, the economies compound.

There are many more example of how the self-generated movement can provide solutions to otherwise intractable problems in vision (Ballard 1991; Blake and Yuille 1992).

The Self-Organization of Model Visual Systems during "Development" Is Enhanced by Eye-Position Signals An additional advantage of interactive vision is its role in the construction of vision systems. Researchers in computer vision often reckon—and bemoan—the cost of "hand" building vision systems, but rarely consider the possibility of growing a visual system. Nature, of course, uses the growing strategy, and relies on genetic instructions to create neurons with the right set of components. In addition, interactions between neurons as well as interactions between the world and neurons, are critical in getting networks of neurons properly wired up. Understanding the development of the brain is perhaps as challenging a problem as that of understanding the function of the brain, but we are beginning to figure out some of the relevant factors, such as position cues, timing of gene expression, and activity-dependent modifications. Genetic programming has been explored as an approach to solving some construction problems, but value of development as an an intermediary between genes and phenotype is only beginning to be appreciated in the computational community (Belew 1993).

Churchland, Ramachandran, and Sejnowski

Most activity-dependent models of development are based on the Hebb rule for synaptic plasticity, according to which the synapse strengthens when the presynaptic activity is correlated with the postsynaptic activity (Sejnowski and Tesauro 1989). Correlation-based models for self-organization of primary visual cortex during development have shown that some properties of cortical cells, such as ocularity, orientation, and disparity, can emerge from simple Hebbian mechanisms for synaptic plasticity (Swindale 1990; Linsker 1986; Miller and Stryker 1990; Berns et al. 1993). Hebbian schemes are typically limited in their computational power to finding the principal components in the input correlations. It has been difficult to extend this approach to a hierarchy of increasingly higher-order response properties, as found in the extrastriate areas of visual cortex. One new approach is based on the observation that development takes place in stages. There are critical periods during which synapses are particularly plastic (Rauschecker 1991), and there are major milestones, such as eye opening, that change the nature of the input correlations (Berns et al. 1993).

Nature exploits additional mechanisms in the developing brain to help organize visual pathways. One important class of mechanisms is based on the interaction between self-generated actions and perception, along the lines already discussed in the previous section. Eye movement information in the visual cortex during development, when combined with Hebbian plasticity, may be capable of extracting higher-order correlations from complex visual inputs. The correlation between eye movements and changes in the image contains information about important visual properties. For example, correlation of saccadic eye movements with the response of a neuron can be used in a Hebbian framework to develop neurons that respond to the direction of motion. At still higher levels of processing, eye movement signals that direct saccades to salient objects can be used as a reward signal to build up representations of significant objects (Montague et al. 1993). This new view provides eye-movement signals with an important function in visual cortex both during development and in the adult.

The plasticity of the visual cortex during the critical period is modulated by inputs from subcortical structures that project diffusely throughout the cortex (Rauschecker 1991). The neurotransmitters used by these systems often diffuse from the release sites and act at receptors on neurons some distance away. These diffuse ascending systems to the cerebral cortex that are used during development to help wire up the brain are also used in the adult for signaling reward and salience. The information carried by neurons in these systems is rather limited: there are relatively few neurons compared to the number in the cortex, they have a low basal firing rate, and changes in their firing rates occur slowly. This is, however, just the sort of information that could be used to organize and regulate information storage throughout the brain, as shown below.

Interactive Perception Simplifies the Learning Problem A difficulty facing conventional reinforcement learning is this: Assuming the brain creates and maintains a picture-perfect visual scene at each moment, how does the brain determine which, among the many features and objects it recently perceived, are the ones relevant to the reward or punishment? An experienced animal will have a pretty good idea, but how does its experience get it to that stage? How does the naive brain determine which "stimulus" in the richly detailed stimulus array gets the main credit when a certain response brings a reward? How does the brain know what synapses to strengthen?

This "relevance problem" is even more vexing as there are increases in the time delay between the stimulus and the reinforcement. For then the stimulus array develops over time, getting richer and richer as time passes. The correlative problem of knowing which movement among many movements made was the relevant one is likewise increasingly difficult as the delay increases between the onset of various movements and the rewarding or punishing outcome.[5] These questions involve considerations that go well beyond the visual system, and include parts of the brain that evaluate sensory inputs.

Suppose that evolution has wired the brain to bias attention as a function of how the species makes its living, and that the neonate is tuned to attend to some basic survival-relevant properties. The evolutionary point legitimates the assumption that an attended feature of the stimulus scene is more likely to be causally implicated in producing the conditions for the reward, and assuming that the items in iconic and working memory are, by and large, items previously attended to, then the number of candidate representations to canvas as "relevant" is far smaller than those embellishing a rich-replica visual world representation. Granted all these assumptions, the credit assignment problem is far more manageable here than in the pure vision theoretical framework (Ballard 1991).

By narrowing the number of visuomotor trajectories that count as salient, attention can bias the choice of synapses strengthen. Selective strengthening of synapses of certain visuomotor representations "spotlit" by attention is a kind of hypothesis the network makes. It is, moreover, an hypothesis the network tests by repeating the visuomotor trajectory. Initially the network will shift attention more or less randomly, save for guidance from startle responses and other reflexive behavior. Given that attention downsizes the options, and that the organism can repeatedly explore the various options, the system learns to direct attention to visual targets that it has learned are "good bets" in the survival game. This, in turn, contributes to further simplifying the learning problem in the future, for on the next encounter, attention will more likely be paid to relevant features than to irrelevant features, and the connections can be up-regulated or down-regulated as a result of reward or lack of same (the above points are from Whitehead and Ballard 1990, 1991; see also Grossberg 1987).

Churchland, Ramachandran, and Sejnowski

LEARNING TO SEE

A robust property of animal learning is that responses reinforced by a reward are likely to be produced again when relevantly similar conditions obtain. This is the starting point for behavioral studies of operant conditioning in psychology (Rescorla and Wagner 1972; Mackintosh 1974), neuroscientific inquiry into the reward systems of the brain (Wise 1982) and engineering exploration of the principles and applications of reinforcement learning theory (Sutton and Barto 1981, 1987). Both neuroscience and computer engineering draw on the vast and informative psychological literature describing the various aspects of reinforcement learning, including such phenomena as blocking, extinction, intermittent versus constant reward, cue ranking, how time is linked to other cues, and so forth. The overarching aim is that the three domains of experimentation will link up and yield a unified account of the scope and limits of the capacity and of its underlying mechanisms (see Whitehead and Ballard 1990; Montague et al. 1994).

Detailed observations of animal foraging patterns under well quantified conditions indicate that animals can display remarkably sophisticated adaptive behavior. For example, birds and bees quickly adopt the most efficient foraging pattern in "two-armed bandit" conditions (a: high-payoff when a "hit" and "hits" are infrequent; b: low payoff when a "hit" and "hits" are frequent) (see Krebs et al. 1978; Gould 1984; Real et al. 1990; Real 1991). Cliff-dwelling rooks learn to bombard nest-marauders with pebbles (Griffin 1984). A bear learns that a bluff of leafy trees in a hill otherwise treed with pines means a gully with a creek, and a creek means rocks under which crawfish are often living, and that means tasty dinner.

The questions posed in the previous section concerning reinforcement learning, along with the dearth of obvious answers, have moved some cognitive psychologists (e.g., Chomsky 1965, 1980; Fodor 1981) to conclude that reinforcement learning cannot be a serious contender for the sophisticated learning typical of cognitive organisms. Further skepticism concerning reinforcement learning as a cognitive contender derives from neural net modeling. Here the results shows that neural nets trained up by available reinforcement procedures scale poorly with the number of dimensions of the input space. In other words, as a net's visual representation approximates a rich replica of the real world, the training phase becomes unrealistically long. Consequently, computer engineers often conclude that reinforcement learning is impractical for most complex task domains. According to some cognitive approaches (Fodor, Pylyshyn, etc.), suitable learning theories must be "essentially cognitive," meaning, roughly, that cognitive learning consists of logic-like transformations over language-like representations. Moreover, the theory continues, such learning is irreducible to neurobiology. (For a fuller characterization and criticism of this view, see Churchland 1986.)

By contrast, our hunch is that much cognitive learning may well turn out to be explainable as reinforcement learning once the encompassing details of the rich-replica assumption no longer inflate the actual magnitude of the "relevance problem."

Natural selection and reinforcement learning share a certain scientific appeal; to wit, neither presupposes an intelligent homunculus, an omniscient designer, or a miraculous force—both are naturalistic, as opposed to supernaturalistic. They also share reductionist agendas. Thus, as a macrolevel phenomenon, reinforcement learning behavior is potentially explainable in terms of micromechanisms at the neuronal level. And we are encouraged to think so because the Hebbian approach to mechanisms for synaptic modification underlying reinforcement learning looks very plausible. These general considerations, in the context of the data discussed earlier, suggest that the skepticism concerning the limits of reinforcement learning should really be relocated to the background assumption—the rich-replica assumption. Consequently, the question guiding the following discussion is this: What simplifications in the learning problem can be achieved by abandoning Pure Vision's rich-replica assumption? How much mileage can we get out of the reinforcement learning paradigm if we embrace the assumption that the perceptual representations are semiworld representations consisting of, let us say, goal-relevant properties? How might that work?

Using an internal evaluation system, the brain can create predictive sequences by rewarding behavior that leads to conditions that in turn permit a further response that will produce an external reward, that is, sequences where one feature is a cue for some other event, which in turn is a cue for a further event, which is itself a cue for a reward. To get the flavor, suppose, for example, a bear cub chances on crawfish under rocks in a creek, whereupon the crawfish/rocks-in-water relationship will be strengthened. Looking under rocks in a lake produces no crawfish, so the crawfish/rocks-in-lake relationship does not get strengthened, but the crawfish/rocks-in-creek relationship does. Finding a creek in a leafy-tree gully allows internal diffusely-projecting modulatory systems, such as the dopamine system, to then reward associations between creeks and leafy-trees-in-gullies, even in the absence of a external reward between creeks and leafy-trees-in-gullies.

Given such an internal reward system, the brain can build a network replete with predictive representations that inform attention as to what is worth looking at given one's interests ("that big dead tall tree will probably have hollows in it, and there will probably be a blue-bird's nest in one cavity, and that nest might have eggs and I will get eggs to eat"). To a first approximation, a given kind of animal comes to have an internal model of its world; that is, of its relevant-to-my-life-style world, as opposed to a world-with-all-its-perceptual-properties. For bears, this means attending to creeks and dead trees when foraging, and not noticing much in anything about rocks at lake edges, or sunflowers in a meadow. All of which then makes subsequent reinforcement tasks and the delimiting of what is relevant that much easier. (To echo the school marm's saw, the more you know about the world, the better the questions you can ask of it and the faster you learn.)

A neural network model of predictive reinforcement learning in the brain roughly based on a diffuse neurotransmitter system has been applied to the adaptive behavior of foraging bumble bees (Montague et al. 1994). This is an especially promising place to test the semiworld hypothesis, for it is an example in which both the sensory input and the motor output can be quantified, the animal gets quantifiable feedback (sugar reward) , and something of the physiology of the reward system, the motor system, and the visual system in the animal's brain has been explored. Furthermore, bee foraging behavior has been carefully studied by several different research groups, and there are lots of data available to constrain a network model.

Bees decide which flowers to visit according to past success at gathering nectar, where nectar volume varies stochastically from flower to flower (Real 1991). The cognitive characterization of the bees' accomplishments involves applications of computational rules over representations of the arithmetic mean of rewards and variance in reward distributions. On the other hand, according to the Dayan–Montague reinforcement hypothesis (Montague et al. 1994), when a bee lands on a flower, the actual reward value of the nectar collected by the bee is compared (more? or less? or right on?) with the reward that its brain had predicted, and the difference is used to improve the prediction of future reward using predictive Hebbian synapses. Dayan and Montague propose that the very same predictive network is used to bias the actions of the bee in choosing flowers. Using this nonhomuncular, nondivine, naturalistic learning procedure, the model network accurately mimics the foraging behavior of real bumble-bees (figure 2.16).

That such a simple, "dumb" organization can account for the apparent statistical cunning of bumblebees is encouraging, for it rewards the hunch that much more can be got out of a reinforcement learning paradigm once the "pure vision" assumption is replaced by the "interactive-vision-cum-predictive-learning" assumption. As we contemplate extending the paradigm from bees to primates, it is also encouraging that similar diffuse neurotransmitter systems are found in primates where there is evidence that some of them are involved in predicting rewards (Ljunberg et al. 1992).

Bees successfully forage, orient, fly, communicate, houseclean and so forth—and do it all with fewer than 10^6 neurons (Sejnowski and Churchland 1992). Human brains, by contrast, are thought to have upward of 10^{12} neurons. Although an impressively long evolutionary distance stretches between insects and mammals, what remains constant is the survival value of learning cues for food, cues for predators and so forth. Consequently, conservation of the diffuse, modulatory, internal reward system makes good biological sense. What is sensitive to the pressure of natural selection is additional processors that permit increasingly subtle, fine-grained, and long-range predictions—always, of course, relevant-to-my-thriving predictions. That in turn may entail making better and better classifications

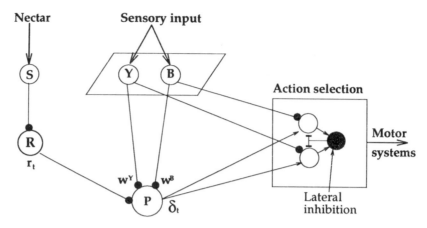

Figure 2.16 Neural architecture for a model of bee foraging. Predictions about future expected reinforcement are made in the brain using a diffuse neurotransmitter system. Sensory input drives the units B and Y representing blue and yellow flowers. These units project to a reinforcement neuron P through a set of plastic weights (filled circles w^B and w^Y) and to an action selection system. S provides input to R and fires while the bee sips the nectar. R projects its output r_t through a fixed weight to P. The plastic weights onto P implement predictions about future reward and P's output is sensitive to temporal changes in its input. The outputs of P influence learning and also the selection of actions such as steering in flight and landing. Lateral inhibition (dark circle) in the action selection layer performs a winner-takes-all. Before encountering a flower and its nectar, the output of P will reflect the temporal difference only between the sensory inputs B and Y. During an encounter with a flower and nectar, the prediction error δ_t is determined by the output of B or Y and R, and learning occurs at connections w^B and w^Y. These strengths are modified according to the correlation between presynaptic activity and the prediction error δ_t produced by neuron P. Before encountering a flower and its nectar, the output of P will reflect the temporal difference only between the sensory inputs B and Y. During an encounter with a flower and nectar, the prediction error δ_t is determined by the output of B or Y and R, and learning occurs at connections w^B and w^Y. These strengths are modified according to the correlation between presynaptic activity and the prediction error δ_t produced by neuron P. Simulations of this model account for a wide range of observations of bee preference, including aversion for risk. (From Montague et al. 1994)

(relative to the animals' lifestyle), as well as more efficient and predictively sound generalizations (relative to the animals' life-style).

To a first approximation, cortical enlargement was driven by the competitive advantage accruing to brains with fancier, good-for-me-and-my-kin predictive prowess, where the structures performing those functions would have to be knit into the reward system. Some brand new representational mechanisms may also have been added, but the increased "intelligence" commonly associated with increased size of the cortical mantle may be a function chiefly of greater predictive-goal-relevant representational power, not to greater representational power per se. Whether some property of the world is visually represented depends on the representation's utility in the predictive game, and for this to work, the cortical representational structures must be plastic and must be robustly tethered to the diffusely projecting systems. World-perfect replicas, unhitched from

Churchland, Ramachandran, and Sejnowski

the basic engines of reward and punishment, are probably more of a liability than an advantage—they are likely to be time-wasters, space-wasters, and energy-wasters.

On this approach, various contextual aspects of visual perception, such as filling in, seeing the dot move behind the occluder, cross-modal effects, and plasticity in exotropia, can be understood as displaying the predictive character of cortical processing.

CONCLUDING REMARKS

A well-developed geocentric astronomy was probably an inevitable forerunner to modern astronomy. One has to start with what seems most secure and build from there. The apparent motionless of the earth, the fixity of the stars, and the retrograde motions of the planets were the accessible and seemingly secure "observations" that grounded theorizing about the nature of the heavens. Such were the first things one saw—saw as systematically and plainly as one saw anything. The geocentric hypothesis also provided a framework for the very observations that eventually caused it to be overhauled.

In something like the same way, the Theory of Pure Vision is probably essential to understanding how we see, even if, as it seems, it is a ladder we must eventually kick out from under us. The accessible connectivity suggests a hierarchy, the most accessible and salient temporal sequence is sensory input to the transducers followed by output from the muscles, the accessible response properties of single cells show simple specificities nearer the periphery and greater complexity the further from the periphery, and so on. Such are the grounding observations for a hierarchical, modular, input–output theory of how we see.

But there are nagging observations suggesting that the brain is only grossly and approximately hierarchical, that input signals from the sensory periphery are only a part of what drives "sensory" neurons, that ostensibly later processing can influence earlier processes; that motor business can influence sensory business, that processing stages are not much like assembly line productions, that connectivity is nontrivially back as well as forward, etc. Some phenomena, marginalized within the Pure Vision framework, may be accorded an important function in the context of a heterarchical, interactive, space-critical and time-critical theory of how we see. Consider, for example, spontaneous activity of neurons, so-called "noise" in neuronal activity, nonclassical receptive field properties, visual system learning, attentional bottlenecks, plasticity of receptive field properties, time-dependent properties, and backprojections.

Obviously visual systems evolved not for the achievement of sophisticated visual perception as an end in itself, but because visual perception can serve motor control, and motor control can serve vision to better serve motor control, and so on. What evolution "cares about" is who survives, and that means, basically, who excels in the four Fs: feeding, fleeing, fighting, and reproducing. How to exploit that evolutionary truism to develop a theoretical framework that is, as it were, "motocentric" rather than "visuocentric" we only dimly perceive. (see also Powers 1973; Bullock et al. 1977; Llinas 1987; Llinas 1991; Churchland 1986). In any event, it may be worth trying to rethink and reinterpret many physiological and anatomical results under the auspices of the idea that perception is driven by the need to learn action sequences to be performed in space and time.

ACKNOWLEDGMENTS

We are especially indebted to Dana Ballard, Read Montague, Peter Dayan, and Alexandre Pouget for their ideas concerning reinforcement learning and vision, and for their indefatigable discussions and enthusiasm. We are grateful to Tony Damasio for valuable discussion on the role of back-projections, the memory-ladenness of perception, and the emotion dependence of planning and deciding, which have become essential elements in how we think about brain function. We are grateful to A. B. Bonds for reading an early version of this chapter, for providing wonderfully direct and invaluable criticism, and for the Bonding—an incomparable extravaganza of neuroscientific fact, engineering cunning, and Tennessee allegory. We thank also Richard Gregory for discussion about the visual system, Christof Koch and Malcolm Young for helpful criticism, and Francis Crick for criticism and valuable daily discussion. Michael Gray provided helpful assistance with figures and references. P. S. Churchland is supported by a MacArthur fellowship; V. S. Ramachandran is supported by the Air Force Office of Scientific Research and the Office of Naval Research; T. J. Sejnowski is an investigator with the Howard Hughes Medical Institute and is also supported by the Office of Naval Research.

NOTES

1. With apologies to Immanuel Kant.

2. For further research along these lines see for example, Ullman and Richards (1984), Poggio et al. (1985), and Horn (1986). For a sample of current research squarely within this tradition, see, for example, a recent issue of Pattern Analysis and Machine Intelligence.

3. Poggio et al. (1985) say: "[Early vision] processes represent conceptually independent modules that can be studied, to a first approximation, in isolation. Information from the different processes, however, has to be combined. Furthermore, different modules may interact early on. Finally, the processing cannot be purely "bottom-up": specific knowledge may trickle down to the point of influencing some of the very first steps in visual information processing" (p. 314). Although we agree that this is a step in the right direction, we shall argue that "trickle" does not begin to do justice to the cascades of interactivity.

4. These brief comments give no hint of the complexities of optic flow cues and their analysis. For discussion, see Cutting (1986).

5. In the case of food-aversion learning the delay between ingested food and nausea may be many hours.

3 Cortical Systems for Retrieval of Concrete Knowledge: The Convergence Zone Framework

Antonio R. Damasio and Hanna Damasio

INTRODUCTION

In this chapter we outline some ideas aimed at understanding the neural processes behind knowledge retrieval in primates. The proposal is neither a model nor a theory. It is a framework. It is large-scale, in both cognitive and neural terms, by which we mean that it deals with psychologically meaningful information, on the one hand, and with neural systems made up of macroscopic units (e.g., cortical regions, nuclei, etc.), and their connectional patterns as studied by current experimental neuroanatomy.

We also want to make clear that, by knowledge, we mean records of interactions between the brain, on the one hand, and the entities and events external to it, on the other. These entities and events exist both outside the organism and inside the organism, in its body proper. Our focus in this chapter is on concrete entities (and their properties) external to the organism.

An additional qualification is in order. Although the entities and events we speak about are real, our framework does not require assuming that they are necessarily as we construct them with our neural machineries. On the contrary, we conceptualize them as neurobiological fabrications, shaped by the organism's dispositions, especially those that pertain to innate neural circuitries in charge of biological regulation for survival.

The framework concentrates on cerebral cortex and has been developed from a background of neuropsychologic studies in humans with lesions in the telencephalon. For that reason we will begin by reviewing pertinent evidence.

BACKGROUND

The findings summarized below are based on both recognition and recall paradigms and on the use of nonverbal as well as verbal stimuli. In general, they show that individuals with lesions in association cortices within the visual, auditory, and somatosensory regions, and within "high-order" temporal cortices, can no longer effectively and reliably conjure up knowledge about some conceptual categories or about unique entities within certain categories. Those individuals have an impaired ability to generate the internal representations on which concept evocation must rely. In other words, they can no longer generate the ephemeral displays of sensory information that, when enhanced by attention, would have become conscious and summed up knowledge pertaining to a concept. At the same time those individuals do not show defective attention, that is, their ability to focus on mental contents and bring them into clear consciousness is not altered. This is important to note since attention is necessary for retrieval of knowledge, and since it is known that patients with lesions in other systems (e.g., in parietal cortices) can perform deficiently on knowledge-retrieval tasks simply because of attention deficits.

The key findings are as follows: First, patients with bilateral damage to the hippocampus proper are unable to learn factual knowledge. Examples of factual knowledge (also known as declarative knowledge) include new entities and events, both as members of a category and as unique exemplars. These patients, however, are able to learn perceptuomotor skills whose retrieval does not require the generation of a conscious internal representation (so-called procedural knowledge). Examples of these skills include the learning of rotor pursuit, mirror tracing, and mirror reading, all of which require the gradual mastering of a task, over several sessions, along a learning curve. They are also able to retrieve previously acquired factual knowledge about varied entities and events, both as members of a category and as unique exemplars. Their perception, language, and motor control are normal (Milner et al. 1968; Corkin 1984; Zola-Morgan et al. 1986; Gabrieli et al. 1988). Patients with Alzheimer's disease, in whom cell-specific and laminar-specific damage compromises hippocampal circuitry (bilaterally) and extensive sectors of high-order association cortices, are similar to patients with damage restricted to hippocampus in that they show defective learning of factual knowledge and normal learning of perceptuomotor skills (Eslinger and Damasio 1986; Van Hoesen and Damasio 1987). They differ from patients with restricted hippocampal damage, however, in that their retrieval of previously acquired factual knowledge is defective, particularly as the disease progresses.

Second, the results of bilateral damage to both the medial temporal region that contains hippocampus, entorhinal cortex, the remainder of parahippocampal gyrus, amygdala, and perirhinal cortex, as well as the nonmedial temporal cortices are remarkably different from those of hippocampal damage alone. The nonmedial temporal region includes area 38 [temporal pole], and areas 21, 20, 36, and part of 37 [the human inferotemporal region; see figure 3.1.]

The patient known as Boswell is exemplary (Damasio et al. 1989). He has a severe impairment in the retrieval of previously acquired factual knowledge, which affects *all unique entities and events*. He cannot narrate any specific episode from the several decades of autobiography that preceded the onset of his lesion. He is unable to recognize family or friends (from

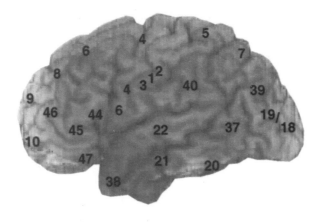

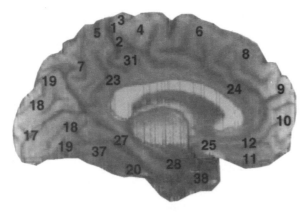

Figure 3.1 Lateral (*top*) and mesial (*bottom*) surfaces of the human brain with the main gyri and cytoarchitectonic areas according to Brodmann. The images were obtained from high resolution MR cuts reconstructed in 3-D by BRAINVOX (Damasio and Frank 1992).

face or voice), unique places or objects (from sight or sound), although he can recognize them at categorical level (as faces, as houses, or cars).

He also shows a remarkable dissociation for nonunique entities. He can always assign any entity to its "supraordinate" taxonomic category. He can recognize different utensils or animals, as being a *utensil* or an *animal*; in other words, he can indicate the most general category to which an entity belongs. On the other hand he can provide only the "middle level" categorization, or the so-called "basic object" level for certain types of entities. As an example, he can provide "basic object" level recognition for houses and tools (*ranch-type* house, *wrench*), but most natural entities baffle him. Confronted with the picture of a camel or a zebra he will appropriately say it is an animal but not go beyond that supraordinate assignment. In brief, when shown the faces of unique persons he was previously familiar with (e.g., Roosevelt or his wife) he was unable to recognize them, but he knows that their faces are human faces. He also knows the meaning of the basic facial expressions shown in those faces (in stills or in motion) even if those expressions are instantiated in faces whose identity he no longer retrieves. He has perfect knowledge of the components of those faces or other objects (shapes, colors, movements). When asked to think about specific faces (or places), he cannot conjure up the face of any one particular person (e.g., he cannot conjure up his wife's face). Yet he can generate an internal representation of a "generic" human face, or of part of a face (e.g., a nose), or of a geometric figure (a square), or of any color. He has *no* perceptual defect (olfaction excepted), no motor impairment, and he can attend effectively to all stimuli he is shown, including the stimuli that he cannot recognize.

The evidence discussed so far suggests that (1) nonmedial temporal cortices (polar, inferotemporal, posterior parahippocampal) are essential for the retrieval of previously acquired factual knowledge, especially for unique exemplars, none of which can be retrieved after substantial bilateral damage to this sector, and (2) neither nonmedial nor medial temporal cortices play a role in the acquisition or retrieval of skill knowledge, basic visual perception, or motor control.

Third, patients with bilateral damage in sectors of the inferior occipital and posterior temporal visual association cortices (areas 18, 19 and posterior 37 in figure 3.1) *cannot* conjure up the unique knowledge pertinent to a unique object when the object is presented visually (i.e., they cannot recognize a unique identity) but can see the object and provide descriptions of its visual details, and cannot generate an internal representation of a unique visual entity when the stimulus is not present (e.g., they are unable to conjure a specific face in the absence of the model). Unlike patient Boswell, however, these patients can retrieve unique knowledge relative to a unique entity when the stimulus is presented through a nonvisual channel. The unique voice of a specific person allows the patient to know the unique identity behind it (Damasio et al. 1990b).

The evidence thus indicates that damage placed anteriorly in the occipitotemporal system precludes retrieval of *any* unique knowledge, regardless of the sensory channel used to trigger the retrieval, and that posterior damage in specific sectors of the visual system precludes retrieval of unique knowledge *only* when the stimulus is visual. The same kind of damage impairs visual imagery but not visual perception.

Fourth, patients with bilateral damage in inferior visual association cortices within the occipital region (especially the medial sector of areas 18 and 19 in figure 3.1) can no longer perceive colors normally, nor can they evoke an internal representation of those colors, such as picturing the color of blood (Damasio et al. 1980). It must be noted that damage to anterior visual and high-order association cortices does *not* preclude color knowledge or color perception. That color imagery is lost with "early" (posterior) cortical damage but not with "late" (anterior) integrative cortical lesions suggests that color is not *re*represented in high-order cortices. When the internal reconstruction of a complex representation requires the evocation of color, the color "content" is generated from "early" visual cortices.

Fifth, damage to human inferotemporal cortices, the areas 21, 20, 36, and 37 that collectively form IT, selectively impairs retrieval of knowledge about certain categories of entities. [Before we go any further an anatomical clarification is in order. Please refer to figure 3.2, comparing the human and monkey temporal lobes. The difference is immense. It is important to note that the region that corresponds, in the human, to the monkey's IT is probably far more posterior and inferior than in the monkey. The results described here relative to the human IT must be considered in the context of this major anatomical difference. The region from which Tanaka and his group (Fujita et al. 1992) have recently recorded, which they call "anterior inferotemporal," is in the middle inferotemporal region of the monkey, and would be in the posteroventral temporooccipital region in humans.] Patients with such damage fail to conjure up knowledge pertaining to some entities but easily conjure up knowledge pertaining to others, along consistent patterns of dissociation (Warrington and Shallice 1984; McCarthy and Warrington 1988; Damasio 1990). The IT region seems necessary to retrieve knowledge about entities that are learned through the visual modality alone and that share physical structures with several other different entities; typical examples are dog-like animals such as fox, raccoon, coyote, wolf, and German shepherd dog, whose physical traits strongly resemble each other. We have designated such entities as visually "ambiguous." Retrieval of knowledge about visual entities with lesser ambiguity does not depend on this region, and that is why animals whose shapes are "outliers" (the elephant is the typical example) pose no problem for recognition. Knowledge of entities that were learned through both the visual and somatosensory modalities (for instance, most manipulable tools and utensils) does not depend on this system either. The evidence uncovers a systematic correspondence between the presence of damage in certain systems and the impaired retrieval of certain types of knowledge. We believe the correspondence arises because of constraints dictated by the physical characteristics of the entities and by neuroanatomical design (Damasio 1989c; Damasio et al. 1990a).

In brief, (1) access to previously acquired nonunique knowledge can be disrupted by lesions in specific neural subsystems, and (2) the defect is not equal for different categories of concrete knowledge. Access to different types of concrete knowledge thus depends on different subsystems. As we will note below, we do not believe that the damaged systems hold records of entity representations per se. We hypothesize that they direct the simultaneous activation of anatomically separate regions whose conjunction defines an entity.

Sixth, damage to left anterior temporal cortices in the temporal pole (area 38) and the anterior part of IT (areas 20 and 21 in figure 3.1) causes a severe defect for naming of concrete entities. Patients cannot access the word forms that belong to unique entities (proper noun), nor can they retrieve the word forms that go with varied nonunique entities. Patients can, however, generate accurate descriptions about all the entities they cannot name.

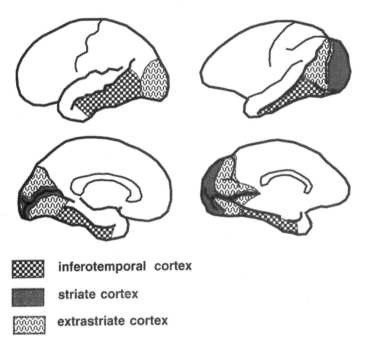

inferotemporal cortex

striate cortex

extrastriate cortex

Figure 3.2 A comparison of human and monkey cortices. Human cortices are on the left and monkey cortices on the right. Note that they are not drawn to scale (the human brain is about 10 times larger than the macaque's). The different position of the inferotemporal cortex, and the different relation that inferotemporal cortex holds to the early visual cortices are obvious.

Only the access to the lexical entries that denote the overall concept is defective. The nonverbal conceptual knowledge of the entities is intact. The patients have no grammatical or phonemic/phonetic defects. We have also discovered that these patients are able to generate the word form that accurately denotes an action or relationship, without the slightest difficulty. They can retrieve word forms (verbs) describing the action of entities whose names they cannot retrieve, or that they may not even know. For instance, shown the picture of a mother duck in a lake, being followed by several ducklings, Boswell said: "The little things [ducks or ducklings, not named] are following her, the mother [duck, not named]. They are all swimming."

Seventh, damage to the cortices in the left temporal *pole* (sparing the anterior part of IT) causes a defect in the retrieval of word forms (e.g., the names of persons or places). Access to word forms for common noun is intact. Access to conceptual knowledge of the entities denoted by both proper and common nouns is also intact (Damasio et al. 1990c; Graff-Radford et al. 1990; Semenza and Zettin 1989). Damage to the same regions in the right hemisphere seems not to compromise lexical access.

These findings suggest that the conjoining of nonverbal and verbal activated representations pertaining to concrete entities depends on a mediator mechanism in left anterior temporal cortices. The mechanism promotes the reconstruction of a word form given the concept, or, conversely, the recon-

Damasio and Damasio

struction of the concept of an object given the word form. The systems that support this mediational mechanism do not contain records for either words or concepts themselves, but rather records of the probable combination between them, that is, the combinations between (1) the many records that subsume the concept of a concrete entity, nonverbally, and (2) the many records that subsume acoustical, somatosensory, and motor patterns with which a given word form can be reconstructed. (It well may be the case that other aspects of lexical characterization, such as the syntactical properties of a given word, will also be activated from these systems.) In brief, these systems promote activity in *other cortices* (and probably basal ganglia), and it is through the resulting activity elsewhere that lexical access or concept access are achieved.

Finally, damage in *left* lateral frontal cortices compromises retrieval of some word forms that denote some classes of verbs, while leaving absolutely intact the retrieval of word forms for nouns. This is interesting evidence for the fact that frontal cortices (and the parietal and mesial cortices that feed into them) are concerned with other aspects of conceptual representation (e.g., space-time trajectories of entities rather than entity structure itself), and neural processing (e.g., attention, governance of response selection, and motor planning). It is intriguing to find retrieval of some word forms for verbs connected with this system given the fact that some verbs do describe actions in space-time rather than structural characteristics of entities (Damasio and Tranel 1993).

REGIONALIZATION OF KNOWLEDGE ACCESS AND CORTICOCORTICAL CONNECTIONS

The evidence summarized here suggests that the access to different levels and types of knowledge depends on different neural systems and is thus regionalized. At first pass, the correspondence between retrieval impairment patterns and site of damage further suggests that

1. Damage to early visual cortices compromises the retrieval of features (e.g., color);

2. Damage to intermediately placed cortices leaves the retrieval of features intact, but may compromise the retrieval of knowledge pertaining to certain categories of concrete knowledge, that is, compromise retrieval of knowledge for some nonunique entities while sparing others;

3. Damage to the anterior-most cortices compromises retrieval of knowledge regarding virtually any unique entity or event (scene) but leaves intact retrieval of features, entity components, and nonunique entities.

Features, nonunique entities, and unique entities are constituted by knowledge of remarkably different ranks in terms of amount of components, relational complexity of those components within the entity, and relational complexity of associations between the entity itself and other entities and events. For instance, to classify an entity as unique, we must know about intrinsic and relational details that are far more complex than those of a nonunique entity. In turn, nonunique entities require more complexity than features.

These findings suggest a tentative principle: access to concrete knowledge of higher hierarchical status requires structures in anteriorly placed temporal cortices, whereas access to concrete knowledge of lower complexity only requires posterior occipital cortices. (It should be noted that when we refer to anterior, or intermediate, or posterior, we do not necessarily include anterior, intermediate, or posterior structure of *both* hemispheres. On the contrary, because of cerebral hemisphere dominance effects, certain types and levels of knowledge may require, say, an anterior, or intermediate cortex of one hemisphere only.)

The cortical regions whose damage we discussed above are anatomically distinguishable in a variety of ways (e.g., cytoarchitecture, subcortical connections, possibly intrinsic circuitry), but we have chosen to focus on their distinction in terms of long-range corticocortical projections. The impaired retrieval of more complex knowledge correlates with damage to the cortices located closest to the apices of feedforward chains culminating in entorhinal cortex, and farthest from the beginning of the feedforward chains in primary sensory cortices. Reciprocating feedback projections from those cortices recapitulate the feedforward projections in reverse direction (Van Hoesen 1982; Felleman and Van Essen 1991). The impaired retrieval of less complex knowledge correlated with damage in cortices located more posteriorly suggests that there is a principled relationship between "rankings" of knowledge access and "rankings" of corticocortical connections.

As we will discuss later, *access to*, leading to *retrieval of*, must be distinguished from *represented at*. The knowledge that can be accessed from anterior temporal cortex is *not* fully represented in anterior temporal cortex in the sense that no "image" is likely to be there. Incidentally, this is the position we take in interpreting the meaning of "face" cells or "hand" cells, or, for that matter, the cells recently described by Tanaka's group (Fujita et al. 1992). Rather, we see these cells as part of the network whose activity may reenact an explicit representation. When they are activated, these cells in high-order cortices contribute to the reenactment of the explicit representation, i.e., they are critical to the neural process on the basis of which we *experience* the representation, but they are neither the "sole basis for" nor the "site of" that neural process. There is no single basis or site for such a process. Finally, we should note that what we call knowledge corresponds to records of interactions between (1) entities and events external to the individual (or internal to the individual but external to the brain), and (2) the brain (such as it is anatomically at the time of the interactions). Although the external entities and events are real, they are not necessarily as we construct them with our neural apparatus.

The relationship suggested by the evidence, then, is between *ranking of knowledge* (as qualified above) and *ranking of corticocortical connections*.

The emphasis on this principle does not mean that we are ignoring the existence and contributory role of many other connections also available at all stations of these hierarchies, that include (1) heterarchical connections between and among parallel cortico-cortical projection streams; (b) local (intrinsic) cortical connections; (c) subcortical connections, direct or re-entrant; and (d) commissural connections. It simply means that we suspect that ranking of cortico-cortical connections is the most distinctive aspect of these regions inasmuch as the substrates of knowledge are concerned. Those other distinctive anatomical aspects of different cortical stations are critical for the continued development and adjustment of the system. Elsewhere (Damasio 1989a,b) and below, we have argued that as the organism interacts with the environment, the selection of circuitries that corresponds to varied interactions with the environment is carried out with the help of those other systems. For instance, subcortical connections hailing from nuclei concerned with biological drives are critical to this process.

KNOWLEDGE REPRESENTATION

Dispositional (Nonmapped) Representations versus Explicit (Mapped) Representations

What did the areas damaged in the patients described above contain or contribute while they were healthy, to the mental representations on which knowledge is based? What is the relation between the knowledge that fails to be retrieved and the station of corticocortical connections destroyed by a lesion? We will entertain two alternatives. In the first alternative, the conceptual knowledge relative to a given entity is contained in the circuitry of the damaged region. (This is not only the classic alternative but the one that remains implicit behind most current work in neuropsychology and neurology.) Both the reactivation of sensory and motor properties defining a concept, as well as the "know-how" needed to reconstitute those properties, would depend on that region alone. Their reinstantiation in consciousness would result from attended neural activity *within* that one region. In that traditional alternative, the failure in knowledge retrieval comes from the absence of the region and of the high-level representations previously contained in it. We must reject this alternative. We believe that the results discussed above are incompatible with this view, because the nature of the losses following lesions at different levels would be different. Rather than respecting the levels of complexity and the relative functional kinship of entities, the lesions would lead to "unprincipled" losses in which all knowledge about entire categories of entities would simply vanish.

In the alternative we favor the sensorimotor properties on which a concept is based would be *retrieved from* early sensory and motor cortices, and the effects of focal attention would be placed in those parts of the distributed network. The retrieval would have been *directed by* the area now damaged, had it been intact. The circuitry of the damaged region previously contained knowledge about *which* separate brain regions should be reactivated to conjure up varied properties, and about *how* those regions should be related temporally and spatially.

The failure in knowledge retrieval comes from the damaged region being unable to conduct the reconstruction, but the reconstruction itself would have depended on circuitry in many other regions. In short, a lesion in a given place modifies the "working habits" of nonlesioned regions that are connectionally related to the lesioned region. Following a lesion, the remainder of the brain works *differently* rather than as before. This may sound trivial but it is not. The results of lesion experiments are often interpreted as if the healthy systems connected to the damaged area continue doing precisely what they did before the lesion.

The alternative we favor implies a relative functional compartmentalization for the normal brain. One large set of systems in early sensory cortices and motor cortices would be the base for "sense" and "action" knowledge, i.e., the highly multiregional substrate for "explicit" representations which are the key to our experience of knowledge. Another set of systems in higher-order cortices would orchestrate time-locked activities in the former, that is, would promote and establish temporal correspondences among separate areas. Yet another set of systems would ensure the attentional enhancement required for the concerted operation of the others. These sets of systems would operate under two major influences: (1) internal biases, expressed in brain core networks concerned with enacting biological drives and instincts required for survival, and (2) the structures and actions in the external environment.

In the overall system we envision, there is neither cartesian dualism between matter and mind, nor homuncular dualism between "images" and a "perceiver" of those images. There is also no infinite regress. On the contrary, neurally speaking, there is a finite number of convergence steps downstream from the system in which the neural activity corresponding to explicit images takes place. But the top steps in the hierarchy are neither a homunculus nor a perceiver. They are merely the most distant convergence points from which divergent retroactivation can be triggered. What is infinite, instead, is the ceaseless production of new activity states, in early sensory cortices and in motor cortices, across time. It is those neural states, one after the other, that can be said to constitute "regresses" for the previous state. But it should be noted that they occur in the *same* set of systems, at *different* times, unlike the classic and much maligned homuncular regress that occurs in different neural sites, ever more removed from the "perceptual" site, in space and in time. It is the perpetually recursive property of corticocortical systems that permits this special form of regress. We have previously discussed evidence in support of this view (Damasio 1989a,b); additional arguments are discussed in Damasio (1994).

Reconstructing Mapped Representations

The reconstruction of pertinent property "representations" is thus accomplished in many separate cortical regions, by means of long-range corticocortical feedback projections that mediate relatively synchronous excitatory activation. In the large scale reconstruction that we envision from higher-order cortices (such as those in anterior temporal lobe), the time scale of the synchronization would be in the order of several hundred msec, and even beyond 1000 msec, the scale required for meaningful, conscious cognition. But at more local levels, for instance, in posterior temporal cortices, the scale would be smaller, in the order of tens of milliseconds. The return projections necessary for the reconstruction are aimed toward layers I and V of the cortex, mainly the former, in which feedforward projections originated (see Rockland and Virga 1989). We have called the neural device from which reconstructions are conducted a *convergence zone*.

Convergence Zones

In essence, a convergence zone is an ensemble of neurons within which many feedforward/feedback loops make contact. Its connectional structure is as follows: a convergence zone receives feedforward projections from cortical regions located in the connectional level immediately below, sends reciprocal feedback projections to the originating cortices; sends feedforward projections to cortical regions in the next connectional level and receives return projections from it; is influenced by a broad class of cortices concerned with attentional control and response selection, such as prefrontal and cingulate directly or indirectly, which are in turn reentrantly connected to basal ganglia; receives projections from heterarchically placed cortices; and receives projections from subcortical nuclei in thalamus, basal forebrain, brain stem, etc. This rich network of extrinsic connections is complemented by a complex network of intrinsic intralaminar and interlaminar connections.

A convergence zone is located within a convergence region. We envision that there are in the order of thousands of convergence zones, which are all microscopic neuron ensembles, located within the macroscopic convergence regions that have been cytoarchitectonically defined and that number about one hundred. Both convergence regions and convergence zones come into existence under genetic control. But epigenetic control, as the organism interacts with the environment, may alter convergence regions, and massively alter convergence zones through synaptic strengthening. As noted above, synaptic strengthening occurs under particular conditions, in which circumstances external to the brain match the survival needs of the organism, its intentions so to speak, as expressed in biological drive networks. It is reasonable, then, to talk about synaptic strengthening as a selective process, in the sense in which Changeux (1976) and Edelman

(1987) have used the concept, although this does not commit us to a particular unit of selection such as a neuron or a neuron group.

A convergence zone is thus a means of establishing, through synaptic strengthening, preferred feedforward/feedback loops that *use* subsets of neurons within the ensemble. A subset of the neurons in the convergence zone would "learn" to activate a large number of spatially distributed neural ensembles, in temporal proximity, by means of feedback projections. The convergence zone could be excited by any (or a subset) of the feedforward projections that were originally paired with the feedbacks coming out of the convergence zone, or by feedback projections of convergence zones from a higher station, or from heterarchical connections.

A convergence zone develops under the influence of (1) temporally close activity in multiple feedforward and feedback lines that are simultaneously active when a number of anatomically separate regions are active and are providing the normal substrate for a given perceptual/thought process, and (2) modulatory action from feedback and feedforward projections from ipsilateral and contralateral cortices, and subcortical nuclei, during (1). The development of a convergence zone also depends on local interactions among neurons (e.g., from their intrinsic collateral arborizations). A convergence zone would be the result of convergence of feedforward inputs, but its feedback projections operate by diverging toward the origin of feedforward projections. Naturally, when we refer to neurons in a convergence zone we refer to the synaptic pools made up of contacts among those neurons.

We have hypothesized that there would be two main types of activity in a convergence zone. In the stable type, the excitation of one or a few neurons feeding into it generates maximal temporally close activity in many feedbacks that participate in the convergence zone. This then generates temporally close activations in several regions that originally projected forward to it. What we envision convergence zones achieving is the recreation of separate sets of neural activity that were grossly simultaneous, that is, that coincided during the time window necessary for us to attend to it and be conscious of it, which means hundreds of milliseconds. However, this may not necessarily translate into simultaneous activity within the convergence zone. In fact, it is much more likely that there would be an extremely fast sequence of activations that would make separate neural regions come on-line in some order imperceptible to consciousness.

Convergence zones also fire forward, through their feedforward projections, into other convergence zones located at a higher level. In turn, feedback firing from that higher level convergence zones would broaden the scope of regions activated in response to the initial stimulus. This first type of activity depends on strong local synaptic linkages among subsets of neurons in a convergence zone. In a second type of activity, less stable modes of firing would activate subsets of feedbacks, leading to novel combinations of activity. This would depend on transient combinations.

In short, knowledge retrieval would be based on relatively simultaneous, attended activity in many early cortical regions, engendered over several recursions in such a system. Separate activities in early cortices would be the basis for reconstructed representations. The level at which knowledge is retrieved (e.g., supraordinate, basic object, subordinate) would depend on the scope of multiregional activation. In turn, this would depend on the level of convergence zone that is activated. Low level convergence zones bind signals relative to entity categories (e.g., the color and shape of a tool), and are placed in association cortices located immediately beyond (downstream from) the cortices whose activity defines featural representations. In humans, in the case of a visual entity, this would include cortices in areas 37 and 39, downstream from the maps in V3, V4, and V5. Higher-level convergence zones bind signals relative to more complex combinations, for instance, the definition of object classes by binding signals relative to its shape, color, sound, temperature, and smell. These convergence zones are placed at a higher level in the corticocortical hierarchy (e.g., within more anterior sectors of 37 and 39, 22, and 20). The convergence zones capable of binding entities into events and describing their categorization are located at the top of the hierarchical streams, in anterior most temporal and frontal regions.

The "firing" knowledge embodied in the convergence zone is the result of previous learning, during which feedforward projections and reciprocating feedback projections were simultaneously active. Both during learning and retrieval, the neurons in a convergence zones are under the control of a variety of cortical and noncortical projections. This includes: projections from thalamus, the nonspecific neurotransmitter nuclei, and other cortical projections from convergence zones in prefrontal cortices, cortices located higher up in the feedforward hierarchy, homologous cortices of the opposite hemisphere, and heterarchical cortices of parallel hierarchical streams. The essence of this framework, then, comprises reconstruction of entities and scenes from component parts and integration of component parts by time correlations. The requisite reactivation is mediated by excitatory projections.

CONCLUDING REMARKS

In closing, we would like to add a few words concerning the evidence now available for this framework, as well as the relation it may have to other theoretical approaches. In addition to the evidence adduced above, from human neuropsychology and from experimental neuroanatomy, there is also supporting evidence in the recent neurophysiological findings of Singer and colleagues (1993), Eckhorn and colleagues (1988), and Fetz et al. (1991), all of which indicate the presence of temporal correlations among anatomically separate cortical regions relative to a single stimulus. These results have been obtained in perceptual or motor experiments, but there is no reason why they cannot be extrapolated for the knowledge retrieval processes we are discussing here. The essence of these results is that geographically separate regions of brain can be active at the same time when their activity is related to the same thing. There is also evidence for the type of functionally segregated neuron ensembles we call convergence zones in the recent work of Fujita et al. (1992), and we interpret the classical "face" and "hand" cells as being part of convergence zones. Concerning other approaches, the massive recurrence of neuroanatomical pathways that we propose for knowledge retrieval is a component of Edelman's model of visual perception (Edelman 1987), in which feedback activity is subsumed by the concept of "reentry." There are, however, several distinctions. Edelman's model does not use a convergence–divergence architecture. The maps are fully and reciprocally interconnected, in both a hierarchical and heterarchical manner. This characteristic seems well suited to the constructive roles that the very early visual cortical regions are likely to play, and for which a convergence zone architecture might not be sufficient. It is supported by neuroanatomical findings (Zeki and Shipp 1988; Rockland and Pandya 1979). Another distinction concerns the fact that reentry's principal means of operation is the "synthesis of signals" within the neuronal populations that get reentered. The framework we have proposed does not include such a means of operation and relies instead on a correlative operation which appears similar to the one Singer and von der Malsburg envision (although in recent work from the same group, reentry also accomplishes a correlative function (see Tononi et al. 1992; Edelman 1992).

ACKNOWLEDGMENTS

We thank Kathleen Rockland, Gary W. Van Hoesen, and Christof Koch for their valuable suggestions. Supported by ONR grant N00014-91-J-1240.

4 The Interaction of Neural Systems for Attention and Memory

Robert Desimone, Earl K. Miller, and Leonardo Chelazzi

INTRODUCTION

The visual recognition of objects depends on a cortical processing pathway that begins in area V1 and continues through areas V2, V3, V4, and the inferior temporal (IT) cortex (Ungerleider and Mishkin 1982; Desimone et al. 1985; Maunsell and Newsome 1987; Desimone and Ungerleider 1989; Felleman and Van Essen 1991). As one proceeds along this pathway, the receptive fields of neurons increase steadily in size and the analysis of object features becomes increasingly complex. Although these large fields will typically contain many different stimuli, we have previously shown that spatial attentional mechanisms limit the amount of information that is processed within them. When one attends to a stimulus at one location within the receptive field of a neuron in V4 or IT cortex, the responses to stimuli at other, ignored, locations are suppressed (Moran and Desimone 1985). Thus, spatially directed attention controls the information processed in extrastriate cortex and regulates access to memory. Generally speaking, we remember what we attend to. The spatial attention system that controls extrastriate processing appears to involve a number of different cortical and subcortical structures, several of which are closely associated with the oculomotor system (Desimone et al. 1990; Posner and Driver 1992; Posner and Petersen 1990; Colby 1992; see chapter 9 by Posner and Rothbart).

Although attention may regulate access to memory, the reverse also occurs. Consider the following scenarios. While driving to work, you pay little attention to all of the surrounding cars traveling in their lanes but react immediately to a car that makes an unexpected change in direction in front of you. At work, you walk into your office and are startled that your familiar desk chair has been replaced by a new one. After studying the new chair, you search for your coffee cup buried among the objects cluttering your desk. When it is found, you switch your attention to the wall of the room, where you know you will find your clock.

Each of the everyday behaviors described above illustrates how memory guides attention. In the case of the car changing direction, the car violated the expectation of its behavior built up over the course of the previous few seconds or minutes, thereby eliciting attentional and orienting responses.

In the case of the new chair, attention was attracted by the mismatch between the new chair's image and the representation of the familiar chair in long-term memory. In both cases, the representation of stimuli in short- and long-term memory contributed as much to their salience as did purely visual properties such as color or brightness. In the case of the coffee cup, it was the representation of the cup in long-term memory that guided the attentional search of the cluttered desk. Even in the case of the clock on the wall, which might be considered a simple case of spatially directed attention, it was the memory of the location of the clock in the room that guided the locus of attention. As we will see below, memory not only influences attention, but is intertwined with the on-going processing of visual information in the cortex.

In this chapter we take a bottom-up approach to developing models of higher brain function. We first describe some new results on the properties of cortical neurons in monkeys performing mnemonic and attentional tasks. With these physiological findings serving as both constraints and inspiration, we begin to sketch out how the neural systems underlying memory and attention interact, resulting in self-directed behavior.

SHORT-TERM MEMORY

Most cognitive scientists and neuroscientists would probably accept the notion that the neural mechanisms of memory include a facility for the temporary storage of information. The neuropsychological evidence for separate mechanisms of long- and short-term memory is dramatic, as amnesic patients with damage to the medial temporal lobe may show normal retention of information for a few seconds or minutes but may be completely unable to hold memories for longer periods of time (see Baddeley 1986, 1990, for reviews of both normal human memory and amnesia). Although one can make numerous additional distinctions among different types or components of both long- and short-term memory, we will use the terms in a generic sense and describe some of the physiological results first before speculating on what aspects of memory they may explain.

Recent physiological results suggest that at least one type of short-term memory may be an intrinsic property of visual cortex. In our work, we record from neurons in anterior IT cortex of rhesus monkeys performing delayed matching to sample tasks (Miller et al. 1991b, 1993). In the standard form of this task, a monkey is shown a "sample" stimulus followed, after a short delay, by a "test" stimulus, and it must indicate whether or not the test stimulus matches the sample. Thus, the task requires a type of stimulus memory, or recognition memory, lasting for the length of a behavioral trial. Several studies have shown that IT neurons respond differently to the test stimulus depending on whether it matches the sample (Gross et al. 1979; Mikami and Kubota 1980; Baylis and Rolls 1987; Vogels and Orban 1990; Riches et al. 1991; Eskandar et al. 1992), and other studies have shown that some IT neurons have elevated activity during the de-

Desimone, Miller, and Chelazzi

lay period, as if they are actively maintaining the memory of the sample (Fuster and Jervey 1981; Mikami and Kubota 1980; Miyashita and Chang 1988; Miyashita 1988; Sakai and Miyashita 1991; Fuster 1990). However, for a neural memory mechanism to be useful, it must have the capacity to retain information over long intervals that are not blank but rather are filled with new stimuli entering the visual system, competing for processing, and presumably activating the same cells involved in the storage of memory traces. To study this interplay between perceptual and mnemonic processing, we record from neurons in anterior ventral IT cortex in a task that requires the monkeys to retain items in memory while concurrently processing new stimuli (Miller et al. 1991b, 1993).

On each trial of the task, from 0 to 5 stimuli intervene between the sample and the final matching test stimulus. The stimuli are complex patterns, such as faces, fruit, or textures, which the monkey has seen before. These stimuli elicit stimulus-selective responses from the large majority of cells, but we do not attempt to understand which features of the stimuli activate a given IT neuron. The basis of object coding in IT cortex is currently not understood—we simply assume that it occurs (see chapter 8 by Poggio and Hurlbert).

How is the memory of the sample maintained? The surprising result is that, for nearly half the cells in the cortex, the memory of the sample is reflected in the responses to the test stimuli. That is, of the cells that respond to a given sample stimulus, the responses of about half of them are a joint function of the current stimulus (i.e., the magnitude of response depends in part on how well the stimulus fits within the cell's "feature domain") and the stimulus in memory. For the large majority of these cells, responses to the test items are suppressed if they match the sample in memory (see also Baylis and Rolls 1987; Riches et al. 1991; Eskandar et al. 1992). The reduction in response is typically proportional to the magnitude of the response to a given stimulus (measured when it is presented as a sample or non-matching item). For example, if a cell prefers red stimuli, these stimuli will show the greatest reduction in response when they match the sample item. Furthermore, the suppression is maintained even if up to five nonmatching stimuli intervene, which is the maximum we have tested. The "memory-span" of this suppressive effect seems to be as long as the monkey can perform the task. A few cells show opposite behavior (enhanced responses with matching), and the remaining cells (about half the population) seem to convey only sensory information and are not affected by memory.

Not only are responses to matching items affected by the specific item in memory, but so are responses to nonmatching items, an effect that has been observed by Eskandar et al. (1992). That is, the response to a given nonmatching item depends on which stimulus had been seen as the sample. There is suggestive evidence that these effects are related to the similarity of a given nonmatching item to the stimulus in memory (Miller et al. 1993). The more similar is the current stimulus to the one in memory, the more the response to the current stimulus appears to be suppressed. Thus, IT cells

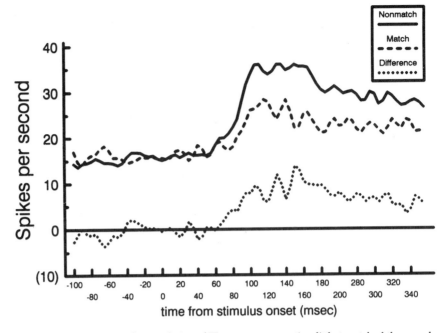

Figure 4.1 Responses of a population of IT neurons to test stimuli that matched the sample stimulus in memory and, for comparison, the response to the same stimuli when they did not match. The difference line plots the difference between the two histograms. The bin width is 10 msec. (Adapted from Miller et al. 1993)

seem to be computing similarity to memory traces, rather than matching per se.

On the basis of these data, we have proposed that a population of IT neurons functions as "adaptive mnemonic filters" whose responses are a joint function of the current stimulus and stored memory traces. A given IT neuron gives its best response to stimuli that contain features within the cell's feature domain (i.e., has the appropriate color, shape, texture, etc.) but that have not been recently seen. Repetition is, in a sense, a type of stimulus feature.

A critical question for constructing a model of short-term memory is whether the mnemonic influences on IT responses are generated within (or before) IT cortex or result from feedback from other structures. An analysis of the time course of the effects in IT argues against certain types of feedback. Figure 4.1 shows the time course of the population response to matching and nonmatching items in IT cortex. The suppression of response to matching items (compared to nonmatching) begins at the onset of the visual response (i.e., within 80 msec of stimulus onset). Thus, it seems highly unlikely that the suppression is due to feedback from memory structures beyond IT cortex that do the actual detection of matching items. It is still possible, however, that critical feedback occurs at the time of storage of the memory trace and persists through the time of retrieval.

The fact that the suppression begins by nearly the first stimulus-evoked

Desimone, Miller, and Chelazzi

action potential in IT cortex also argues against the idea that the effect depends on lengthy temporal processing in this region. We have tested whether the match-nonmatch status of a stimulus is coded by specific temporal variations in response (i.e., responses that differ in their time-course but have the same average rate) by using the principal components of the spike trains of IT cells to classify stimuli as matching or nonmatching, on a trial-by-trial basis. In contrast to recent findings of Eskandar et al. (1992), we find no advantage in using the first three principal components of the responses to classify stimuli as matching or nonmatching, compared to using just the average firing rate. Since Eskandar et al. found significant temporal variations in IT responses in monkeys performing a matching to sample task with short delays and without intervening items, the two results together suggest that successive stimulus presentations do cause significant temporal variations in neuronal responses but that do not span long delays or intervening items (although other explanations are possible, including differences in recording sites in IT cortex). These results do not, of course, rule out "fast" temporal mechanisms such as synchronized firing among IT cells, which could, in principle, cause virtually instantaneous changes in firing rates among coupled IT cells (see chapter 5 by Koch and Crick and chapter 10 by Singer).

We have also examined activity during the delay intervals of the matching task, to see whether cells might maintain a representation of the sample in memory through maintained activity in the retention interval (Fuster and Jervey 1981; Mikami and Kubota 1980; Miyashita and Chang 1988; Miyashita 1988; Fuster 1990; Sakai and Miyashita 1991). Although a quarter of the cells show stimulus-specific activity in the delay interval immediately following the sample stimulus, this activity appears to be "reset" by the first intervening item (Miller et al. 1993). That is, after the first intervening test item, the amount of activity in the subsequent delay interval seems to be determined more by the intervening item than by the sample stimulus in memory. Thus, it is unlikely that maintained IT activity in the delay mediates the memory of the sample in this particular short-term memory task. Nonetheless, as we will see later in the chapter, maintained activity during delay periods is a prominent feature of neuronal activity in some tasks. Further, we have evidence that the delay activity can be switched on and off under the monkey's voluntary control (Chelazzi and Desimone, unpublished data). Maintained activity during retention intervals, easily disrupted by intervening stimuli, may nonetheless serve as a type of visual "rehearsal," that helps solidify memory traces (for example, its effect on memory may be equivalent to that of a longer stimulus presentation time), or as a kind of visual "sketchpad" (Baddeley 1986) for comparing stimuli presented in different modalities or different spatial locations (see below).

The sensitivity of IT neurons to repetition suggests an analogy to figure-ground separation in the spatial domain. Many cells at all levels of the visual system respond best to contrast of some sort. For these cells, the greater the similarity between the stimulus in the receptive field and those in the silent surround (stimulation of which does not elicit responses by itself), the more the response to the receptive field stimulus is suppressed. As one advances through the visual system, the stimulus features that are contrasted may become more sophisticated and the spatial areas over which the interactions occur may become larger. Based on these properties, it has been conjectured that one of the functions of visual cortex is to separate figures from background (Allman et al. 1985a; Desimone et al. 1985).

The properties of IT neurons suggest that figure-ground separation occurs in the temporal domain as well. As a result, stimuli that have not been recently seen or are unexpected may tend to pop out from an array of repeated items. In a sense, the past functions as the surround, which is compared with the present stimulus in the receptive field. This temporal figure-ground extraction may occur automatically, as repetition effects in IT cortex have been found both in passively fixating and in anesthetized animals (Miller et al. 1991a).

If the analogy with spatial receptive fields is valid, temporal figure-ground extraction is probably not unique to IT cortex but may be an intrinsic property of visual cortex. Figure-ground separation in the spatial domain appears to build incrementally as one moves through the visual system, and there may be a comparable build up of temporal processing. Nelson (1991), for example, finds that cells in striate cortex of the cat show orientation-specific suppression lasting a few hundred milliseconds. Orientation-specific temporal interactions are also found in area V4 (Haenny et al. 1988; Maunsell et al. 1991). Our own preliminary evidence in V4 in monkeys performing the same task and viewing the same stimuli that we used in IT cortex suggests that suppression with repetition occurs in V4 (Miller, Li, and Desimone, unpublished data). Presumably, the temporal interval over which stimuli are compared is much smaller (and less able to span intervening stimuli) in earlier visual areas, and the features that are compared are less complex than in IT cortex. The results of the experiments described below indicate that, in IT cortex at least, the temporal interval over which stimuli are compared can span periods comparable to long-term memory.

The notion of temporal figure-ground may also have applications to artificial visual systems. Current artificial systems typically separate visual processing from visual memory, whereas the biological visual system seems to integrate both at relatively early stages, building on them incrementally. This integration and incremental build-up may not only result in processing efficiencies and in a natural interface to attentional systems, but may also simplify visual recognition, which would not have to be accomplished in one final stage.

REPRESENTATION OF FAMILIARITY

Physiological studies of short-term memory, like those described above, typically use stimuli that are highly familiar to the monkey, and mnemonic

Desimone, Miller, and Chelazzi

effects are confined to a single trial of the monkey's task. However, by testing the responses to novel stimuli that the monkey has never seen before, one can observe adaptive memory filtering spanning time periods consistent with long-term memory formation for stimuli in IT cortex (Miller et al. 1991b; Li et al. 1993). As with short-term memory, we use the term "long-term memory" here in a generic sense, without (initially) claiming any specific type or component. The behavioral task we use to study long-term memory is basically the same as in the short-term memory study, but we monitor changes in the response to the initially novel sample stimuli over the session, as the animal gradually becomes familiar with them. Thus, we study the memories that are incidentally acquired during task performance. For a third of the cells, the response to novel stimuli systematically declines as the stimuli become familiar to the animal over the course of an hour-long recording session. This decline is stimulus-specific and is found even when more than 150 other stimuli (the maximum tested) intervene between repeated presentations. Virtually all of these cells also show selectivity for particular visual stimuli and, thus, are not "novelty detectors" in the sense of cells that respond to any novel stimulus. Rather, IT cells combine sensitivity to novelty and familiarity with sensitivity to other objects features such as shape and color. Furthermore, as shown in figure 4.2, these same cells show suppression to matching stimuli within a trial, which is added to the suppression caused by familiarity. Thus, these cells appear to communicate both types of information—both recency and familiarity—and the effects of memory appear to be additive. As in the short-term memory experiment, at least half the cells convey sensory information only and are unaffected by the contents of memory.

To assess the possible role of feedback, we measure the time course of the response suppression to familiar stimuli in the population of cells. We compare the average population response to novel sample stimuli and to the same stimuli after they had been seen just once before. The response waveform to the stimuli seen once before does not show suppression until 170–180 msec after stimulus onset. This is clearly enough time for the suppression to be mediated by feedback to IT. However, after the stimuli had been seen as samples just one additional time, suppression is evident from the very onset of the visual response, 80 msec after stimulus onset. Thus, after the first couple of presentations of a stimulus, IT networks appear to detect familiar stimuli on their own.

Because fewer IT cells respond strongly to a stimulus after it has become familiar, familiarity may cause a "focusing," or narrowing, of activity across IT cell populations, resulting in a sparser representation of the stimulus (see chapter 1 by Barlow and chapter 8 by Poggio and Hurlbert). Cells that poorly represent the important features of a given familiar stimulus may be winnowed out of the population of strongly activated cells through the adaptive filtering mechanism, in much the same way that an excess of cells and connections are pruned during development. If the remaining activated cells have the appropriate association with cells cod-

The Interaction of Neural Systems for Attention and Memory

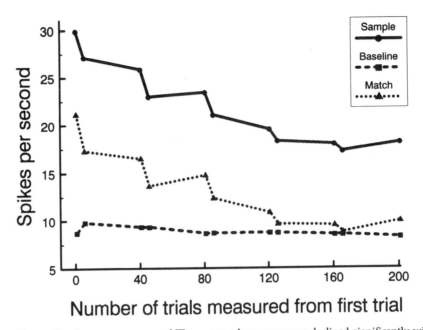

Figure 4.2 Average responses of IT neurons whose responses declined significantly with increasing stimulus familiarity over the recording session. The curves show the average response to 20 initially novel stimuli tested for each cell (each cell was tested with a new set). The solid line indicates the responses to the stimuli when they were samples, the dotted line indicates responses to the same stimuli when they were matching test items at the end of each trial, and the dashed line gives the baseline (prestimulus) firing rate. Trial number is measured from the first trial of a given stimulus. The staircase appearance results from two different numbers of intervening trials between successive presentations of a given novel stimulus as the sample. Three or 35 trials, in alternation, intervened before a given stimulus was repeated as the sample on another trial. Responses declined most when only 3 trials intervened but the decrement was retained ("remembered") even when 35 trials intervened. The greater decrement after 3 trials demonstrates that the decrement is stimulus-specific and cannot be explained by simple fatigue. If the cells were simply becoming fatigued, the decrement should have been greater after 35 intervening trials, as the cells were stimulated by far more stimuli during that interval than when only three trials intervened. (Adapted from Li et al. 1993)

ing contextual information (the circumstances in which the stimuli were seen, for example), they might mediate, in part, an explicit memory of the familiar stimulus.

On the other hand, the adaptive filtering mechanism might mediate priming phenomena, which reflect implicit memory. In a typical priming task for visual patterns, subjects are first shown a list of drawings, without any instruction to remember them. Later, they are given a picture recognition task, and their performance is usually faster or better for the stimuli that had been seen before, even if they have no conscious memory of having seen them (Schacter et al. 1990, 1991; Squire 1992). It is commonly believed that priming is due to a tendency of neuronal populations to be more easily activated if they have been activated previously (see chapter 9 by Posner and Rothbart). For individual cells, we have shown that this is

Desimone, Miller, and Chelazzi

not the case, at least in IT cortex. However, as we suggest above, it is the elimination of certain cells from the activated population that may be important in forming the underlying neuronal representation of a stimulus. Repetition may speed the construction of this critical population, resulting in faster and better recognition when a stimulus is repeated. Results from a recent PET study of priming are consistent with the idea that priming is associated with a reduction of activity in the cortex. Subjects performing a word-stem completion task showed less activation of temporal cortex when they had previously seen the words (Squire et al. 1992).

ADAPTIVE FILTERING CELLS AND BEHAVIOR

Could the responses of IT cells actually support the animal's behavioral performance in recency memory tasks or in tasks requiring judgments of novelty and familiarity? We have used both simulated neural networks and statistical models (discriminant analysis) to classify stimuli as matching or nonmatching based on the trial-by-trial responses of individual IT neurons (Miller et al. 1993). The networks or models are trained on half the data and then applied to the other half, for cross-validation. Although no individual cell performs as well as the animal as a whole, we find that we can, in principle, achieve the animal's behavioral level of performance by averaging the responses of only 25 neurons. We would expect to obtain similar results from the novelty-familiarity data. Thus, mnemonic information equivalent to the animal as a whole is apparently distributed down to the level of small neural populations. Comparable results have been found in area MT for motion information (Britten et al. 1992).

The classification model we used assumes that the decision networks which interpret the outputs of IT cells "know" (or have been trained on) the match–nonmatch response distributions of the cells for specific stimuli. One can avoid this assumption if the decision network is supplied with information from both the adaptive cells in IT as well as from the cells that provide purely sensory information. The responses of these latter cells, which make up about 50% of the total population of cells in IT cortex, would provide a stable sensory "referent" for comparison with responses of the adaptive cells. As shown in figure 4.3, the difference in response of the two populations would be proportional to the similarity of the current stimulus to stimuli held in either short- or long-term memory. If the responses of the minority of cells showing match enhancement were added to the sensory pool, the magnitude of the response difference would be even larger, resulting in a better signal-to-noise ratio.

Beyond its role in memory tasks, we believe that the activation level of adaptive cells in IT cortex (or the mismatch signal between the adaptive cells and the sensory cells) may be an important drive on attentional and orienting systems. Although no individual cells in IT cortex appear to be either "novelty detectors" or "unexpected stimulus detectors," the summed activity of adaptive cells in IT cortex could provide a signal to

The Interaction of Neural Systems for Attention and Memory

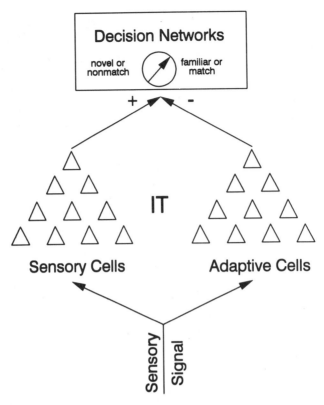

Figure 4.3 Model of how sensory cells and adaptive mnemonic filter cells in IT cortex contribute to novelty or match-nonmatch decisions. The plus and minus are meant to indicate that the outputs of the adaptive and sensory cells work in opposition, and are not meant to imply excitation or inhibition. (Adapted from Miller et al. 1993)

other systems that the current stimulus is new and deserving of attention. Conversely, it could be regarded as a mechanism for discounting the familiar and the expected. As Barlow points out in chapter 1, discounting the expected features of the incoming sensory messages seems to be an essential element of effective learning. As the organism orients and attends to a new stimulus, activated IT cell populations shrink to the critical set necessary for representing the stimulus. This shrinkage reduces the overall activity in IT cortex, reducing the drive on the orienting system and freeing the organism's attention for other, competing, stimuli. Thus, it is behavior that completes the loop. One could view this as a memory-guided system for self-directed behavior, whose goal is to incorporate knowledge about new stimuli into the structure of the cortex (see figure 4.4). Carpenter and Grossberg's (1987) Adaptive Resonance Theory (ART) makes use of a similar sort of feedback between memory and attention as a means for adaptively adjusting the categories in which items are stored in memory.

The adaptive filtering mechanism suggests that attention will be automatically biased toward stimuli that are novel or have not been recently seen. However, it is frequently necessary to voluntarily attend to familiar

Desimone, Miller, and Chelazzi

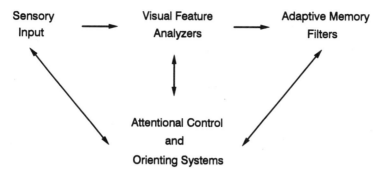

Figure 4.4 Interaction between systems for memory and attention. In this scheme, a stimulus that is novel or has not been recently seen will activate adaptive memory filter cells in IT cortex, which in turn will drive attentional and orienting systems. This will lead to increased attention and contact with the new stimulus, causing adaptation of synaptic weights in IT cortex, reducing the activation of the cells. When the novelty of the new stimulus and the activation of IT cortex is sufficiently diminished, the system will be ready to process other competing stimuli. (Adapted from Li et al. 1993)

or expected stimuli and even suppress attention to novel stimuli (e.g., when looking for a pariticular item or when trying to "pay attention"). How this might be accomplished neurally is considered in the next section.

ACTIVE MEMORY MECHANISMS

It is commonly assumed that the standard matching to sample task that we and others have used is solved by holding the sample, A, "in mind" (i.e., in some short term storage buffer) during the delay and comparing each of the test items to it. That is, the memory mechanism is thought to require an active process, or "working memory," linked specifically to the sample stimulus, much the way one might rehearse a new phone number. However, in the standard form of the task, the sample stimulus (e.g., "A") is the only stimulus repeated within a given trial (e.g., "A B C D A"). Conceivably, then, the task might as well be solved by a neural mechanism that automatically detects any stimulus repetition (e.g., "A A"), without requiring active maintenance of the sample stimulus memory. Does the adaptive filtering mechanism function automatically, or does it require active storage of the sample item? To distinguish between these possibilities, we record from IT neurons in two versions of the matching to sample task (Miller and Desimone 1994). One version is the same as used previously, which may be solved by detecting any repetition within a trial ("A B C A"). In the other version, intervening nonmatch items during the delay period may match each other, but the animal must ignore these and respond only to the one item that matches the sample (e.g., "A B B A," where only A is a match). We will refer to these as the standard and ABBA versions, respectively.

Interestingly, animals initially taught the standard task respond to the repeated nonmatch stimuli when first presented with ABBA trials (i.e., they respond to the second occurrence of "B"). The animals therefore apparently learn to solve the task using a simple repetition rule. After additional training, animals eventually learn to match items to only the sample stimulus.

After the monkeys have learned the ABBA task, we still find adaptive filtering cells whose responses are suppressed to the matching stimulus. However, the ABBA trials reveal that the responses of these cells are also suppressed to the intervening stimuli that match each other. Thus, the adaptive filtering mechanism seems to be sensitive to simple repetition. It should be noted that human observers easily notice the repeated intervening stimuli when performing the ABBA task, at the same time that they are able to detect the one stimulus that matches the sample.

By contrast, another class of IT cells gives enhanced responses to matching stimuli (which was rarely found prior to the training on the ABBA trials) but only to the one stimulus that matches the sample item on a given trial, that is, the one stimulus the animal is actively holding in memory. Simple stimulus repetition has no effect on these cells, as repeated intervening items show no enhancement. These cells apparently can be dynamically "biased" to respond to a specific stimulus, and this bias can span many seconds and at least several intervening stimuli. Such a bias is presumably mediated by "back projections" to IT cortex (see chapter 2 by Churchland, Ramachandran, and Sejnowski and chapter 12 by Ullman), possibly from prefrontal cortex. The fact that the bias is apparently under the animal's control suggests a relation to the active components of "working memory." In this case, IT cortex appears to be the site of dual short-term memory mechanisms operating in parallel: a suppressive mechanism underlying the automatic detection of repetition and an enhancement mechanism underlying working memory. In the next section, we consider how these different memory mechanisms function when viewing a typical crowded scene.

VISUAL SEARCH

In visual search for a particular object, the representation of the object in memory is used to guide the search of the external scene (e.g., searching for a face in a crowd). This type of search is often distinguished from "preattentive" or "pop-out" tasks, where a subject finds a stimulus that stands out from its background on the basis of a strong featural cue such as color or luminance. Most agree that such stimuli are extracted from their background based on a parallel process operating over the full display. However, there are two competing notions of how search of the former sort takes place. We will describe the two extremes, but hybrid models are possible. According to the serial search account, the scene is searched element by element by a "spotlight" of attention (Bergen and Julesz 1983; Treisman 1988). The element selected by attention is evaluated by a recognition memory process, which is terminated when the target is found. As elements are added to the scene, it takes longer and longer to find the target. By contrast, according to parallel search accounts, all elements of the

Desimone, Miller, and Chelazzi

scene are processed in parallel and compete for access to decisional mechanisms, attentional mechanisms, and so on (Duncan and Humphreys 1989; Bundesen 1992). The mnemonic template of the searched-for object biases the competition in favor of the neurons coding that particular object, much the same way that a strong color difference may bias the competition towards a unique element in a "pop-out" display. As elements are added to the scene, it takes longer to find the target because of the reduction in signal-to-noise ratio. Attentional scrutiny may follow the localization of the target according to this view, but is not essential to find it. We have recently begun to investigate the neural basis of this type of visual search in anterior IT cortex (Chelazzi et al. 1993). Although we are not yet able to provide a full account of the mechanism of search, the results show how memory and attention interact in IT cortex.

Monkeys are presented with a complex picture (the cue) at the center of gaze to hold in memory. The cue is always either a "good" stimulus that elicits a strong response from the cell or a "poor" stimulus that elicits little or no response. Both good and poor cues are highly familiar to the monkey. After a delay, the good stimulus and the poor stimulus are both presented simultaneously as choice stimuli, at extrafoveal locations in the contralateral visual field. Because of the large receptive fields of IT neurons, the two choice stimulus locations are typically both within the receptive field. The animal is trained to make a saccadic eye movement to the target stimulus that matches the cue, ignoring the nonmatching stimulus (the "distractor"). Thus, unlike in our previous short-term memory experiments, the animal does not have to indicate whether a matching stimulus was present (one was always present in the choice array), but rather must find the matching stimulus, or separate it from the nonmatching stimulus in the array. Any suppression of response to the matching stimulus in this task might place it at a competitive disadvantage to the nonmatching stimulus, possibly interfering with its ability to capture attention. A further difference is that the cue and the matching choice stimulus are never presented at the same retinal location, that is, the cue and matching target never activate the same retinal elements. It is possible, therefore, that the animal must compare the choice stimuli with a more abstract representation of the cue.

We find that the cue typically initiates activity that persists through the following delay period among the neurons that are tuned to the cue's features. Although the frequency of this maintained activity is relatively low (average = 7.9 Hz following best cue and 5.6 Hz following worst cue) it is much more common than in our previous memory experiments and lasts for a considerable time (at least 3 sec, the maximum delay tested).

In addition to information about the cue, IT cells communicate information about the target. Relative to the preceding delay activity, the initial population response to the array is about the same regardless of which choice stimulus is the target. By contrast, the late phase of the response changes dramatically depending on whether the animal is about to make an eye movement to the good or poor stimulus. If the target is the good

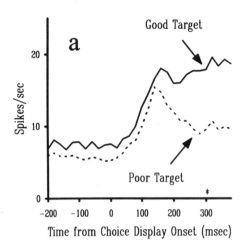
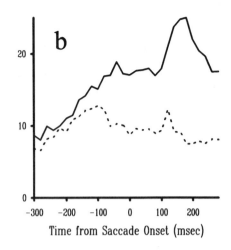

Figure 4.5 Average firing rates of IT cells to choice arrays in a visual search task, in which either the good (solid lines) or poor (dashed lines) stimulus was the target. (*a*) Responses time-locked to array onset. Average time of saccade onset, 306 msec after array onset, is indicated by an asterisk. (*b*) Same data as in (*a*), but responses are time-locked to saccade onset. Data in all graphs are from trials in which the target and distractor appeared in the hemifield contralateral to the recorded cell, whether the target was in the upper quadrant and the distractor in the lower, or vice versa. (Adapted from Chelazzi et al. 1993)

stimulus, the response remains high, but if the target is the poor stimulus, the response to the good distractor stimulus is suppressed even though it is still within the receptive field (figure 4.5). This suppression of response begins about 200 msec after the onset of the choice array, or about 90–120 msec before the start of the eye movement. The cells respond as though 100 msec or so before the eye movement, the target stimulus "captures" the response of the cells, so that neuronal activity in IT would reflects only the target's properties. This target information in IT cortex is available to drive attentional as well as oculomotor systems, resulting in eye movements to the chosen object (Glimcher and Sparks 1992). More generally, networks in IT cortex may select the visual objects that are acted on by attentional and motor systems when selection is guided by object features.

These effects are found when target and distractor are both within the hemifield contralateral to the recorded neuron. By contrast, there is little effect on responses prior to the eye movement when the stimuli are separated by the vertical meridian. Moran and Desimone (unpublished data) and Sato (1988) also found reduced effects of spatially directed attention in IT cortex when attended and ignored stimuli were located in opposite hemifields, suggesting that stimuli in the two hemifields may be processed largely independently. Although IT neurons often have bilateral fields, the response to stimuli in the contralateral hemifield is typically much larger than to stimuli in the ipsilateral hemifield. When both choice stimuli are in the contralateral hemifield, a further surprising result is that many IT cells respond differently depending on the relative spatial locations of target and distractor. Thus, some spatial (retinal) information appears to be re-

Desimone, Miller, and Chelazzi

Cells in IT cortex

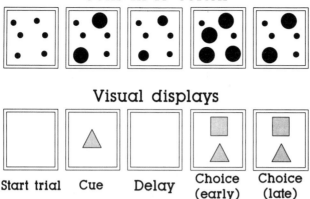

Visual displays

| Start trial | Cue | Delay | Choice (early) | Choice (late) |

Figure 4.6 Upper diagrams are schematic representations of activity in a population of IT neurons during performance of the task. Each dot represents an individual cell, and the size of the dot indicates relative firing rate. Lower diagrams illustrate the visual displays during the relevant portions of the task. A specific cue activates the subpopulation of IT cells tuned to any of the various features of the cue. During the delay period, this subpopulation remains more active than other cells. When the choice array first appears, cells are initially activated by whichever stimulus they prefer in the array, regardless of which is the target. In some cases the initial responses are identical but in others the response to the array with the good target starts off at a higher rate because of persisting activity from the delay. Later, within 90–120 msec of saccade onset, the cells tuned to the properties of the target stimulus remain active, whereas cells tuned to the properties of the distractor are suppressed. Whether this final divergence in activation results from competitive interactions within IT cortex (e.g., through mutual inhibition between cells selective for target and distractor) or from interactions between IT cortex and an attentional control system is not yet known. (Adapted from Chelazzi et al. 1993)

tained in IT cortex, despite the large receptive fields. Retinal location may simply be another coarsely coded feature in IT cortex and thus be available to oculomotor systems for directing the eyes.

It seems as though we have narrowed the selection component of visual search to a 200-msec period following the onset of a peripheral search array. By the end of this time, neural responses begin to communicate almost exclusively the properties of the chosen target. Within this critical 200-msec period, we have observed the same mnemonic phenomena we observed in IT cortex in the memory tasks with a single stimulus in the field, namely suppressed responses when the target array contains a stimulus that matches the previously seen cue (adaptive filtering cells), enhanced responses (for other cells) when the good stimulus is the target, and elevated activity in the delay interval when the good stimulus is the cue/target. Any or all of these effects on responses may bias competition within IT networks such that only those cells coding the target are responding by the end of 200 msec (figure 4.6).

It is particularly intriguing that the maintained activity during the delay interval is much more pronounced than in the previous memory experiments and, in some conditions, it persists into the initial response to the choice array. If the monkey is given the good stimulus as a cue, for example, the cell often shows higher activity during the delay, increasing the initial response to the array with the good stimulus as the target. This heightened activity may give those cells tuned to the properties of the target a competitive advantage within IT, causing them to win a competition against the cells tuned to the properties of the distractor. Such a mechanism would be consistent with the "parallel search" model described above. Increasing the steady-state firing rate of a population of cells coding a particular stimulus may be a way of segregating, or binding, those cells that will contribute the most to a later perceptual decision. Consistent with this idea, we have recently found evidence for sustained firing of V4 cells when attention is directed to the center of the cell's receptive field (Luck et al. 1993). Alternatively, the higher steady-state activity might turn out to be simply a reflection of greater synchronous firing among the relevant population of neurons, in which case the synchronicity itself could be the binding mechanism, not the maintained activity (see chapter 5 by Koch and Crick and chapter 10 by Singer).

Finally, although we have described the responses of IT cells as though all of the critical interactions take place within IT cortex itself, it is possible that the ultimate suppression of IT responses to distractors is due to interactions with attentional systems outside of IT cortex. A spatial attention system could, for example, gate inputs into IT from one stimulus in the array at a time. The appropriate modulation of IT responses by the joint action of the selected stimulus and the stimulus held in memory would constitute "recognition," terminating the scan. Such a mechanism would be consistent with the "serial scan" model described above. Anatomical considerations dictate that such a mechanism could operate in either of two ways (Desimone 1992): either by gating the inputs into IT cells (Anderson and Van Essen 1987) or by gating the cells themselves on and off (Crick and Koch 1990b; Niebur, Koch, and Rosin 1993). Different approaches to attentional gating are described in chapter 5 by Koch and Crick and chapter 13 by Van Essen, Anderson, and Olshausen.

CONCLUSIONS

It is natural to think of attention as the gateway to memory, as we typically remember only those things that we attend to. Correspondingly, ignored stimuli are filtered from the receptive fields of extrastriate neurons. Information cannot be remembered if it has been removed from the visual system. However, as we have seen, the contents of memory also guide our attention. Memory for objects is reflected both in the maintained activity of IT cells in the absence of any stimuli as well as in the responses of cells to current stimuli. New or not-recently-seen stimuli cause the greatest activation of adaptive filtering cells in IT cortex, and the difference in overall activity between these cells and sensory cells unaffected by memory may be one of the signals that drives attentional mechanisms. This temporal "figure-ground" mechanism most likely builds up incrementally within the visual cortex. As the organism attends to the new inputs, activity in the adaptive filtering cells declines, reducing the drive on attentional systems. Many times, though, we need to search for a particular familiar object and suppress orienting to novel ones. In this case, top-down mechanisms are able to bias those cells that code the searched-for item, resulting in enhanced activation when the stimulus occurs. An even more pronounced case of memory-guided attention occurs in visual search, where the representation of a target item in memory is used to guide the search of an external scene containing many stimuli. Within 200 msec, the interaction between the neural representation of the external array and the memory trace of the target results in the target "capturing" the responses of IT cells. Information about nontargets is almost completely suppressed. Thus, interactions between memory and attention in IT cortex result in the selection of objects that are foveated and acted on.

ACKNOWLEDGMENTS

We gratefully acknowledge the advice and support of Mortimer Mishkin in all phases of the work. Lin Li and John Duncan collaborated on the studies of memory and visual search, respectively. Andreas Lueschow and Leslie Ungerleider provided helpful comments on the manuscript. The work was supported in part by the Human Frontiers Science Program Organization and the Office of Naval Research.

5 Some Further Ideas Regarding the Neuronal Basis of Awareness

Christof Koch and Francis Crick

INTRODUCTION

What goes on in our head while we perceive a seagull gliding effortlessly through the air? And what happens if we drive down the freeway, concentrating on an upcoming lecture and barely aware of our visual surroundings? While our "mind" is occupied with perceiving the seagull or contemplating some future event, our brain continues to perform a series of quite complex functions such as climbing over rocks on a beach or maneuvering a car. What causes one event to capture our attention (i.e., to have access to an privileged internal state that is central to our conscious experience), while the myriads of other sensory events that we are continously being bombarded with never reach awareness? Because we are neuroscientists, we are couching this question in a radical reductionist framework: What is the correlate of awareness (and ultimately consciousness) at the level of nerve cells? Expressed differently: What telltale signs should the electrophysiologist look for when searching for the neuronal correlate of awareness? A particular type of neuron in a particular part of the brain? A special form of neuronal electrical activity?

With very few exceptions, almost no modern neuroscientist has asked this sort of question, let alone provided an answer. The problem of "awareness" is either felt to be purely philosophical or too elusive to study experimentally. In our opinion, such timidity in the face of one of the most puzzling questions that we can ask is ridiculous! Given the pre-Copernican state of the field, it is too early for any definite theory of the neuronal basis of awareness. Yet we believe that the time is ripe to provide at least a theoretical framework to allow us to seek answers using the well-proven tools of the neuroscience, in particular electrophysiology!

We have already (Crick and Koch 1990a,b, 1992) described our general approach to the problem of visual awareness. In brief, we believe the next important step is to find experimentally the *neural correlates* of various aspects of visual awareness, that is, how best to explain our subjective mental experience in terms of the behavior of large groups of nerve cells. At this early stage in our investigation we will not worry too much about many fascinating but at the moment unrewarding aspects of the problem, such as the exact function of visual awareness, what species do and what species do not have awareness, different forms of awareness (such as dreams and visual imagination), and the deep problem of *qualia*. We here restrict our attention mainly to results on man and on the macaque monkey, since their visual systems appear to be somewhat similar and, at the moment, we cannot obtain all the information we need from either of them separately. When this information is lacking we will refer to related results on the cat.

Our main assumption is that, at any moment, the firing of some but not all the neurons in what we call the visual cortical system (which includes the neocortex and the hippocampus as well as a number of directly associated structures, such as the visual parts of the thalamus and possibly the claustrum) correlates with visual awareness. Yet, visual awareness is highly unlikely to be caused by the firing of *all* neurons in this system that happen to respond above their background rate at any particular moment. If at any given point in time only 1% of all the neurons in cortex fire significantly, about one billion cells in sensory, motor, and association cortices would be active and we would never be able to distinguish any particular event out of this vast sea of active nerve cells. We strongly expect that the majority of neurons will be involved in doing computations, while only a much smaller number will express the results of these computations. It is probably that we become aware of only the latter. There is already preliminary evidence from the study of the firing of neurons during binocular rivalry that in area MT of the macaque monkey only a fraction of neurons follows the monkey's percept (Logothetis and Schall 1989). We can thus usefully ask the question: What are the essential differences between those neurons whose firing does correlate with the visual percept and those whose firing does not? Are the "awareness" neurons of any particular cell type? Exactly where are they located, how are they connected, and is there anything special about their patterns of firing?

To look for such neurons it may be useful to have some tentative ideas as to where they are and how they might behave, if only to guide experiments. The following suggestions are offered in that spirit.

In our previous papers (Crick and Koch 1990a, 1992) we outlined the ideas of psychologists that consciousness involves a form of short-lasting memory, possibly including both iconic as well as short-term memory, and that it appears to be greatly enriched by selective attention. Accordingly, we need to search for a mechanism that mediates selective visual attention as well as a short form of visual memory.

We also explained that there does not appear to be one single cortical area whose activity corresponds to what we see. The necessary visual information is distributed over many cortical areas. Thus, there has to be a process that in some way binds together the neural activity in many different places to form our unitary view of an object or event in the visual scene. How this is done is often referred to as the *binding problem*.

It is important to distinguish at least three different forms of binding (Crick and Koch 1990a). A simple cell, responding best to motion perpendicular to its optimal orientation, can be thought to exemplify a low-level type of binding (at least three features, spatial location, orientation and direction of motion are combined in one such cell). The second type of binding is acquired by learning a small class of ecological very important stimuli, such as faces (or letters for humans). Due to repeated exposure to these patterns, a set of neurons has become uniquely responsive to just these stimuli. Because both of these mechanisms have only a fairly limited capacity, a third type of very rapid and transient form of binding with a practically infinite capacity is proposed. It seems probable that an attentional mechanism is usually necessary for this rapid and transient binding to occur. This rapid form of binding, can, by overlearning, eventually be carried out by a specialized set of neurons (i.e., by the second form of binding).

We also hypothesized that in addition to vivid visual awareness there may be an extremely transient form of awareness that symbolizes only rather primitive features without binding them together. We will return to this in the last section.

OUR THEORETICAL FRAMEWORK

It may be useful to state our 'fundamental' assumptions at the beginning. We assume the following:

1. To be aware of an object or an event the brain has to construct an *explicit, multilevel, symbolic interpretation* of part of the visual scene. By *explicit* we mean that one such neuron (or a few closely associated ones) must be firing above background at that particular time in response to the feature they symbolize. The pattern of color dots on a TV screen, for instance, contains an "implicit" representation of, say, a person's face, but only the dots and their locations are made explicit here; an explicit face representation would correspond to a light that is wired up in such a manner that it responds whenever a face appears somewhere on the TV screen. By *multilevel* we mean, in psychological terms, different levels such as those that correspond, for example, to lines, eyes, or faces. In neurological terms we mean, loosely, the different levels in the visual hierarchy (see Felleman and Van Essen 1991). By *symbolic*, as applied to a neuron, we mean that neuron's firing is strongly correlated with some "feature" of the visual world and thus symbolizes it (this use of the word "symbol" should not be taken to imply the existence of a homunculus who is looking at the symbol). The *meaning* of such a symbol depends not only on the neuron's receptive field (i.e., what visual features the neuron responds to) but also on what other neurons it projects to (its projective field). Whether a neural symbol is best thought of as a scalar (one neuron) or a vector (a group of closely associated neurons as in population coding in the superior colliculus; Lee et al. 1988) is a difficult question that we shall not discuss here.

2. Awareness results from the firing of a coordinated subset of cortical (and possible thalamic) neurons that fire in some special manner for a certain

length of time, probably for at least 100 or 200 msec. This firing needs to activate some type of short-term memory by either strengthening certain synapses or maintaining an elevated firing rate or both. Experimental studies involving short-term memory tasks in the temporal lobe of the monkey (Fuster and Jervey 1981) have provided evidence of elevated firing rates for the duration of the interval during which an item needs to be remembered. It is at present not possible to assess empirically to what extent synapses undergo a short-term change during a memory task in the animal. We are assuming that the semiglobal activity that corresponds to awareness has to last for some minimum time (of the order of 100 msec) and that *events within that time window are treated by the brain as approximately simultaneous*. An example would be the flashing for 20 msec of a red light followed immediately by 20 msec of a green light in the same position. The observer sees a transient yellow light (corresponding to the mixture red and green) and *not* a red light changing into a green light (Efron 1973). Other psychophysical evidence shows that visual stimuli of less than 120–130 msec produce perceptions having a subjective duration identical to those produced by stimuli of 120–130 msec (Efron 1970a,b).

3. Unless a neuron has an elevated firing rate and unless it fires as a member of such an (usually temporary) assembly, its firing will not *directly* symbolize some feature of awareness at that moment.

These ideas, taken together, place restrictions on what sort of *changes* can reach awareness. An example would be the awareness of movement in the visual scene. Both physiological and psychophysical studies have shown that movement is extracted early in the visual system as a primitive (by the so-called short-range motion system; Braddick 1980). We can be aware that something has moved (but not *what* has moved) because there are neurons whose firing symbolizes movement as such, being activated by certain changes in luminance. To know *what* has moved (as opposed to a mere change of luminance) there must be active neurons somewhere in the brain that symbolize, by their firing, that there has been a change of *that* particular character. In any instance, such neurons *may* be present in the brain but it cannot be assumed that they *must* be there.

3.1 As a corollary, we formulate our *activity principle*: Underlying every direct perception is a group of neurons strongly firing in response to that stimulus that come to *symbolize* it. An example is the "Kanizsa triangle" illusion, in which three Pacmen are situated at the corners of an triangle, with their open mouths facing each other. Human observers see a white triangle with illusory lines, even though the intensity is constant between the Pacmen. As reported by von der Heydt et al. (1984), cells in V2 of the awake money strongly respond to such illusory lines. Another case is the filling-in of the blind spot in the retina (Fiorani et al. 1992). Since we do not have neurons that explicitly represent the blind spot and events within it, we are not aware of small objects whose image projects onto them and can only infer such objects indirectly.

A semiglobal activity that corresponds to awareness does not itself symbolize a *change* within that short period of awareness unless such a change is made explicitly by some neurons whose firing makes up the semiglobal activity (for what else but another group of neurons can express the notion that a change has occurred?). These ideas are very counterintuitive and are not easy to grasp on first reading, since the "fallacy of the homunculus" slips in all too easily if one does not watch out for it.

3.2 It follows that active neurons in the cortical system that do *not* take part in the semiglobal activity at the moment can still lead to behavioral changes but without being associated with awareness. These neurons are responsible for the large class of phenomena that bypass awareness in normal subjects, such as automatic processes, priming, subliminal perception, learning without awareness, and others (Tulving and Schacter 1990; Kihlstrom 1987) or take part in the computations leading up to awareness. In fact, we suspect that the majority of neurons in the cortical system at any given time are not directly associated with awareness!

The elevated firing activity of these neurons also, of course, explain blindsight and similar clinical phenomena where patients with cortical blindness can point fairly accurately to the position of objects in their blind visual field (or detect motion or color) while strenuously denying that they see anything (Weiskrantz 1986; Störig and Cowey 1991).

We have argued (from the experiments on binocular rivalry) that the firing of some cortical neurons does *not* correlate with the percept. It is conceivable that *all* cortical neurons may be capable of participating in the representation of one percept or another, though not necessarily doing so for *all* percepts. The secret of visual awareness would then be the type of activity of a *temporary subset* of them, consisting of all those cortical neurons that represent that particular percept at that moment. An alternative hypothesis is that there are special sets of "awareness" neurons somewhere in cortex (for instance, layer 5 bursting cells; see below). Awareness would then result from the activity of these special neurons.

In any case, it is crucial to ask what exactly is the detailed nature of the particular type of neuronal activity giving rise to awareness? Does it involve oscillatory activity, synchronized firing, or bursts of high-frequency activity? And how can this activity be decoded?

THE NEURONAL SPIKE CODE OF AWARENESS

Earlier, we postulated that awareness is mediated by a coordinated subset of cortical (and possibly also thalamic) neurons firing in some special manner for a certain length of time. How can this coordinated subset of cells be formed—and disbanded—quickly? Given that the only rapid mode of communication among cortical cells involves action potentials, at least four possibilities for defining membership in this assembly using spikes come to mind: high-frequency activity (rate code), oscillations in the 40 Hz range, bursting, and synchronized firing activity. We do not discuss here the later idea, that all cortical neurons whose detailed spike patterns are temporally synchronized to each other, firing action potentials at about the same time, constitute the neuronal assembly coding for awareness, but refer the interested reader instead to chapter 10 by Singer on this topic (see also Abeles 1991). Singer also discusses the relationship between oscillations in the γ range and synchronization.

It is possible, of course, that the brain uses much more complex spatiotemporal neuronal activity patterns than discussed here to encode aspects of awareness. The principal component analysis of spike trains in the awake monkey (Richmond and Optican 1990, 1992) suggests such a possible coding. We feel that at this preliminary stage in our investigations it is best to first investigate the more obvious possibilities.

When discussing these different scenarios, it is important to keep in mind that at the psychological level, awareness of an event or object appears to involve attending to this object *and* placing it into short-term memory. We must therefore ultimately find a link between the different forms of constituting a neuronal assembly and short-term or working memory (Crick and Koch 1990a).

Rate Coding

The simplest encoding uses mean firing frequency. Visual awareness of objects is correlated with all neurons, say in inferior temporal (IT) cortex, that fire above a certain threshold, irrespective of the temporal character of the spike train. This assumes that all IT neurons corresponding to nonattended features are suppressed, as suggested by the attentional experiment of Moran and Desimone (1985) in monkey IT. In other words, at any given time only those neurons whose features correspond to the current content of awareness are highly active, while the firing of the vast majority of neurons is not significantly elevated above their background rate. Here the binding problem is solved trivially by virtue of the fact that only those neurons corresponding to the attended event or object are active and the problem of incorrect binding does not arise.

The beautiful experiments carried out by Newsome and his colleagues (Newsome et al. 1989; Britten et al. 1992) support the notion that the firing rate of MT cells directly encode motion perception. Analyzing the total number of spikes discharged by a cell during the 2 sec long experiment allows an "ideal" observer to mimic the observed psychophysical performance of the animal to a remarkable extent. In these experiments, the direction and speed of the random dot motion stimulus used were optimized for each cell recorded from.

As shown by the Logothetis and Schall (1989) experiment, only a minority of neurons in MT follow the changing percept associated with the binocular rivalrous stimulus. Thus, it may well be possible that awareness only correlates with high-frequency activity in parts of cortex further removed from the sensory periphery, such as V4, IT, or prefrontal cortex due to relevant differences in circuitry or biophysics (only enabling neurons in these areas, for instance, to store short-term memory). In our opinion it is unlikely that cells behave in such an all-of-none manner in higher cortical areas, that is firing only if they belong to the neuronal assembly encoding awareness.

The major computational disadvantage of rate encoding for solving the binding problem is the associated loss in bandwidth. Using a sort of temporal code allows the superposition of information coded via the firing frequency (i.e., orientation, speed, hue, etc.), and information coded in time, here that the neurons are part of a particular assembly (von der Malsburg 1981; von der Malsburg and Schneider 1986).

Neuronal Oscillations

In our original publication on this topic (Crick and Koch 1990a), we hypothesized that all neurons corresponding to various aspects of the object the observer is currently directing attention to fire in an oscillatory and semisynchronous manner, binding them together. Chapter 10 by Singer discusses the background and current status of these oscillations (see also Koch 1993). We here briefly highlight their status with respect to cat and monkey cortex.

Oscillations in the Cat Semisynchronous oscillations in local field potentials as well as multi- and single-unit activity in the 30–70 Hz range were reported in the first visual (V1) area of the lightly anesthetized cat (Eckhorn et al. 1988; Gray and Singer 1989; Gray et al. 1989). These oscillations were subsequently shown to be present in awake kittens and more recently in alert cats.

The 40-Hz oscillations are most clearly seen in the local field potentials, rather than in individual neuron, although recently, Jagadeesh et al. (1992) using in vivo patch-clamping, reported that the intracellular membrane potential of cells in cat visual cortex oscillate in the 40 Hz range. About 10% of simple cells, but more than half of all recorded complex cells show oscillations with a mean value around 50 Hz. The oscillations may last for relatively short periods (such as 200 msec) and vary in frequency between trials, so that averaging over longer periods may make them almost invisible. The oscillations themselves are not locked to the stimulus. The site of origin of these oscillations is still controversial. Two intracellular studies have revealed depolarization-dependent subthreshold oscillations in cortical cells in the 40 Hz range (Llinas et al. 1991; Nunez et al. 1992). Yet, strong 50-Hz oscillations have been seen in cat geniculate cells and may originate already in the retina (Ghose and Freeman 1992); furthermore, subthreshold oscillations in the 20–40 Hz band can be evoked by intracellular current injections in cat thalamocortical relay cells (Steriade et al. 1991). It appears that 40-Hz oscillations can be generated at a number of sites in the nervous system.

When recording from two distinct sites in cortex, the oscillations can be phase-locked with a phase-shift of ±3 msec around the origin (Engel et al. 1990; see also chapter 10 by Singer). When the distance between the two sites is small, the cross-correlation depends little on the preferred orientation of the units being recorded from. However, at distances up to 10 mm and across the vertical midline (i.e., when recording from the two cortical hemispheres; Engel et al. 1991b) the cross-correlation shows a significant peak only if the two stimuli are spatially aligned with each other. It can be shown that stimulation of the two sites by an elongated single bar leads to strong synchronization among the firing activities at the two sites, while stimulation with two separate, but still aligned, bars leads to a reduced cross-correlogram (Gray et al. 1990). This has yet to be shown for the alert animal.

The fact that the 40-Hz oscillations are seen in lightly anesthetized cats is not by itself fatal to our theory, as the anesthetic may remove or reduce some other essential aspect of visual awareness, such as short-term memory.

Oscillations in the Monkey   There has, so far, been much less work on the oscillations in monkey visual cortex. Livingstone (1991) has seen strong oscillations in both single-unit as well as local field potentials in V1 in the anesthetized monkey (see also Freeman and van Dijk 1987). Recording from two sites with two electrodes, she often observes phase-locked oscillations using either a single bar or two bars. Kreiter and Singer (1992) report brief periods of highly oscillatory activity in multiunit data in the awake and fixating monkey in area MT. Based on their criteria, 58% of units show significant oscillatory activity in the 30–60 Hz range. Oscillatory response episodes are often of short duration (< 300 msec), do not occur on each trial, and can vary in their oscillation frequency both within and between trials (Kreiter and Singer 1992). Thus, averaging cell responses over long time will render the oscillations invisible. Nakamura et al. (1991, 1992) record single neurons in the temporal lobe of monkey during a short-term memory task. About one quarter of their neurons show a stimulus-dependent, sustained firing following the short display of a particular figure. The associated autocorrelograms showed pronounced oscillations whose frequency varied from one stimulus to the next between 3 and 28 Hz. However, the majority of oscillations had frequencies less than 8 Hz.

A thorough search for oscillatory neuronal responses in the monkey was carried out by Young et al. (1992). The autocorrelation function associated with multiunit activity showed oscillations in the 12–13 Hz (α range) at about 10% of recording sites in areas V1 and MT and no oscillations in inferior temporal cortex of the anesthetized monkey. In the behaving monkey (performing a face discrimination task), only 2 of 50 recording sites in IT showed oscillations in the 40–50 Hz range (Young et al. 1992; see also Tovee and Rolls 1992). Bair et al. (1994) analyzed the response of 212 cells in extrastriate cortex (area MT) in the macaque monkey while the monkey performed a very demanding motion discrimination task (Newsome et al. 1989). Applying the same criteria as Kreiter and Singer (1992) in their study of multiunit activity in MT, Bair et al. (1993) find only a single cell that shows strong oscillatory activity. This result, obtained using random-dot stimuli, is quite distinct from the high percentage of oscillatory cells reported to be found in the same cortical area when using high-contrast bar stimuli (Kreiter and Singer 1992).

Two groups find oscillations in somatosensory and motor regions of the alert monkey. Murthy and Fetz (1992) report the existence of large, 25–35 Hz, oscillations in both local field potentials and single-unit activity in pre- and postcentral cortex of two awake rhesus monkeys. These oscillations are synchronized across up to 20 mm. Sanes and Donoghue (1993) record local field potentials from up to 12 sites in motor and premotor cortical areas while the monkey is waiting for a visual cue to carry out a hand movement. These potentials show oscillations in the 20–50 Hz range that are phase-locked across different sites. Once the visual cue appears and the movement is executed, both the oscillations as well as the synchronization of neuronal activity abruptly ceases.

Our original hypothesis (Crick and Koch 1990a) was that the phase-locked firing of a set of neurons at 40 Hz was the neural correlate of visual awareness. Such a set would correspond to the semiglobal activity referred to earlier. So far the experimental evidence has lent rather little support to this hypothesis, though it may still be true that the 40 Hz oscillations are used as part of the processes leading up to visual awareness, such as figure-ground segregation, as first suggested by Milner (1974) and discussed at length by von der Malsburg (1981; von der Malsburg and Schneider 1986). More experimental work on the natural history of the 40-Hz oscillations, where and when they can be evoked, etc., is urgently required, especially in the alert macaque monkey.

Bursting and the Lower-Layers Hypothesis

Although it is possible that *all* cortical neurons can, at one time or another, be part of the neural correlate of awareness, it is sensible to explore the idea that only a limited subset of cortical neurons has this property. Our lower-layer hypothesis states that the neural correlates of visual awareness occur mainly in the lower layers 5 and 6 of the cortex. The input layer as well as neurons in the upper layers 2 and 3 are assumed to be mainly concerned with *unconscious* processing. We were led to this hypothesis for several reasons.

What Becomes Conscious   Cognitive scientists, such as Johnson-Laird (1988), have suggested that the content of consciousness consists of the *results* of neural computation while the interim results associated with the computations leading up to these results are themselves largely unconscious. The only cortical layer that has neurons that project right out of the cortical system (that is, neither to other cortical areas, nor to the thalamus nor the claustrum) are in layer 5. For instance, layer 5 pyramidal cells in the early visual cortices, including V1, V2, V3, and MT (Ungerleider et al. 1983) project to the superficial layers of the superior colliculus and to the pontine nuclei. The corticospinal pyramidal tract, with one milion axons the largest descending fiber tract from the human brain, originates in layer 5 of primary motor, supplementary motor, and premotor cortical areas and projects onto interneurons and motoneurons in the spinal cord. Similarly, the massive projection system linking virtually the entire neocortex with the striatum in the monkey originates in layer 5 (Jones et al. 1977). It could be argued that what the cortex sends elsewhere in the brain are likely to be the *results* of its computations.

Sleep and the Lower Layers   A second reason that attracted our attention to the lower layers comes from the study by Livingstone and Hubel (1981) of the same neuron in cat striate cortex, both when the animal was awake and when it was in slow-wave sleep. Their main result was that the general character of the neuronal responses was similar in these two states, though the signal-to-noise ratio was improved in the awake animal. In addition, some neurons fired more strongly when the animal was awake. Such neurons were found predominantly (but not exclusively) in the lower cortical layer. They confirmed this result by using double-labeled deoxyglucose studies. These showed that the average neuronal activity was greater in the lower cortical layers of V1 when the animal was awake.

   We would like to note in this context that Livingstone (1991) reports high-frequency oscillations only in the superficial layers and in layer 4, but *not* in layers 5 and 6 of primary visual cortex in the anesthetized monkey. It would be quite intriguing to know to what extent oscillations can be found in the deep layers of V2.

Bursting Cells   There exists a subclass of pyramidal cells in the lower layers that behaves differently from other pyramidal or spiny stellate cells. Intracellular current injections into cells in rodent slices of sensorimotor cortex has revealed two types of pyramidal cells (McCormick et al. 1985; Connors and Gutnick 1990; Agmon and Connors 1992). The majority of these in vitro cells respond to a sustained, intracellular current by a train of action potentials, which adapt within 50–100 msec to a more moderate discharge rate ("regular spiking" cells; figure 5.1A) and correspond to pyramidal cells throughout all layers, including layer 5 (figure 5.1B). A second class of neurons responds to the depolarization by generating a short burst of 2–4 spikes, followed by a long hyperpolarization. This cycle of burst and hyperpolarization persists for as long as the current stimulus persists ("intrinsically bursting" cells; figure 5.1C). The latter cells appear to be confined (at least in rat and guinea pig slice) to layer 5 (Agmon and Connors 1992). Bursting cells are quite large and have apical dendrites that extend up to layer 1 and arborize there (Chagnac-Amitai et al. 1990; Larkman and Mason 1990; figure 5.1D). In rat and cat visual cortex, the

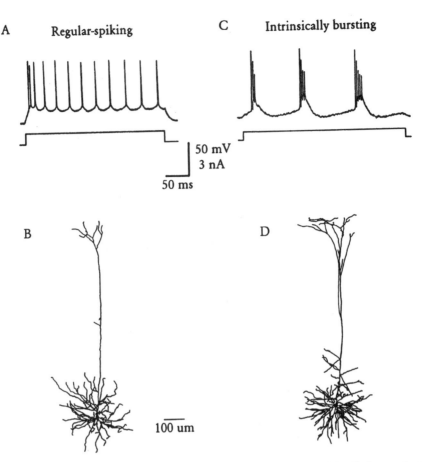

A Regular-spiking   C Intrinsically bursting

50 mV
3 nA
50 ms

B   D

100 um

Figure 5.1 "Regularly spiking" and "intrinsically bursting" pyramidal cells from rodent somatosensory slice. (*A*) The majority of pyramidal cells in all layers respond to a rectangular current pulse injected at the soma by a train of adapting action potentials. They are found throughout layers 2–6. (*B*) The morphology of a such a "regular spiking" biocytin-filled layer 5 pyramidal cell. Notice the sparse dendritic tuft in layer 1. Frequently, the apical tree of "regularly spiking" cells does not extend past layers 2/3. (*C*) Burst cells, limited to layer 5, respond to the same intracellularly delivered current step with a typical pattern of bursts. (*D*) Their morphology reveals extensive dendritic branching in layer 1. Layer 5 bursty cells project to subcortical targets. (Drawings are modified from Agmon and Connors 1992 and B. W. Connors, personal communication)

axons of these cells project subcortically, here to the ipsilateral superior colliculus, while pyramidal cells with short dendrites not reaching into layer 1 project into other parts of cortex (Hübener et al. 1990; Kasper et al. 1991). In guinea pig sensorimotor and visual cortex maintained in vitro, layer 5 bursting cells project to the superior colliculus and to the pontine nuclei (Wang and McCormick 1993).

Layer 5 burst-generating neurons typically exhibit rhythmic burst firing in the frequency range of 0.2–10 Hz, depending on the level of somatic depolarization (Wang and McCormick 1992), while Silva et al. (1991) demonstrated that layer 5 slice neurons can generate brief periods of oscillatory field potentials in the 5–10 Hz band in response to activation of excitatory afferents.

This classification of pyramidal cells into "regular spiking" and "intrinsically bursting" cells has not be carried out in primate cortex. However, extracellularly recorded cells in the awake monkey frequently show a burst-like pattern (figure 5.2). In their statistical analysis of firing properties of neurons in area MT in the behaving monkey, Bair and colleagues (1994) found that two thirds of the recorded cells frequently fire in bursts of 2–4 spikes within 2–6 msec and show a small peak in the 25–50 Hz band in the associated power spectrum. The amplitude of the peak of spectrum in this frequency band relates directly to their propensity to fire bursts. The statistical properties of this cell class can be fitted by the assumption of Poisson-distributed bursts with a burst-dependent refractory period. The remaining third of their cells have an autocorrelation function and an interspike interval distribution compatible with the notion that spikes are Poisson distributed with a refractory period (figure 5.2). Cells are either of the bursting or of the nonbursting type and do not change from one type to the other. It is not known whether these bursting cells correspond to the intracellularly defined "intrinsic bursters."

In their study of the relationship between single unit firing properties and the behavior of the animal, Newsome and colleagues (Newsome et al. 1989; Britten et al. 1992) recorded from MT cells while the monkey performed a near-threshold direction-of-motion discrimination task. Using signal-detection theory (ROC analysis) based on the total number of spikes occurring during the trial (here 2 sec), they obtain a neuronal threshold and compared it against the more conventional psychometric threshold of the animal, finding that the two are very similar. Bair et al. (1994) show that the sensitivity of the ROC analysis can be improved, in a few cases by a factor two, if a "burst" of spikes is treated as a single event, rather than as consisting out of a variable number of single action potentials. This argues for the idea that "bursts" are events that are treated differently by the nervous system than isolated spikes (see also Bonds 1992).

Bursting and Short-Term Memory   Why should the nervous system have two types of neurons, one signaling isolated spikes and the other predominantly responding in bursts of spikes? Could these two cell types convey fundamental different types of information (Crick 1984)? One biophysical plausible argument is that bursting neurons are much more efficient at accumulating calcium in their axonal terminals than cells that fire isolated spikes (that is, four spikes within a 10-msec interval cause a much larger increase in intracellular calcium at the end of the last spike than four spikes arriving within a 40-msec interval). Because intracellular calcium accumulation in the presynaptic terminal is thought to be mainly responsible for various forms of short-term potentiation (in particular facilitation and augmentation; Magleby 1987), it may well be that the primary function of layer 5 bursting neurons is to induce this non-Hebbian (that is, nonassociative)

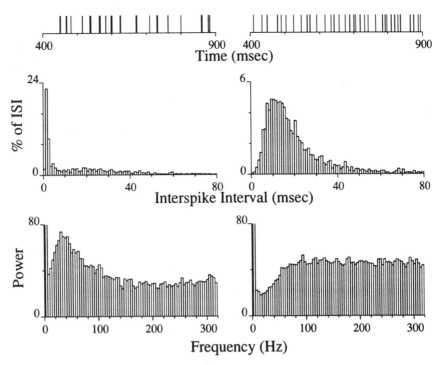

Figure 5.2 Spike train statistics from two extracellularly recorded cells in cortical area MT in the behaving monkey responding to randomly moving dots. The top row illustrates a typical 500-msec segment of spike occurrence times. The distribution of intervals between two consecutive spikes (ISI) is shown in the middle row (note the different scales). The power spectra of the two cells are shown in the bottom row. Both cells fire at roughly similar average rates (40–60 spikes/sec). One-third of all cells are of the nonbursty type, with an ISI and a power spectrum expected from a Poisson process with a refractory period. Statistics from one such cell are shown in the right column (averaged over 30 trials). About two-thirds of cells are of the bursty type, with a significant fraction—if not the majority—of the interspike intervals falling into the first three bins of the ISI and a peak in the power spectrum in the 20–60 Hz range. Statistics from such a bursty cell, based on 15 trials, are shown in the left column. The relationship—if any—between these bursting and the "intrinsically bursting" pyramidal cells from slices (see figure 5.1) is not known. (Drawings are modified from Bair et al. 1993)

type of synaptic plasticity at its postsynaptic targets outside of the cortical system. Recurrent spiking activity within one or several seconds at these synapses will then cause a greater postsynaptic signal than without the "priming" by the previous burst. In essence, the burst of spikes acts to turn on short-term memory, which then decays over several seconds (an interesting theoretical question is whether or not very short-term memory needs to be associative).

Thus a very simplistic answer to the question "Which neurons fire in such a way that they correlate with awareness?" would be "The large pyramidal cells in layer 5 that fire in bursts and project outside the cortical system!" It would be marvelous if this were true but the answer is unlikely to be as simple as that.

Some Further Ideas Regarding the Neuronal Basis of Awareness

The Lower Layers and the Thalamus   Pyramidal cells in both layer 5 and layer 6 project to the various thalamic nuclei, including the lateral geniculate nucleus (LGN), as well as the inferior, lateral, and medial pulvinar nuclei. In monkey, only layer 6 of area V1 projects back to the LGN, while higher cortical areas project to—and receive from—the different pulvinar nuclei. In primate area V1, cells in layer 5 as well as the deep part of layer 6 project to the pulvinar (Conley and Raczkowski 1990), while higher cortical areas project from the deep layers into the different pulvinar nuclei (with the general rule that as one goes from occipital lobe to more anterior cortical areas, the thalamic target areas move from inferior to lateral to medial pulvinar). The precise layer of origin of this corticothalamic projection is not know.

In cat, about half of all pyramidal cells in layer 6 project back to the LGN while others project to the claustrum. This corticogeniculate projection is so massive that at least 10 times more fibers project down than project from the LGN in V1 (Sherman and Koch 1986). It is known in cat that the propagation delay of these cells is unusually long and heterogeneous, ranging from 2 to 20 or more msec (Tsumoto et al. 1978), in agreement with their unmyelinated nature. These could conceivably form reverberating circuits that hold activity in very short-term memory. Furthermore, the circuit LGN → layer 6 → LGN is composed of neurons whose axons have very few *horizontal* collaterals. This may prevent the "reverberation" from spreading too easily to adjacent neurons. Under an anesthetic such possible reverberations may be too weak to become established (incidentally, almost all work on the function of the corticogeniculate pathway has been done only on anesthetized animals and therefore its main function may have been missed; Sherman and Koch 1986).

A final observation of possible relevance to the distinction between lower and upper layers is that visual-induced activity can be blocked by NMDA antagonists in the superficial, but not in the deep layers of cat visual cortex (Fox et al. 1989, 1990). Iontophoretic and radioligand binding studies from a number of different labs argue for high densities of NMDA receptors in superficial and low densities in the input and the deep layers of cortex (summarized in Tsumoto 1990).

HYPOTHESES ABOUT CONNECTIONS

So much for the lower cortical layers. Are there any signs of connections *between* cortical areas that might relate to visual awareness? We have been able to think of three different clues.

1. *Uses of content of "consciousness."* The first asks the question, What is conscious information used for? In neuronal terms one might expect that the information is not only exported subcortically but is also made available to at least the hippocampal system and the higher, planning levels of the motor system, probably located in or near the anterior portion of the cingulate sulcus. We will not describe these connections in detail as so far

we have not been able to turn up any neuroanatomical data that might help. For example, exactly *which* neurons project from the higher levels of the visual system to those to the motor system? We will not detail this idea further though we shall continue to keep an eye on it.

2. *Visual processing and backprojections.* The second idea was put forward some time ago in a general way by Milner (1974). He proposed that an essential feature of visual processing would turn out to be the back projections to V1 (or conceivably to V1 and/or V2) as these areas are the only ones with *detailed* information about precise visual location. Supporting evidence comes from a study of the somatosensory system that uses a combination of current source analysis and somatosensory-evoked potentials in the awake monkey (Cauller and Kulics 1991a,b). These authors argue that the backward projections from S2 to S1 (that are targeted specifically to the superficial layers 1 and 2) are involved in the conscious process of touch sensation as measured by the evoked potentials.

Looking at the diagram of Felleman and Van Essen (1991) for the connections between cortical areas in the visual system of the macaque monkey, it would appear that while areas MT and V4 do send projections back to V1, the inferotemporal regions do not. However, more recent evidence (K. Rockland, personal communication) suggests that projections between cortical areas may frequently be nonreciprocal. While the inferotemporal regions do not *receive* a direct projection from V1, Rockland claims that they do send backprojections to V1. Of course there may be several kinds of backprojections. Clearly this idea of Milner's needs to be kept in mind as further results come in.

3. *Reentrant projections.* The third idea about connections is based on "reentrant" or backward connections, so strongly emphasized by Edelman (1987). The basic idea is that consciousness is, in some sense, the brain reflecting on itself and that this needs reentrant connections—,that is, connections that form a circuit that finishes where it began.

An obvious case of massive reentrant connections is those formed by the hippocampal system in the medial temporal lobe. Its input comes mainly from the entorhinal cortex and its output returns there, though to a different cortical layer. However, as we know from such patients as H.M. and Boswell, the complete removal of this system by certain types of brain damage in humans does not affect immediate awareness, though the patient cannot remember any recent incident that took place more than a minute or so before (Squire and Zola-Morgan 1991). There are very many other reentrant connections in the brain, so it is not easy to discover which ones might be intimately involved in visual awareness. Are there any that have some unusual character?

DIFFERENT FORMS OF AWARENESS

Before closing, let us return to a topic we discussed only tangentially. In our original manuscript (Crick and Koch 1990a), we postulated the existence of two forms of awareness: a brief and very transient one and a form associated with selective, visual attention. It is the presence of the latter, coupled with short-term memory, that we believe mediates vivid awareness. However, in the absence of the former, our visual environment would have the appearance of a tunnel, in which the currently attended location appears in vivid detail with its associated perceptual attributes while everything else is invisible or hazy. We were therefore led to postulate another form of *fleeting awareness*, enough to mediate the perceptual richness we take for granted when looking at the world.

Due to its lack of an attentional mechanism, we believe that fleeting awareness is not associated with solving the transient type of binding problem but to encode only perceptual features that are bound within single neurons due to epigenetic factors or overlearning.

However, given the complexities of the brain, it may well be possible that many more different forms of awareness coexist at all times, each with different functional abilities. For instance, it may well be that each corticothalamic–cortical loop instantiates it own form of short-term memory and awareness, each with its own representation and time scale.

Recent psychophysical experiments by Braun and Sagi (1992) and Braun (1994) support this point of view. The visual attention of subjects was "tied down" to a particular spot on a monitor by asking them to carry out a difficult discrimination experiment. With focal attention thus distracted, Braun projected a number of objects in an annulus around the focus of attention and asked subjects to respond if an odd-man-out was present in these displays. His results show that subjects could well detect a large object among many small objects, a red among many grays, a low-spatial frequency grating among many high-frequency ones, a circle among many triangles, or a triangle among many distracting circles. However, subjects were frequently unable to detect a small object among large ones, a gray object among many distracting red ones, etc. In the absence of attention, subjects could even reliably report the hue of two bright spots, one above and one below the location of attention. Thus, in this case, awareness of these objects is mediated in the absence of focal attention, supporting the idea that different forms of "vivid awareness" might exist, each one with its own set of properties.

It is important that neurobiological theories of these phenomena do not treat short-term memory, visual perception, or awareness as single, monolithic entities with but a single neuronal implementation. They may rather be the end product of a large number of highly interactive neuronal mechanisms (Minsky 1985).

A SUMMARY OF OUR SPECULATIONS

To assist the reader we now list very briefly the various speculative ideas put forward in this chapter. These speculations do not all make a coherent set, though certain combinations of them do.

Koch and Crick

1. The brain constructs an explicit, multilevel, symbolic interpretation of parts of its environment.

1.1. To do this it usually needs some form of attentional mechanism.

2. The form of awareness associated with focal attention is caused by the firing of a temporally coordinated assembly of neurons firing in some special manner for at least 100 or 200 msec.

2.1. This special form of neuronal activity induces short-term memory.

3. If neurons are not part of this transient subset, they can still influence behavior but do not contribute toward awareness.

3.1. Underlying every direct perception is a group of neurons strongly firing and participating in the temporally coordinated neuronal assembly.

4. Semisynchronous, neuronal oscillations in the 25–55 Hz band could cause neurons to be coordinated, giving rise to short-term memory and thus to awareness.

5. The neural correlate of awareness occurs mainly in the lower layers.

5.1. The neural correlate of awareness is associated with the bursting neurons in layer 5, some of which project outside the cortical system.

5.2. The loop between deep layers in cortex, the different thalamic nuclei, and back to cortex may implement short-term memory.

5.3. The neurons in the upper cortical layers are mainly concerned with unconscious processing.

6. Various types of neural connections may be associated with some forms of visual awareness. Possible examples are:

6.1. Connections to the hippocampal system and the higher planning levels of the motor system, direct backprojections to V1 (and possibly V2), and reentrant connections within layer 4 or between cortical areas at the same level in the anatomical hierarchy.

ACKNOWLEDGMENTS

Christof Koch is supported by the Air Force Office of Scientific Research and the Office of Naval Research. Francis Crick is supported by the J.W. Kieckhefer Foundation.

6 Perception as an Oneiric-like State Modulated by the Senses

Rodolfo R. Llinás and Urs Ribary

An issue clearly fundamental to the understanding of central nervous system function lies in the similarities and differences between wakefulness and dreaming. Indeed, from the standpoint of the thalamocortical system it has been shown that, as will be described in detail later, these two states have a common intrinsic implementation mechanism and so may be considered, in that sense, as fundamentally equivalent (Llinás and Paré 1991). The implications of such a hypothesis could be far-reaching if wakefulness is demonstrated to be, as is dreaming, a closed intrinsic functional state. If this were the case, the central difference between the two would be the degree of their modulation by sensory input.

WAKEFULNESS AND REM SLEEP

Paradoxical sleep is characterized by the repeated occurrence of rapid eye movement (REM)—from which the alternative designation "REM sleep" was derived—and by muscular atonia. One of the most salient differences between the wakefulness and dreaming states resides in the fact that sensory input does not generate the expected cognitive consequences that it does in the awake state. With respect to other sleep states, REM sleep differs in that sensory thresholds for awakening are the highest in REM sleep, except for stage IV (Rechtschaffen et al. 1966; Williams et al. 1964), and that subjects awakened during REM sleep often report having been dreaming.

Of central interest here is the finding that the averaged evoked potentials (AEPs) recorded from the scalp in response to sensory stimulation during waking and REM sleep are very similar, but they differ strikingly from those recorded during non-REM sleep. For instance, the early components of the auditory evoked potential in humans (<10 msec) (Moller and Burgess 1986) do not display state-dependent fluctuations during the sleep-waking cycle (Campbell and Bartoli 1986; Giard et al. 1988; Picton and Hillyard 1974). However, those middle-latency components (10–80 msec) that seem to reflect early thalamocortical activity decreased in amplitude from waking to stage IV but returned to normal (Chen and Buchwald 1986) or surpassed waking values in REM sleep (Deiber et al. 1989; Mendel and Goldstein 1971; Mendel and Kuperman 1974).

Likewise, short-, middle-, and long-latency components may also be distinguished in somatosensory evoked potentials. Among the early components, only the positivity at 15 msec ($P \sim 15$) does not display state-dependent fluctuations (Yamada et al. 1988). The amplitude of the other components decreases markedly from waking to stage IV but partially recovered in REM sleep (Yamada et al. 1988). The latency of the P20 component (which presumably reflects the primary cortical response) increases from waking to stage IV but returned close to waking values in REM sleep.

The Central Paradox

Since the brain's response to sensory stimulation is very similar during REM sleep and wakefulness, the threshold for awakening should be lowest in REM sleep. As stated above, however, this is not the case in humans or in other mammals where the auditory threshold for awakening is clearly higher in REM sleep than in deep slow-wave sleep (Jouvet and Michel 1959). These studies point to a central paradox of REM sleep: stimuli that are perceived in the waking state do not awaken subjects in REM sleep even though the amplitude of the primary evoked cortical responses is generally similar to, or higher than, those in the waking state. In other words, although the thalamocortical network is as excitable during REM sleep as in the waking state, the input is mostly ignored.

The resolution of this paradox probably lies in the nature of brain function in a most fundamental sense. In particular, the fact that the late potentials (P100, P200, P300) following sensory stimuli are abolished in REM sleep (Goff et al. 1966; Velasco et al. 1980) suggests that the ongoing activity that generates cognition during dreaming prevents the early thalamocortical activation from being incorporated into the intrinsic cognitive world. Perhaps then, an altered state of attention is the most likely origin for the high threshold for awakening from REM sleep (Llinás and Paré 1991).

Is "Cognition" during REM Sleep Similar to That in Wakefulness?

One tool available to study the functional state of the brain during REM sleep is a comparison of the dreams of control subjects with those of patients suffering from various central or peripheral nervous system dysfunctions.

The decline of higher cognitive abilities following circumscribed lesions of the temporal and parietal associative areas is also reflected in dream content. For instance, patients afflicted with unilateral neglect resulting from right parietal lobe damage, in which the opposite half of the visual field is not perceived, report similar lack of perception in their dreams (Sacks 1991; M. Mesulam, personal communication). Similarly, people inhabiting the dreams of prosopagnosic subjects are faceless (A. Damasio, personal communication). Interestingly, when awake these patients perceive facial features but they cannot use such features to recognize individual faces.

These observations indicate that mentation during dreaming operates on the same anatomical substrate as does perception during the waking state.

From the fact that similar deficits are formed in wakefulness and dreaming, it may be concluded that a possible approach to understanding the nature of wakefulness is to consider it as one element in a category of intrinsic brain functions, in which REM sleep is another element. The difference between these two states would be that in REM sleep, the sensory specification of the functionalities carried out by the brain is fundamentally altered. That is, REM sleep can be considered as an intrinsic state in which "attention" is turned away from sensory input (Llinás and Paré 1991).

In proposing that wakefulness is nothing other than a dreamlike state modulated by the presence of specific sensory inputs (Llinás and Paré 1991), the following must be considered. The thalamus is classically regarded as the functional and morphological gate to the forebrain (Steriade et al. 1990). Indeed, with the exception of the olfactory system, all sensory messages reach the cerebral cortex through the thalamus (Jones 1985). Yet, synapses established by specific thalamocortical fibers comprise a minority of cortical contacts. For example, in the primary somatosensory and visual cortices, the axons of ventroposterior thalamic and dorsal LGN neurons account for, respectively, 28% and 20% of the synapses in layer IV and adjacent parts of layer III (LeVay and Gilbert 1976), where most thalamocortical axons project. Even in primary sensory cortical areas, most of the connectivity does not represent sensory input transmitted by the thalamus, but rather input from cortical and nonthalamic CNS nuclei. Indeed, corticostriatal, corticocortical and corticothalamic pyramidal neurons receive, respectively, 0.3–0.9%, 1.5–6.8%, and 6.7–20% of their synapses from specific thalamocortical fibers, and less than 4% of the synaptic contacts on multipolar aspiny neurons in layer IV originate in the thalamus (White and Hersch 1981, 1982).

Moreover, the connectivity between the thalamus and the cortex is bidirectional. Indeed, layer 6 pyramidal cells project back to that area of the thalamus where their specific input arises (Jones 1984), and layer 5 cells project to the nonspecific thalamus. The number of corticothalamic fibers is about one order of magnitude larger than the number of thalamocortical axons (Wilson et al. 1984). Looking at the peripheral input, the number of optic nerve axons projecting to the LGN is much smaller than the number of corticothalamic axons projecting to the LGN (Wilson et al. 1984).

Clearly, the sensory input arising from the thalamus is necessary for perception; in the absence of specific inputs, there is no externally guided sensory function. However, the specific thalamocortical input accounts for a minority of the synaptic contacts in the cortex.

Let us briefly discuss the nature of the interaction between this set of innate mechanisms and the sensory world. At the outset, it must be recognized that sensory events are nothing other than simplifications determined by the physical properties of our sensory organs. Similarly, the internal representation derived from the sensory specification is constrained by the "computational" capabilities of the brain. According to this view, the model of the world emerging during ontogeny is governed by innate predispositions of the brain to categorize and integrate the sensory world in certain ways. Although the particular computational world model derived by a given individual is a function of his sensory exposure, the resulting functional accommodation is genetically determined. As a result, sensory inputs presented during adult life would convey only the parameters required to specify the dimensions relevant to the cognitive domains which stemmed from this evolutionary process. These cognitive domains could be used to recreate world-analogues during dreaming or, once specified by sensory inputs, to generate an adaptive representation of the environment.

Thus, we may consider a closely related problem, that of the open (extrinsic) or closed (intrinsic) nature of nervous system function. One view stipulates that the brain states that represent the external world are point-to-point representations, having as their basic currency a set of elaborate reflexes. This view may be traced back, in modern times, to William James (1890). An opposite point of view is that the brain is basically a recurrent or closed system. Support for the latter proposal comes from electrophysiological studies indicating that the intrinsic membrane properties of neurons allow them to oscillate or resonate at different frequencies (Llinás 1988a) and that such intrinsic activity, by supporting rhythmic oscillatory events, may play a fundamental role in CNS function (Llinás 1988a; see chapter 10 by Singer). It can be argued that the insertion of such elements into complex synaptic networks allows the brain to generate dynamic oscillatory states that deeply influence the brain activity evoked by sensory stimuli.

PERCEPTION AS GENERATED BY A CLOSED SYSTEM

Several factors suggest that the brain is essentially a closed system capable of self-generated oscillatory activity that determines the functionality of events specified by the sensory stimuli. First, as stated above, only a minor part of the thalamocortical connectivity is devoted to the reception and transfer of sensory input. Second, the number of cortical fibers projecting to the specific thalamic nuclei is much larger than the number of fibers conveying the sensory information to the thalamus (Wilson et al. 1984). Thus, a large part of the thalamocortical connectivity is organized in what is presently known as reentrant activity (Edelman 1987) or previously viewed as reverberating activity (Lorente de Nó 1932). Third, the insertion of neurons with intrinsic oscillatory capabilities into this complex synaptic network allows the brain to generate dynamic oscillatory states which shape the computational events evoked by sensory stimuli. In this context, functional states such as wakefulness (or REM sleep and other sleep stages) appear to be particular examples of the multiple variations provided by the self-generated brain activity.

Much neuropsychological evidence also supports this view of the brain as a closed system in which sensory input plays an extraordinarily important but, nevertheless, mainly modulatory role. The cases of prosopagnosic patients dreaming of faceless characters indicate that the significance of sensory cues is largely dependent on their incorporation into larger cognitive entities and on the functional state of the brain. In other words, sensory cues gain their significance by virtue of triggering a preexisting disposition of the brain to be active in a particular way.

That for the most part connectivities present at birth in humans are modified only in detail during normal maturation has been suspected from the inception of neurological research (Cajal 1929; Harris 1987). The localization of function in the brain began with the identification of a cortical speech center by Broca and was followed by the discovery of point-to-point somatotopic maps in the motor and sensory cortices (Penfield and Rasmussen 1950), and in the thalamus (Mountcastle and Hennemann 1949, 1952).

A totally different type of functional geometry (Pellionisz and Llinás 1982) suggests the existence of temporal mapping. This has been far more difficult to conceptualize, since its study requires an understanding of simultaneity in brain function not usually considered in neuroscience.

40-HZ ACTIVITY AND COGNITIVE CONJUNCTION: THE CASE FOR TEMPORAL MAPPING

Synchronous activation has recently been seen in the mammalian cerebral cortex. Visual stimulation with light bars of optimal dimensions, orientation, and velocity may synchronously activate cells in a given column in the visual cortex (Eckhorn et al. 1988; Gray et al. 1989; Gray and Singer 1989). Moreover the components of a visual stimulus that relate to a singular cognitive object (such as a line in a visual field) produce coherent 40-Hz oscillations in regions of the cortex that may be separated by as much as 7 mm (Gray et al. 1989; Gray and Singer 1989). And a high correlation coefficient has been found for 40-Hz oscillatory activity between related cortical columns.

These findings have inspired a number of theoretical papers with the view that temporal mapping is very important in nervous system function. The central tenet can be summarized simply. Spatial mapping allows a limited number of possible representations. However, the addition of a second component (serving to form transient functional states by means of simultaneity) generates an indefinitely large number of functional states, as the categorization is accomplished by the conjunction of spatial and temporal mapping.

Magnetoencephalographic recordings performed in awake humans (Llinás and Ribary 1992) revealed the presence of continuous and coherent 40-Hz oscillations over the entire cortical mantle. The presentation of auditory stimuli produced a clear resetting of this 40-Hz activity. Phase comparison of the oscillatory activity recorded from different cortical regions revealed the presence of a 12- to 13-msec phase shift between the rostral and caudal pole of the brain (Llinás and Ribary 1992).

The high degree of spatial organization displayed by this 40-Hz oscillation suggests that it may be a candidate mechanism for the production of temporal conjunction of rhythmic activity over a large ensemble of neurons. Furthermore, it has been shown that the sparsely spiny layer IV neurons of the cortex (Llinás et al. 1991) are capable of 40-Hz activity (figure 6.1). This inhibitory input would produce rebound sequences (probably dependent on persistent sodium conductances) in thalamically projecting pyramidal neurons. These cells would then generate a 40-Hz inhibitory rebound oscillation in cells of the reticularis (RE) thalamic nucleus, a group of GABAergic neurons projecting to most relay nuclei of the thalamus (Steriade et al. 1984). More recently, it has been demonstrated that, in addition to oscillations due to cortical circuit properties, thalamic neurons in vivo can also oscillate intrinsically at 40-Hz, using ionic mechanisms similar to those of the spiny-layer neurons (Steriade et al. 1991). Consequently, spe-

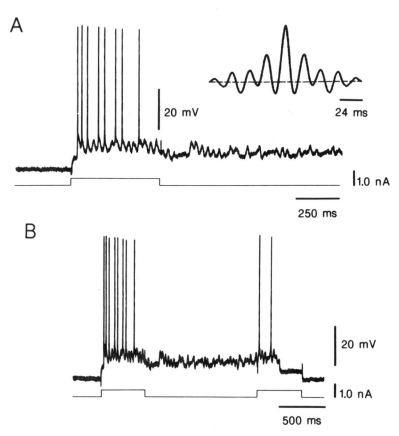

Figure 6.1 In vitro intracellular recording from a sparsely spinous neuron of the fourth layer of the frontal cortex of guinea pig. (*A*) The characteristic response obtained in the cell following direct depolarization, consisting of a sustained subthreshold oscillatory activity on which single spikes can be observed. The intrinsic oscillatory frequency was 42 Hz, as demonstrated by the autocorrelogram shown in the upper right corner. (*B*) The same record as in (*A*) but at slower sweep speed, demonstrating how the response outlasts the first stimuli but comes to an abrupt cessation in the middle of a second stimulus. (Modified from Llinás et al. 1991)

cific corticothalamocortical pathways could be led to resonant oscillation at 40-Hz. According to this hypothesis, RE cells would be responsible for the synchronization of the 40-Hz oscillations in distant thalamic and cortical sites. Indeed it has been shown that neighboring RE cells are linked by dendrodendritic and intranuclear axon collaterals (Deschênes et al. 1985; Yen et al. 1985).

BRAINSTEM INFLUENCE ON THALAMIC FIRING MODE

While the firing mode of thalamocortical cells is related to the expression of intrinsic membrane properties, the state-dependent fluctuations in membrane potential seem to result from extrinsic synaptic influences. Thus, during REM sleep temporal associations that generate subjectivity may not coincide with the temporal maps, and only strong sensory inputs are capable of resetting such temporal conditions. In short, if the sensory input coming to the brain is not put in the context of thalamocortical reality by being correlated temporally with ongoing activity, the stimulus does not exist as a functionally meaningful event.

If this is the case, we may conclude that the perception of external reality is an intrinsic function of the CNS, developed and honed by the same evolutionary pressures that generated other specializations. Moreover this implies that secondary qualities of our senses such as colors, identified smells, tastes, and sounds are inventions of our CNS that allow the brain to interact with the external world in a predictive manner (Llinás 1988a). The degree to which our perception of reality and "actual" reality overlap is inconsequential as long as the predictive properties of the computational states generated by the brain meet the requirements of successful interaction with the external world.

If we assume that the phase shift observed in these preliminary studies is related to the presence of coherent waves that scan our brain at 40 Hz, we can conclude that consciousness is not a continuous event. Rather it is determined by the simultaneity of activity in the thalamocortical system modulated by the brainstem, and fed—when one is awake, by sensory input and when one is asleep, by circuits that support memories.

THALAMOCORTICAL RESONANCE AS THE FUNCTIONAL BASIS FOR CONSCIOUSNESS

From the above, it follows that the major development in the evolution of the brain of higher primates, including man, is the enrichment of the corticothalamic system. This is supported by evolutionary studies if one considers the increase in corticalization in mammals. The increase in the surface area of the neocortex in man is approximately three times that of higher apes (Lande 1979).

How can this thalamocorticothalamic functional state generate the unique experience we all recognize as existence of self or existence of the here and now? In principle, the activity generated via thalamocortical interactions may mimic the responsiveness generated during the waking state (i.e., reality-emulating states, such as hallucinations, may be generated). The implications of this proposal are of some consequence, for this means that if consciousness is a product of thalamocortical activity, it is the dialogue between the thalamus and the cortex that generates subjectivity.

EXPERIMENTS SUPPORTING THE SIMILARITIES OF REM SLEEP AND WAKEFULNESS

If, as stated above, 40-Hz thalamocortical resonance is responsible for the global temporal mapping that generates cognition, such global conjunction should be present during the dreaming state. In fact, it has been recently reported that 40-Hz activity occurs in an organized fashion and demonstrates a rostrocaudal phase shift during REM sleep (Llinás and Ribary 1993).

Magnetoencephalography (MEG) was utilized in that study. Three sets of studies addressed issues concerning (1) the presence of 40-Hz activity during sleep, (2) the possible differences between 40-Hz resetting in different sleep/wakefulness states, and (3) the question of 40-Hz scan during REM sleep.

To this effect, spontaneous magnetic activity was continuously recorded and filtered at 35–45 Hz during wakefulness, delta sleep, and REM sleep, using a 37-channel sensor array. Because Fourier analysis of the spontaneous, broadly filtered rhythmicity (1–200 Hz) demonstrated a large peak of activity at 40-Hz over much of the cortex, we feel that such filtering is permissible. Large coherent signals with a very high signal-to-noise ratio were easily recorded from all 37 sensors, corresponding to activity in different regions of the cortex (figure 6.2B). This single 0.6-sec epoch illustrates the global spontaneous oscillation in an awake individual.

A second set of experiments examined the responsiveness of the 40-Hz oscillation to stimuli during these three different functional states. As shown previously, 40-Hz oscillation may be reset by sensory stimuli (Llinás and Ribary 1993; Galambos et al. 1981; Pantev et al. 1991). This is clearly observed following auditory stimulation. In these experiments, the auditory stimulus consisted of frequency-modulated 500-msec tone bins, triggered 100 msec after the onset of the 600-msec recording epoch, randomly sampled over a time period of approximately 10 minutes. The stimuli were delivered to the subject during conditions of wakefulness (figure 6.2C), delta sleep (figure 6.2D), and REM sleep (figure 6.2E). In agreement with previous findings (Llinás and Ribary 1993; Galambos et al. 1981; Pantev et al. 1991), auditory stimuli (arrowhead) produced well-defined 40-Hz oscillation (Ribary et al. 1991). When a similar set of stimuli was delivered during delta sleep, no resetting was observed in this or any of six other subjects where this experiment was performed, resetting was not observed during REM sleep, as shown in figure 6.2E.

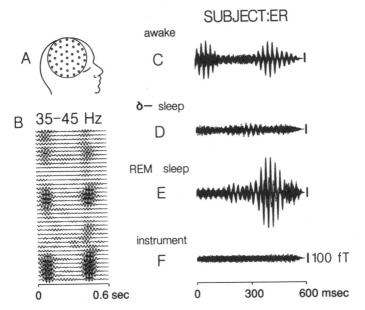

Figure 6.2 Forty-Hertz oscillation in wakefulness and a lack of 40-Hz reset in delta sleep, and REM sleep. Recording using a 37-channel MEG. (*A*) Diagram of sensor distribution over the head; in (*B*) the spontaneous magnetic recordings from the 37 sensors during wakefulness are shown immediately below (filtered at 35–45 Hz). In (*C–F*) averaged oscillatory responses (300 epochs) following auditory stimulus. In (*C*), the subject is awake and the stimulus is followed by a reset of 40-Hz activity. In (*D*) and (*E*), the stimulus produced no resetting of the rhythm. (*F*) The noise of the system in femtotesla (fT). (Modified from Llinás and Ribary 1993)

The level of coherence present at all recording points was illustrated by superimposing the 37 traces recorded during a 600-msec epoch (figure 6.2C). It is clear from such recording that while there is coherence among the different recording sites there is also a phase shift of the oscillation along the different sites (Llinás and Ribary 1992).

These findings indicated that while electrically the awake and REM sleep states are similar with respect to the presence of 40-Hz oscillations, the central difference between these states is the lack of sensory reset of the REM 40-Hz activity. By contrast, during delta sleep, the amplitude of these oscillators differs from that of wakefulness and REM sleep, but as in REM sleep, there is no 40-Hz sensory response.

The findings indicate therefore that during wakefulness and REM sleep a very specific 40-Hz thalamocortical resonance is active and has very similar global properties. Moreover, while both states can generate cognitive experiences, the recordings indicate, as is commonly known, that the external environment is, for the most part, excluded from the imaging of the oneiric states. This further substantiates a recent proposal (Llinás and Paré 1991) that the dream state is characterized by an increased attentiveness to an intrinsic state in the sense that external stimuli do not perturb the intrinsic activity.

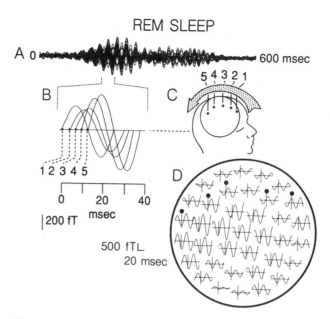

Figure 6.3 Rostrocaudal phase shift of 40-Hz during REM sleep as measured using MEG (see also figure 6.2). The upper trace (*A*) shows synchronous activation in all 37 channels during a 600-msec period. The oscillation in the left part of trace (*A*) has been expanded in trace (*B*) to show five different recording sites over the head. The five recording sites of trace (*B*) are displayed in (*C*) for a single epoch to demonstrate the phase shift for the different 40-Hz waves during REM sleep. The direction of the phase shift is illustrated by an arrow above (*C*). The actual traces and their site of recordings for a single epoch are illustrated in (*D*) for all 37 channels. fT, femtotesla. (Modified from Llinás and Ribary 1993)

In a third set of experiments the issue of the front-to-back phase shift of the 40-Hz activity over the head during REM sleep was addressed. Spontaneous 40-Hz activity during a single 0.6-sec epoch in REM sleep (figure 6.3A and B) and an expanded portion of this burst (figure 6.3B) show the well-organized 12-msec phase shift for the 40-Hz oscillation observed from recording sites 1 to 5, as illustrated schematically in figure 6.3C. The actual recording sites are illustrated for the epoch shown in A in figure 6.3D. A similar 12-msec phase shift was also observed in the same individual in the awake state with the exception that, during REM sleep, the rostrocaudal sweep is better organized and more repeatable, probably since the sweep is not continually reset by incoming sensory stimuli.

The significant new finding here is the fact that during the period corresponding to REM sleep (in which a subject, if awakened, reports having been dreaming), 40-Hz oscillation similar in distribution phase and amplitude to that observed during wakefulness is observed. In the five individuals in whom these recordings were made, the overall speed of the rostrocaudal scan, which averaged approximately 12.5 msec, corresponded quite closely to half a 40-Hz period. This number is the same as that calculated by Kristofferson (1984) for a quantum of consciousness in his psychophysical studies in the auditory system.

A second significant finding related to the fact that during the dreaming state, 40-Hz oscillations are not reset by sensory input although clear evoked potential responses indicate that the thalamoneocortical system is accessible to sensory input (Llinás and Paré 1991; Steriade 1991). This we consider to be the central difference between dreaming and wakefulness. The recordings suggest that we do not perceive the external world during REM sleep because the intrinsic activity of the nervous system does not place sensory input in the context of the functional state being generated by the brain at that time (Llinás and Paré 1991). That is, that the dreaming condition is a state of hyperattentiveness in which sensory input cannot address the machinery that generates conscious experience. Relating to the morphophysiological basis for this scanning property, a very attractive hypothesis could be that the "nonspecific" thalamic system—in particular, the intralaminar complex—may be an important part of this process. Indeed, the intralaminar complex represents a cellular mass that projects to the most superficial layers of all cortical areas, to include primary sensory cortices (Jones 1985) in a spatially continuous manner. The cells in this group may also have the necessary interconnectivity to sustain a propagation wave within the nucleus, which could result in the 40-Hz phase shift observed at the cortical level that is generating the rostrocaudal 12.5-msec phase shift. This possibility is particularly attractive given that damage of the intralaminar system results in lethargy or coma (Facon et al. 1958; Castaigne et al. 1962) and that the electrophysiological properties of single neurons, especially during REM sleep, burst in firing with a 30–40-Hz periodicity (Steriade et al. 1993) as is in keeping with the macroscopic magnetic recordings observed in this study.

BINDING BY SPECIFIC AND NONSPECIFIC 40-HZ RESONANT CONJUNCTIONS

The results reported above and other recent findings indicate that 40-Hz oscillation is present at many levels in the CNS. Indeed, such a property is found in sites as peripheral as the retina (Ghose and Freeman 1992), and olfactory bulb (Bressler and Freeman 1980), in the thalamus, specific and nonspecific (Steriade et al. 1993a), in the thalamic reticular nucleus (Pinault and Deschenes 1992a), and in the neocortex (Llinás et al. 1991). Moreover, it has been shown that some of the 40-Hz recorded in the visual cortex is correlated with retinal 40-Hz (Ghose and Freeman 1992). Thus, 40-Hz oscillation involves not only the cortical but also the thalamocortical interactions. Such a possibility is indicated in the diagrams in figure 6.4. Forty-Hertz oscillation of specific thalamocortical neurons (Steriade et al. 1991) can establish (as shown in figure 6.4, left) thalamocortical resonance via fourth-layer inputs, which resonates with inhibitory interneurons at that level (Llinás et al. 1991). Such oscillation can reenter the thalamus via the layer 4 pyramidal cells (Steriade et al. 1990) and resonate with both the nucleus reticularis and in the specific thalamic nuclei (Pantev et

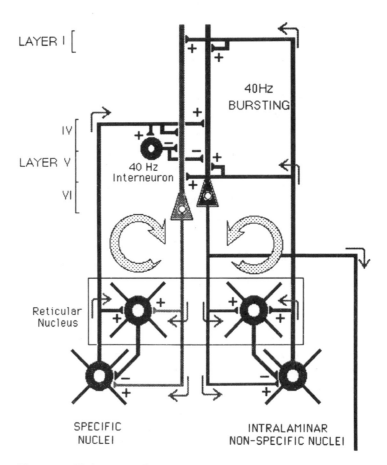

LAYER I

LAYER V
IV
VI

40Hz
BURSTING

40 Hz
Interneuron

Reticular
Nucleus

SPECIFIC
NUCLEI

INTRALAMINAR
NON-SPECIFIC NUCLEI

Figure 6.4 Thalamocortical circuits proposed to subserve temporal binding. Diagram of two thalamocortical systems. (*Left*) Specific sensory or motor nuclei project to layer 4 of the cortex, producing cortical oscillation by direct activation and feedforward inhibition via 40-Hz inhibitory interneurons. Collaterals of these projections produce thalamic feedback inhibition via the nucleus reticularis. The return pathway (circular arrow on the left) reenters this oscillation to specific and reticularis thalamic nuclei via layer 6 pyramidal cells. (*Right*) Second loop shows nonspecific intralaminary nuclei projecting to the most superficial layer of the cortex and giving collaterals to the reticular nucleus. Layer 5 pyramidal cells return oscillation to the reticular and the nonspecific thalamic nuclei, establishing a second resonant loop. The conjunction of the specific and nonspecific loops is proposed to generate temporal binding. (Modified from Llinás and Ribary 1993)

al. 1991). This view is therefore different from the binding hypothesis proposed by Crick and Koch where the authors proposed cortical binding due to the activation of specific thalamic inputs (Crick and Koch 1990a; see also chapter 5).

On the other hand, a second system (figure 6.4, right) is represented by the intralaminary cortical input to layer 1 of the cortex and its return-pathway projection via fifth and sixth layer pyramidal systems to the intralaminary nucleus, directly and indirectly, via collaterals to the nucleus reticularis (Jones 1985). The cells in this system have been shown to oscillate in 40-Hz bursts (Steriade et al. 1993a), and to be organized in space as a toroidal mass having the possibility of recursive activation (Krieg 1966), which could result in the recurrent activity ultimately responsible for the rostrocaudal cortical activation found in the present MEG recordings.

Finally, it is also evident from the literature that neither of these two circuits alone can generate cognition. Indeed, as stated above, damage of the nonspecific thalamus produces deep disturbances of consciousness while damage of specific systems produces loss of the particular modality.

From the above, a very tentative hypothesis may be proposed relating to the overall organization of brain function in very gross and oversimplified terms. Indeed, the "specific" thalamocortical system (to be understood not only as that relating to the primary sensory modalities but rather to nuclei that project mainly, if not exclusively, to layer 4 in the cortex, whether sensorimotor or associative) is viewed as encoding specific sensory and motor "information" by the resonant thalamocortical system specialized to receive such inputs (e.g., the LGN and visual cortex).

If this were to be the case and optimal activation of any such loop would tend to oscillate at close to 40-Hz, activity in the "specific" thalamocortical system could then be easily "recognized" over the cortex by this oscillatory characteristic. In such a scheme then, areas of cortical sites "peaking" at 40-Hz would represent the different components of the cognitive world that have reached optimal activity at that point in time. The problem now would be that of the conjunction of such a fractured description into a single cognitive event. This could be done, we propose, by the summation of 40-Hz activity along the radial dendritic axis of the cortical elements, which would occur when the specific and nonspecific 40-Hz activity is superposed in time, and on the same set of neurons.

In short, the system would work by bringing central neurons to optimal firing patterns via dendritic integrations based on passive and active dendritic conduction along the apical dendritic core conductors. In this manner, the time-coherent activity of the specific and unspecific oscillatory inputs, by summing distal and proximal activity in given dendritic elements, would serve to enhance de facto 40-Hz cortical coherence by their multimodal character, and serve as one mechanism for global binding.

In this manner the specific system would provide the content, and the nonspecific system the temporal conjunction of such content, into a single cognitive experience.

SO, WHY DO WE DREAM?

At this time nothing other than hypotheses can be offered with respect to the final physiological role of dreaming in brain function. An excellent summary of these different points of view has been published recently. (Hobson 1988).

The categories discussed most often today relate to either (1) the Freudian view concerning a subconscious drive or (2) the "mesencephalic" origin, in which dreaming is considered as the forebrain interpretation of otherwise meaningless "brainstem noise." We have a different view that is based on the fact that more often than not one either dreams about recent events or about ongoing problems. On this basis one may consider that dreaming may be the necessary consequence of the parallel nature of the neuronal organization in the CNS. So, given a particular question to be resolved, the CNS generally embarks on simultaneous but diverse possible solutions to such problems, in a parallel fashion. Given this strategy, chances are that a given solution is arrived at before other alternatives.

This, however, does not mean that the alternatives are not considered further. In fact, it often happens that having come to a solution considered adequate, a second may "pop up" in one's mind at a later time. We may further consider that at the end of the day we may have many such partial computations being performed prior to our falling asleep. The possibility is there that in dreaming we "download" the other possible solutions and thus prevent the overloading of circuits with the accumulation of an ever-increasing set of ongoing partial solutions as new problems are considered. This particular point of view may be supported in part by the fact that excellent solutions to problems may arise in dreams.

Something quite similar may be said with regard to slow-wave sleep. In this case the very slow oscillatory nature of the neuronal rhythmicity observed during this functional state (Steriade et al. 1993c) is probably closer to the grooming functions that most animals perform upon finishing a task—whether it is the smoothing of ruffled feathers after flight or the grooming of vibrissa following ingestion of food. What all of this may have in common is that the time spent to generate repeating and rather simple movements (i.e., grooming) or repeating and rather simple patterns of brain activity facilitates the return of neuronal circuits to a readiness state that follows the end of grooming of peripheral receptors and effectors.

The use of such metaphors is but a first approximation to what may ultimately be shown to be the real function of sleeping and dreaming. However, when one considers the fact that sleeping is of such importance that no higher nervous system has evolved away from this time-consuming activity, we must assume that its presence is vital to normal brain function.

7 Neuronal Architectures for Pattern-theoretic Problems

David Mumford

As is abundantly clear from the other chapters of this book, there are many *levels* at which one can attack the problem of modeling the computations of the cortex. For example, at one extreme, one can model how the action potentials received at each synapse are combined in the dendritic tree, or, at the other, one can develop a functional theory of the different cortical areas. But, in addition to choosing a level, modeling requires you to choose some description for the *class of problems* that you expect the cortex is solving, or the *class of signals* that you expect the cortex to be processing. Folk psychology provided the labels for the original cortical area theory of Gall, and cognitive psychology continues to provide a more sophisticated framework for assigning task and function labels to cortical areas (cf. Luria 1962; Fodor 1983; Kosslyn and Koenig 1992). Neurologists use the results of a limited battery of tests, supplemented by their own ability to empathize with the mental state of their patients, as the evidence to be correlated with the nature of the brain damage. For several decades, visual neurophysiologists have relied on the presentation of moving edges and bars and sine wave gratings: the implicit assumption is that distinctive patterns of response to these embody the basic elements of low level visual processing.

The point of departure of this chapter is the proposition that the computational analysis of vision—and speech, tactile sensing, motor control, etc.—(the theory of the computation as Marr called it [Marr 1982]) is reaching a point where it can provide a clearer and deeper description of the essential tasks of vision as well as a wide range of other cognitive tasks. For instance, the development of algorithms for character recognition or for face recognition or for road tracking from a moving vehicle (three problems that have been much studied on account of their potential applications) forces the researcher to deal with noisy, complex real world data. In doing this, one's initial ideas about what parts of the problem are difficult, what parts are simple, may turn out to be quite wrong. Quite often, a step that one thinks of as a simple preprocessing clean up operation turns out to be very difficult and pinpoints for you a new class of problems that had been ignored. *Introspection turns out often to be a very poor guide to the complexity of a problem.* The reason for this, we believe, is our subjective impression of perceiving instantaneously and effortlessly the significance of sensory patterns (e.g., the word being spoken or which face is being seen). Many psychological experiments, however, have shown that what we perceive is not the true sensory signal, but a rational reconstruction of what the signal should be. This means that the messy ambiguous raw signal never makes it to our consciousness but gets overlaid with a clearly and precisely patterned version that could never have been computed without the extensive use of memories, expectations, and logic. Only when you attempt to duplicate such a skill by computer do you discover all the hidden complexity in the computation.

We believe that this analysis, which we call "Pattern Theory" (a term introduced in the pioneering work of Grenander some 15 years ago), leads not merely to a few broad guidelines on the problems faced by a brain, but to a rather specific set of computational tasks, and to a flow chart of how the pieces should be put together. This analysis is very different from most of the orthodox analyses of cognitive problems: it is very distinct from the standard AI view, which takes formal logic and the formal linguists' analysis of language into atomic units and air tight rules, as the universal language of cognition. As we shall see, it fits naturally, instead, with such nonlogical data structures as probabilities, fuzzy sets, and population coding. Moreover, it is very distinct from the pure feedforward analyses such as Marr's analysis of vision (Marr 1982), in that it is based in an essential way on a relaxation between feedforward and feedback processes. Having this analysis, we can go directly to neuroanatomy and neurophysiology and ask if there are structures in the brain that suggest being designed to implement one or more of these basic computational building blocks. If these computations do indeed represent fundamental cognitive operations, one hopes that the basic circuitry is not hidden, but clearly expressed in the anatomy of the cortex, especially in its layers, pathways, and cell types. The method to follow, we believe, is to seek the simplest mechanisms compatible with present knowledge of the anatomy and physiology of cortex, seeking direct analogies between the computational architecture and the neural architecture.

In the next section, we outline the ideas of Pattern Theory and introduce three basic ideas of this theory. There follow sections in which each of these ideas is detailed and its connections with neuroanatomy and neurophysiology are described. We suggest, where possible, the most specific predictions these theories make and propose experimental tests in several cases. The biological ideas in this paper are developments of those described in our earlier two-part paper (Mumford 1991, 1992). The formalism of Pattern Theory presented here is developed at greater length in Mumford (1993).

WHAT IS PATTERN THEORY?

The starting point of Pattern Theory is the idea that sensory signals are coded versions of what is really going on in the world, and that the task of sensory information processing is to reconstruct as much as possible a full description of the state of the world. We may define the goals of the field as

the analysis of the patterns generated by the world in any modality, with all their naturally occurring complexity and ambiguity, with the goal of reconstructing the processes, objects, and events that produced them.

For example, these patterns may be those of visual signals, that is, 2D arrays of intensity and color measurements as received by the rods and cones in the retina. Or they may be the patterns of auditory signals, that is, the time-varying vibration patterns of the inner hair cells generated by the complex cochlear filter. In the visual example, one seeks first to reconstruct the pattern of discrete objects in the world, their distances from the observer, surface markings, and how they are illuminated so as to produce the observed signal. In the case of speech, the first step is to reconstruct the events in the throat and mouth of the speaker and then to label these as the events associated to specific phonemes in a specific language, plus pitch and stress data to be used in further processing.

But Pattern Theory goes further and asserts that a parallel analysis can be applied to higher cognitive levels as well. Consider a medical expert system—or a physician. Both of these educated devices accept as input a description of the symptoms, test results, and a partial history of a specific patient. This data can be viewed as a coded signal generated by the processes at work in the patient's body. The task of medical expert system or the physician is to reconstruct a full description of these hidden processes. Many cognitive tasks can be analyzed in this way. The world contains unknown processes, objects, and events—hidden random variables in the language of the probabilist. But they are not totally hidden, as partial encoded information about them comes to the observer through various sensory channels or lower level analyses. The goal is to estimate the world variables.

How does Pattern Theory propose to carry out this reconstruction? There are three characteristic ideas in Pattern Theory. The first idea is that to successfully reconstruct the world variables, one must learn to *synthesize* the coded signals that one observes, so that tentative reconstructions of the world variables can be checked by comparing the actual observed signal with synthesized signals. This means that the architecture is *not* purely feedforward, bottom-up, but fundamentally recursive combining feedforward actions with feedback, top-down processing with bottom-up. The second idea is that the encoding processes, which transform the state of the world into the received sensory signal, are not completely arbitrary (e.g., the logician's general recursive functions), but processes of several restricted sorts—*deformations* is Grenander's word—that reoccur in all sensory channels and in higher cognitive problems. This means that the architecture can be customized to *decode* these specific types of deformations to reconstruct the state of the world. The third idea is that this reconstruction can (and must) be learned from experience, that one learns both which

Neuronal Architectures for Pattern-theoretic Problems

hidden variables best describe the patterns in the signals, hence the world itself, and the *priors* on these variables to be able to best compute them. In the rest of this chapter, we want to discuss these three ideas.

THE ANALYSIS-SYNTHESIS LOOP AND CORTICAL FEEDBACK PATHWAYS

Two Different Flow Charts

The first basic idea of Pattern Theory is that to analyze some class of signals, you must learn to synthesize these signals given typical values of the world variables. To recognize some class of objects visually, you must know how to synthesize images of them; to recognize words, you must know how to synthesize the actual sound patterns; to diagnose a disease, you must be able to describe its typical presenting symptoms.

Although this sounds like common sense, it distinguishes Pattern Theory from the majority of computational and modeling theories, because it implies that top-down feedback processes are just as important as bottom-up feedforward processes. Consider how many classification algorithms are purely feedforward: feature-based winner-take-all ("Pandemonium") algorithms, feedforward neural nets (even with backpropagation, in which feedback is used for learning, but not in practice), tree-based classifiers like CART, and parametric statistical modeling. None of these handles gracefully a new and unexpected stimulus, because they have not *explicitly modeled* the stimuli they have been trained on, and therefore cannot recognize novelty. At best, they can incorporate significance levels, and flag suspicious stimuli if none of their categories fits with overwhelming significance. Unfortunately, this often miscarries with borderline cases. One reason is that, because of the distortions caused by "interruptions" (i.e., overlapping objects, events or processes—see below), correct instances of a category are often present but with part of their characteristic pattern missing (e.g., a letter partially covered by an ink blot). In this case, part of the stimulus will fit the category very well, part not at all, and a feedforward classifier may mistake them for a different category. In contrast, incorrect instances, like a letter from a foreign alphabet, may roughly resemble one of expected categories, say an english letter, and therefore be mistaken for it by a feedforward classifier. The moral is that it is much more significant for a part of the stimulus to match closely the prototype of a category, than for all of it to match slightly. This kind of distinction cannot be made unless a top-down synthesis stage is part of the recognition algorithm.

The simplest type of pattern synthesis consists in simply storing prototypes or templates for each category to be recognized. Note that this is not the same thing as storing prototype feature vectors (e.g., mean values of the features for all instances of signals from a given category). This is because there is usually no way to reconstruct the signal itself from its features. In contrast, a template (as the word is used in traditional pattern recognition) is a particular signal that can be directly compared with the incoming signal. Such templates are also incorporated in the pattern completion operation of various neural nets such as Kohonen's and in the seeking of "energy minima" in the attractive neural nets of Hopfield. In a simple world such templates might suffice but, because the many different signals belonging to a single category (e.g., all varieties of the letter *A*) differ by complex transformations such as domain warping (see below), a single template will rarely match the actual signal at all well. Too many factors affect every real world stimuli for a simple Sears-Roebuck catalog of patterns to be useful. Each instance of a category can be positively identified only by actively synthesizing it: combining the templates of those objects or processes present on all scales, distorting them in the correct ways, and removing parts that are absent. This is why Pattern Theory presupposes an analysis-synthesis *loop* in which feature extraction and feedforward style classification is combined with a feedback step in which the system attempts to duplicate the stimulus by combining and transforming its basic prototypes.

Figure 7.1 contrasts the flow charts of traditional bottom-up recognition systems with that of Pattern Theory. Note that Pattern Theory proposes that analysis and synthesis should be carried out *iteratively*. Thus, at the first stage, if there is no expected pattern, the features of the actual signal are extracted exactly as in the traditional flow chart and passed to a recognizer. *However*, next the recognizer draws on its database of prototypes to synthesize a standard instantiation of the hypothetical object being seen. In subsequent iterations, the hypothesis will be refined: details on size, orientation, shading if present, and missing and/or extra parts will be computed by comparing the synthesized image with the true image and computing features of the residual or difference between these. That does not mean that the true image is thrown away. But a steady state would mean that the synthesized image agrees, *up to acceptable error*, with the true image and the features of the residual are too small to modify the hypothesis further. There is no need to send any more feedforward signals when the feedback pathway already predicts the input signal. (This is like driving home on a well-known road and not needing to pay attention to anything that you see because it always agrees with what you expect, hence never generates a residual.)

What is an acceptable error in synthesizing the signal is something that must also be modeled explicitly and differently for each category of signal. Thus modeling the detailed contour of the nose is quite significant for face recognition, but modeling the shape of a stapler is not significant when performing office tasks. Modeling the details of the grain of an oak floor is not significant, but the exact shape of the stripes or spots on the back of a large member of the cat family is. This is a major difference between Pattern Theory and Barlow's theory (see chapter 1). In Barlow's theory, modeling patterns allows you to distinguish that part of the signal that is familiar and has predictable structure from the novel information in the

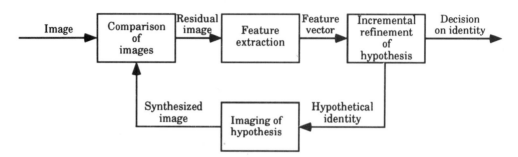

Figure 7.1 (*Top*) The traditional bottom-up approach to recognition in which a feature vector is computed first and this compared with prototype vectors, one for each category. (*Bottom*) The alternative proposed by Pattern Theory in which a bottom-up/top-down relaxation explicitly models the image by comparing it with images synthesized from high-level descriptions.

signal—which resembles noise. Pattern Theory, however, distinguishes *two* parts to this "information": the high-level description from which the signal is being synthesized and the residual error that is hard or impossible to model. The former is truly informative and is passed on to higher levels, and the latter is discarded as being truly noise.

Note that the flow chart of Pattern Theory is also different from that proposed by Poggio (e.g., in chapter 8 by Poggio and Hurlbert). They propose a very specific mechanism for combining multiple instances of a specific category by comparing each with the true signal and *interpolating*. But this comparison is feedforward and is hard-wired by radial basis functions, so that if further kinds of variability are encountered, one must multiply the sets of stored instances, allowing for all combinations of each type of variability. In contrast, Pattern Theory is feedback, so it can synthesize dynamically every new signal and thus potentially model a much larger class of deformations. How this can be done neurally will be discussed below.

This feedback stage is not unlike *mental imagery*, which, as Kosslyn has discovered, is a complex synthesis and reconstruction of something that has all the qualities of actual stimuli from the external world. As he suggests and both MRI and PET scans seem now to confirm, this something may be low-level activity in the sensory areas of the brain, even V1, just like what we propose for our feedback (see Le Bihan et al. 1992; Kosslyn et al. 1993).

We may summarize our argument by saying:

$$\text{Feedback} = \frac{\text{Synthesis of signals from memory}}{} = \frac{\text{Use of (flexible) templates}}{} = \frac{\text{Mental imagery}}{}$$

To give these ideas a more concrete flavor, we want to take a particular image: the old man on a bench shown in figure 7.2a. We assume that you instantly recognize the content of the image. But how did you do this? A blow up of his face (at the same resolution) is shown in figure 7.2c: his ear is the only vaguely recognizable part of his face and his hand blends into his face, creating the two utterly misleading spots of light where you see past his face. Figure 7.2b shows what a state-of-the-art edge detector (Canny's) produces (such detectors require various parameters to be set by the user and we have selected those that seemed more or less optimal): not only are the edges of his face not found, but even the outline of his coat is fragmented. Finally, note that the most salient "object" in the image is his cap, which, by itself, could be virtually anything. How do feedback loops help you analyze this man? There are two stages here: in the low-level feedback loops, low-level templates and low-level segmentation (= clustering into distinct objects) take place, while in the high-level feedback loops, models of objects such as bodies, heads, and benches are fit to the image. To make this plausible, let me point out how much could, in principle, be done in low-level fitting operations: first, the pieces of the bench on each side of the man can be grouped, using an interupted line template. Next, a textured, fragmented contour along the back of his coat can be assembled into a model of a backlit, wrinkled, and rounded object. And his cap comes forward because it occludes the background and his face and simultaneously the fact that the black triangle over his eyes is a shadow can be deduced. All of these deductions involve fitting simple models of scene fragments. At this point, there is finally a chance for high-level models to find the right parts of the scene to fit and we already know enough about the lighting to know what would be in shadow and what would be brightly lit (e.g., the back of his head).

Besides arguing for the flow chart in figure 7.1, this example is also useful in contrasting Pattern Theory with the feedback theory of Ullman (chapter 12). Our analysis of the old man example requires multiple independent and concurrent loops, *low-level and high-level*, some modeling shading, some modeling depth planes, some modeling clothed bodies, and some modeling faces. This suggests that Ullman's theory with a single bottom-up search and single top-down search could not easily solve the old man puzzle. Postulating multiple independent feedback loops, instead of one global feedback from stored knowledge to the sensorium, is also helpful in comparing Pattern Theory with Marr's theory of vision (Marr 1982). Marr was very influenced by several examples in which top-down information was either not needed or ignored in accomplishing some feedforward computational task (e.g., fusing random-dot stereograms or

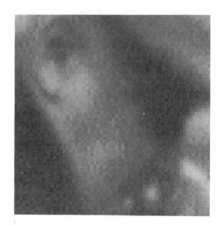

Figure 7.2 An image that illustrates some difficulties in recognition. (*Top*) The image. (*Bottom left*) Canny's edge detector applied to the image. (*Bottom right*) A blow-up of the face showing the lack of recognizable features.

construction of 3D models from unorthodox 2D views by victims of agnosia). This led him to propose a purely feedforward theory of vision. We would argue that all his examples are evidence against *strong feedback models*, like Ullman's, in which high-level knowledge is fed back all the way to low-level stages, and that none of his examples contradicts the hypothesis that multiple, more local, feedback loops are being used.

Evidence from Neuroanatomy

We now turn to the cortex itself and ask whether we can find a confirmation in its structures of the theory that bottom-up pattern analysis cannot be done independently of top-down pattern synthesis. Indeed, one of the main themes in neuroanatomy in the last several decades has been the discovery that the cortex is naturally divided into distinct *areas* that are *reciprocally* connected by pathways created by the axons of their pyramidal neurons. Pattern theory strongly suggests that these pairs of pathways should instantiate the dual computational processes of analysis and synthesis. This proposal is strongly supported by the still emerging picture of the *cortical layers* connected by these pathways. Some of these pathways terminate principally in layer 4, the standard "input" layer for bottom-up cortical processing, the route from raw sensory input to higher association areas: it is natural to propose that these pathways carry out pattern analysis. Other pathways terminate mostly in layers 1 and 6, the top and bottom of the cortical plate, and are typically dual to the first set (i.e., if area A is connected by the first type of pathway to area B, then one of the second type connects area B back to area A). Pattern Theory suggests that these pathways should carry out pattern synthesis.

These cortical feedback pathways are, perhaps, the most complex piece of wiring in the brain and it is astonishing that evolution has been able to create them. Does their evolution support our proposal that all cortico-cortical pathways should belong to two separate systems, a bottom-up processing pathway and a top-down processing pathway? The homologies between mammalian neocortex and reptilian telencephalic structures are not obvious and there has been much debate on them. One set of homologies is the so-called *dual origin* hypothesis, which goes back to the pioneering work of Marin-Padilla (1978). This theory has been developed by Karten and most recently by Deacon (see Karten and Shimizu 1989; Deacon 1990) and has been gaining adherents. It proposes that the six-layered mammalian neocortex is not homologous to a single structure in the reptile, but that two structures, separate in the reptile, have become merged in the mammal. More specifically, (1) the top and bottom layers of the mammalian neocortex when originally formed in the embryo are homologous to the two-layered *dorsal cortical plate*, or pallium, of the reptile, and (2) that the population of neurons that migrates during mammalian embryogenesis to form the inner layers of the neocortex is homologous to the neurons of the *dorsal ventricular ridge* in the reptile.

This theory is shown, in simplified form, in figure 7.3. What Deacon has pointed out is that this theory explains beautifully the existence of reciprocal pathways and their most common laminar patterns (Deacon 1990, pp. 686–691, especially last paragraph). Note that in the reptile, there are no directly reciprocal pathways, all loops being longer and more indirect. But the original pallium carries its own internal connections, labeled "A" in figure 7.3, many of which emanate from the olfactory and limbic cortex and proceed caudally. Moreover, the dorsal ventricular ridge (DVR) has its internal pathways labeled "B," which proceed rostrally. When in the mammalian embryo the homologous structure to the latter migrates inside the homologous structure to the former, Deacon proposes, because of the conservatism of evolution, that homologous connections will still be established: the pathways A, descending from limbic areas and synapsing on layers 1 and 6, the residues of the dorsal cortical plate, are still laid down and become the top-down pathways of the mammal; and the pathways B,

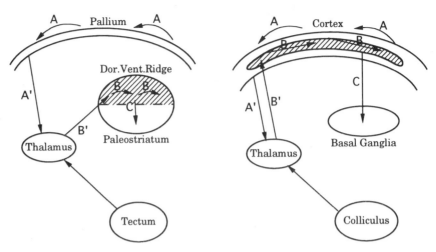

Figure 7.3 A comparison of the main structures in the reptilian (*left*) and mammalian (*right*) brains, illustrating Marin-Padilla and Deacon's theories of the dual origin of the neocortex and its reciprocal pathways from the pallium and dorsal ventricular ridge.

ascending from sensory areas in the DVR, synapsing in the middle layers, become the bottom-up pathways of the mammal. Moreover, the thalamo-cortical reciprocal pathways arise in a similar way, from the *thalamus* → *DVR* pathway B', and the *pallium* → *thalamus* pathway A'. (We have simplified the picture somewhat by excluding the geniculocortical pathway and its precursor.)

One can make a suggestive link of Pattern Theory with the 40- to 60-Hz cortical oscillations that have been observed in the last decade in so many structures in so many distinct recording modes (cf. Singer, chapter 10). The link is the proposal that this oscillation is a reflection of the basic cycle of computation in which bottom-up features are compared with top-down memories and expectations, of the iterative operation of the loop in figure 7.1 (bottom). The strongest evidence for this is the observation that these oscillations lock when the cells are responding to linked parts of the stimulus, both in different parts of V1 and between V1 and V2. It is important to realize that if successive cycles of this oscillation represent successive iterations in a computation, one would not expect exactly the same cells to participate in each cycle. Therefore, the oscillation would be much stronger in field potentials than in single cell recordings. This is exactly what is found. For instance, field potential oscillations were discovered by Freeman in the 1970s (Freeman 1975) in the olfactory bulb and cortex. It is interesting to note that one form they take here is repeated sweeps of rostral-to-caudal excitation, as though the two poles of the olfactory bulb are like two neocortical areas communicating and oscillating via long axons (compare the model of Wilson-Bower 1992). The oscillation even shows up on the entire cortex in human MEG recordings: see Llinas et al. (1991), which shows a 40-Hz oscillation sweeping over the whole cortex and Ribary et al. (1991), which shows the oscillation between cortical and thalamic activity.

If we make a crude connectivity model of the type of circuit that emerges from this analysis, what does it look like? Is it like the "blackbox" computational models that have long been the staples of the computer metaphor (figure 7.4, top)? These diagrams stem from 50 years of development of the computer, starting from Von Neumann. The major computational steps are carefully dissected and put in separate boxes, necessary data flow paths are added, and the whole thing operates like a chemical factory. This point of view is highly developed in the books of Fodor and Kosslyn; its computational foundation has been beautifully expounded in the book by Abelson and Sussman (1985). But this is not what's there! In the cortex, roughly 65% of all cells are pyramidal cells that send their output to distant cortical areas, as well as locally via their axon collaterals. This means that there is *no hiding of local information*, no "local variables" or protected data. A better picture is figure 7.4, bottom. Instead of black boxes with opaque walls, we have apartments in a cheap housing complex with very thin walls! All your neighbors hear everything that is going on in your home. Instead of "hiding local variables," a device central to all modular programming, every little whimsy that occurs to you goes out instantly to all and sundry.

It seems to me that the computational metaphor itself is flawed. Pattern Theory has a clear explanation: these tightly coupled cortical areas are exactly the higher and lower level areas of pattern theory that seek, by a sort of relaxation algorithm, to come to a mutual understanding in which the lower area's more concrete data are fit with a known, more abstract, category expressed by the higher area's activity. This is a fundamental shift in focus from the computational metaphor. Just as, for instance, Edelman has proposed Darwinian, evolutionary metaphors as the right ones for modeling brain function (cf. Edelman 1987), similarly pattern theory implies a new paradigm: that of many different parts of the brain attempting to reconcile their states, their implicit descriptions of part of reality, with the states of other areas, either through bottom-up assertions of facts that have to be dealt with or top-down memories of expected patterns. This is related to Minsky's idea of the brain consisting of many agents, in "Society of Mind" (Minsky 1985).

Does all this speculation mean anything for the experimenter? Does it have any predictive force? To begin with, it implies that there will be more correlation between single-cell responses in different areas than would be expected if the areas were black boxes, hiding their characteristic internal computations from each other. For instance, we see this in the tremendous overlap of the characteristics of single cells in the various visual areas, which has prevented assigning any clear functional role to V3 or V4 (aside from generalities like being concerned with shape or color). What we think is the most important implication, however, depends on a refinement of multiple cell recording techniques: Pollen has proposed a technique for preparing an animal with electrodes recording from cells in two areas that show significant cross-correlation in their spiking (cf. preliminary work

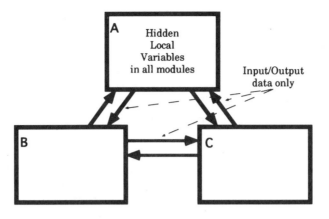

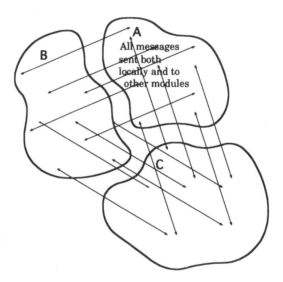

Figure 7.4 *(Top)* The modular approach to cognition and computation, in which individual steps are carried out "privately" and only final results are broadcast. *(Bottom)* The relaxation approach of the cortex, in which two-thirds of all neurons send their output both locally and to distant areas.

by Liu et al. 1992). At this point, instead of looking at the responses of the two cells in isolation, *one can separate for analysis the correlated spikes from the full spike trains.* The theory suggests that this set of spikes may be much less stochastic, carrying the information transmitted between areas, and hopefully correlated much more precisely and predictably with identifiable aspects as the stimulus. To be more specific, we must turn to what the theory conjectures about the content and nature of the representation in individual areas and, using this, its description of the data transmitted back and forth between areas.

THE FOUR BASIC DEFORMATIONS

What They Are

The second basic idea of Pattern Theory is that the processes that transform the world variables into the observed variables are not arbitrarily complicated, varying widely from one channel to another. Instead, four basic transformations, or deformations as Grenander called them, can be found at work in every channel. These are the following:

1. *Noise and blur.* These effects are the basis of standard signal processing, caused, for instance, by sampling error, background noise, and imperfections in your measuring instrument such as imperfect lenses, veins in front of the retina, dust, and rust. Typically, the full real world signal is measured only at discrete sample points; its value at each point gets averaged with its neighbors—this is blur—and corrupted by the addition of some unknown noisy factors. In more cognitive applications, like the medical expert systems, errors in tests, the inadequacy of language in conveying the nature of some pain or symptom, confusing extraneous factors, all belong to this class.

2. *Multiscale superposition.* Signals typically reveal one set of structures caused by one set of phenomena in the world when analyzed locally, at high precision, and other structures and phenomena when analyzed globally and coarsely, at low precision. For instance, in images, local properties include sharp edges, texture details, and local irregularities of shapes, which coexist with global properties like slowly varying shading or texture statistic gradients and the overall shape of an object. In speech, information is conveyed by the highest frequency formants, by the lower frequency vibration of the vocal cords and the even slower modulation of stress. These spatial or temporal frequency bands may be combined additively (as in Fourier analysis or wavelets), multiplicatively (as in AM coding) or by more complex nonlinear rules. In higher order processing, the analog of this decomposition into the "overall" shape versus fine local detail is the hierarchical model of concepts embodied in semantic nets. These models describe a situation partly by its general properties, the very inclusive superordinate categories (in the terminology of Rosch 1978) to which it belongs, and partly by its details, the subordinate categories of Rosch. Thus a patient is in simplest terms "very ill"; in more precise terms the patient has pneumonia, is contagious, should be hospitalized, and in very precise terms is infected by such and such a bacteria, has a temperature of 103, etc.

3. *Domain warping.* Two signals generated by the same object or event in different contexts typically differ because of expansions or contractions of their domains, possibly at varying rates: phonemes may be pronounced faster or slower, the image of a face is stretched or shrunk by varying expression and viewing angle. In speech, this is called "time warping" and in vision this is modeled by "flexible templates." In both cases, there is a

mapping from the domain of one signal to the domain of the other, either a map of time intervals or a map between two-dimensional domains, which carries the salient parts of one signal to the corresponding parts of the other. The cognitive version of this type of distortion is thinking with *analogies*. In an analogy, some or all the elements in two situations can be mapped to each other, preserving many of their interrelations, just as the same elements occur in two faces, with nearly the same spatial relationships. In all cases, the map may be incomplete, in that some parts of one situation may not have corresponding parts in the other. Thus one face may be partially obscured by hair or a bandage, and only the unoccluded parts match up.

4. *Interruptions.* Natural signals are usually analyzed best after being broken up into pieces consisting of their restrictions to subdomains. This is because the world itself is made up of many objects and events and different parts of the signal are caused by different objects or events. For instance, an image typically shows different objects partially occluding each other at their edges. In speech, the phonemes naturally break up the signal and, on a larger scale, one speaker or unexpected sound may interrupt another. Obviously, in the cognitive realm too, several processes may be at work at once, as in a patient who has several medical problems at once. To infer the correct values of the hidden variables, the effects of the different processes must be separated from each other. A general term for isolating the effects of one process, object, or event from all the myriad others going on simultaneously is figure/ground separation.

This part of Pattern Theory has a great deal to say to neuronal models. If these four transformations are universal coding mechanisms, which must be decoded by a brain, there should be mechanisms for all of them if you look in the right way. If they are truly universal, these mechanisms should be general circuits that occur in all areas of cortex. This is the challenge of Pattern Theory. We will discuss in separate sections below possible neuronal correlates of deformations 1, 3, and 4.

Deformation 2, multiscale superposition, has often been discussed for vision as the "pyramid" data structure and associated algorithms often using a moving window of attention. It was only at this meeting, however, that we heard Van Essen and Anderson propose how such a pyramid could be laid out cortically using the *three* areas V1, V2, and V4 (see chapter 13). We will not discuss the decoding of this deformation except to mention that one of the major computations using a pyramid is the discovery of the "part-of" relations between blobs of different sizes (for instance, as a step to recognition of complex objects, e.g., Hong and Rosenfeld 1984). Striking evidence that this is done by the recognition of small and large blobs *in parallel*, with hard-wired "part-of" connections, was recently found by Jeremy Wolf (unpublished), who found that (1) red houses with yellow windows pop-out in a field of differently colored houses and windows, while (2) duplex half red and half yellow houses do not pop-out! I believe this strongly supports Anderson and Van Essen's theory, because it can be explained by the concurrent recognition of the red houses in V2 and yellow windows in V1 with reciprocal V1, V2 pathways marking "part-of" rapidly strengthening the activation to threshold.

Nonlinear Filtering and the Thalamocortical Loop

Let us look at the lowest level loops in the circuitry of the cortex and its immediate neighbors. The most basic of these are the loops connecting various cortical areas with various nuclei in the thalamus, especially the loop between visual area V1 and the LGN. In many cases, these give primary sensory input to the cortex and a natural idea, in the context of pattern theory, is that these would be concerned with correcting for the most basic "deformation" of the sensory signal—noise and blur. For instance, Grossberg has often pointed out that the visual signal coming from the retina must be distorted by the presence of veins on the inner surface of the retina, not to mention the blind spot itself. Ever since Yarbus (1967), it has been known that within each fixation, the eye is far from still, but drifts irregularly, with a constant tremor of several minutes of arc (enough to move sharp edges across several adjacent cones in the fovea). In addition, the light signal, as it strikes the eye, is already the result of conflicting processes that obscure its origin: the "accidental" markings on textured surfaces obscure their shape, and lighting effects are complicated by local self-shadowing and mutual reflections. Although part of the rich complexity of the world, they act like noise and blur if you are attempting to reconstruct the outlines of the major objects in view.

For many years, engineers have proposed appropriate filtering as the universal solution to the problem of compensating for noise and blur. But pattern theory would propose that, like the other types of deformations, they must be corrected for, not by a blind bottom-up filter, but by an adaptive feedback process. This is a logical role to propose for the thalamocortical loop. Specifically, the reciprocal LGN $\rightleftharpoons$ V1 pathways should implement an image processing algorithm, which "cleans up" and disambiguates the visual signal. Typical functions of image processing are noise removal and edge enhancement. No wonder single cell recordings could never find any role for the V1 $\rightarrow$ LGN feedback: the squeaky clean laboratory signals, with edges, bars, and sine wave gratings do not need any image processing! Experimental tests for this hypothesis are easy to draft, once one is committed to presenting more complex and realistic stimuli, for which the response cannot be summarized by linear approximations, like the impulse transfer function. Several such proposals are presented in Mumford (1991, 1992).

How are these image processing tasks accomplished? We assume that the complex cells, whose response, to a first approximation, is like a power Gabor filter with a preferred scale and orientation, attempt to find the salient edges and bars in an image. But typically, many of these will be responding simultaneously in each local region and one must find how to reconcile them (e.g., one such cell "sees" a strong long line, the other an edgelet that is part of texture; or one marks the end of a bar, the other its sides). Before a consistent interpretation is found for each part of an image, many conflicting local organizations may be detected and there is a need for some kind of decision mechanism such as a "winner-take-all" circuit.

There are several hints of such decision mechanisms in the cortico-thalamic projection. Several groups (cf. McGuire et al. 1984 in cat, White and Keller 1987 in mouse) have reported that the axon collaterals of the layer 6 V1 pyramidal cells and especially the corticothalamic projection cells appear to synapse largely on aspinous interneurons, presumably inhibitory cells. This has the look of a winner-take-all network, an organization long predicted in the neural net literature, but never clearly identified in the cortex to our knowledge. Alternately, the inhibitory cells in the LGN could provide a voting mechanism. In other words, if these were absent, the various feedback signals from cortex would simply be averaged in the dendritic trees of LGN "relay" cells. But if some of them synapse on inhibitory cells, they can effectively suppress other feedback and feedforward signals.

Shifter Circuits, Flexible Templates, and Population Coding

A more radical part of the pattern theory analysis is the proposal that domain warping is a universal deformation. This means that *in analyzing signals, and matching signals against patterns in memory, the pattern of activity on the cortex must be displaced (in the two-dimensional coordinates of the cortical surface)*. Such operations have been proposed under the name of "shifter circuits," most recently in Anderson and Van Essen (1987). Although argued for by theorists for some time, only recently has evidence appeared for their existence in cortex. In a beautiful paper on recordings in the parietal lobe, Duhamel et al. (1992) found that activity correlated to the visual location of different objects in front of an awake monkey is shifted on the parietal lobe surface in anticipation of a saccade that will shift the visual sensory signal. In a totally different part of the cortex, Georgopoulos and his group have found that activity in the primary motor area M1 is shifted as the precise coordinates of an intended arm movement are computed. Note that this example is not sensory but motor-planning related: here the activity pattern for one standard arm movement is first recreated in M1, and then it is modified over a 100-msec period by domain warping until it is appropriate for the specific movement presently desired.

The simplest example where there is a need for this mechanism is in the computations of stereo vision, in the correlation of the left and right eye movement. This example was used by Anderson and Van Essen and by Poggio. As they point out, what makes it especially compelling is the existence of tremors in eye position of up to 10 min of arc during a period of fixation: without active compensation for this, stabilizing the image, it is very hard to imagine how stereo cells in V1 can respond robustly to left/right eye feature disparities of only several minutes, let alone account for the psychophysical evidence of disparity hyperacuity of less than a minute of arc. Anderson and Van Essen propose that, in the primate, this is carried out by shifter circuits in V1 that has developed a highly specialized layer 4, making it unique for its cell density among mammalian cortical areas. In the less specialized case of the cat, we would propose that this stabilization results from the action of the LGN $\rightleftharpoons$ V1 loop, rather than a hard-wired shifter circuit in V1 (and that this circuit is reprentative of the general mechanism used to implement domain warping).

What neural circuitry could accomplish this? In figure 7.5, we make a simple proposal. We suggest that (in the cat) each retinal ganglion cells' axons synapse on multiple LGN "relay" cells and that populations of such cells synapse in overlapping ways. Thus one LGN cell receives input from multiple retinal cells, but on distinct branches of its dendritic tree. Normally one of these is the strongest and that retinal cell takes charge of that particular LGN cell. But under cortical influence, both excitatory and inhibitory, some of these synapses can be strengthened and some weakened by local postsynaptic potentials on the different branches of its dendritic arbor. This could be done by a variety of mechanisms, including NMDA channels. In the figure, we have drawn one possibility using inhibitory effects, caused by the dendrodendritic triadic synapses with inhibitory glomerular interneurons. Following Sherman and Koch (1990, pp. 256–266), we have assumed that this interaction takes place on spines of the "relay" cell, where the retinal and glomerular inputs are combined in a synaptic triad functioning like an "x and NOT y" gate. The effect is that each LGN cell is driven by a different retinal cell and the pattern of activation is shifted in the LGN. Note that such shifts must be vertical as well as horizontal, as evidence (cf. Motter and Poggio 1984) shows that the two eyes are usually misaligned vertically by 5 to 10 min of arc. This shifting can accomplish two things at once: it can compensate for tremor and misalignment and it can create a simulated vergence movement to align more closely the left and right eye images, thus reducing the disparity of the signal received by V1 so that the exquisitely sensitive "tuned excitatory cells" of V1 can measure extremely fine residual disparities. One prediction that this makes is that the left and right eye layers of the LGN should interact *through cortical feedback*. Varela and Singer (1987) show that this does happen, and, even more interesting, if the left and right eyes are stimulated with radically conflicting signals, which cannot be put in binocular registration, then the LGN "relay" signals decrease markedly after about 1 sec.

At all levels of the cortex, there is a need to shift patterns of activation in order to find matches between memories and expectations and the particularities of the present situation: a very concrete example is the need to recognize a familiar face with any of the millions of possible combinations of viewpoint, lighting, and expression that can occur. Shifter circuits can accomplish this and *we propose that this shifting is accomplished in general by*

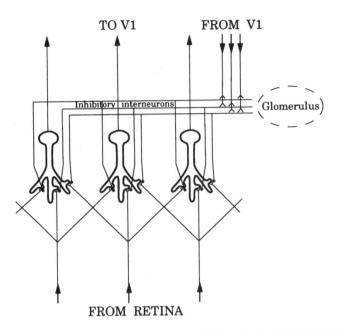

TO V1   **FROM V1**

Inhibitory interneurons   (Glomerulus)

FROM RETINA

Figure 7.5 A possible implementation of shifter circuits in the LGN: V1 feedback excites inhibitory glomerular interneurons that combine with retinal input in "x and NOT y" trisynaptic connections on the LGN relay cells.

the extensive arborization of the feedback pathways, selectively exciting and inhibiting the collateral spread of activity in a given cortical area. This is the natural generalization of the circuits in figure 7.5. Rockland's beautiful tracings of the axons of recurrent axons have shown how amazingly diverse and extensive their arborizations can be (cf. Rockland and Virga 1989, 1993).

From an evolutionary perspective, we can contrast this with what happens in the reptile. The reptile has a more or less rigid body and its tectum contains maps of its visual, auditory, and somatosensory systems in, more or less, hard-wired registration. In such a structure, the sensory systems are forced to combine their data with very little flexibility. In contrast, mammalian cortex has a unique flexibility due to the separation and duplication of cortical mappings. It should be noted that the existence of multiple sensory maps is not particular to higher mammals, but is universal in mammalian neocortex, even in the evolutionary side branches of marsupials and edentates (e.g., essentially all mammals have a homolog of both V1 and V2 [Kaas 1989]). To some extent, this may be a response to the increased flexibility of the trunk, especially the neck, and the development of limbs, which require that the animal have the capability of combining visual, auditory, and tactile sensory data in flexible ways. But it also affords new computational capabilities: in particular, the sensory maps in distinct areas can be dynamically interleaved, creating the domain warping needed for much more sophisticated pattern matching.

An objection to these ideas is that only in primary and (to a lesser extent) secondary sensory and motor areas can one find a coherent meaning to the two-dimensional cortical layout of activation. In higher cognitive areas, it is very hard to imagine why abstract thoughts should have any two-dimensional structure or why shifting patterns of activation on the cortical surface would be useful! We think the answer to this paradox lies in another biological principle, which is strongly at odds with traditional cognitive modeling. This is population coding: many experiments reveal that the brain does not store facts cleanly and discretely, with one neuron firing for one possibility, another for a second, etc. Instead, there is a graded pattern of firing for each possibility, but with shifting strengths (possibly with coherent pulse timing too) for each situation. It seems to me that this is directly connected to the linguistic fact that the meaning of individual words in human languages is not simple and clean either: words always cover a great variety of different related situations. This is exactly what you would expect if language reflects the way neurons fire, and if higher level concepts are population coded like sensory and motor signals. *But a corollary of population coding is that the set of higher level concepts will automatically have geometric structure.* This is because two concepts can, at one extreme, excite nearly identical populations with a small change in the degree of excitation and the marginally excited neurons; and, at the other extreme, can excite totally disjoint populations. We may thus define the *distance* between two concepts by the degree of overlap of their representations, or the correlation of the vectors of neuron-by-neuron excitations that each concept arouses.

Chapter 4 by Desimone et al. describes experiments in area IT that fit nicely into this theory. Their data suggest that perhaps in many cortical areas, there is a tendency to form more and more localized responses to *exactly* repeating stimuli (this is shown negatively by the large numbers of cells whose responses decrease to repetitions). In other words, the cells of a certain class tend to specialize in responding to very precise patterns. If a category is formed by a cluster of such precise instances, we will naturally get a graded population response to new instances of this category, because it will resemble to a greater and lesser degree each of the previously learned instances.

One construct that has often been suggested in this connection is to make a *graph* out of the set of all concepts, or the set of all English words. The edges of the graph represent the most closely connected concepts. Such a graph was proposed, for instance, in Quillian (1967) under the name of an associative net. Perhaps the earliest attempt to do this with a whole language was the Thesaurus of Roget, which is precisely such a graph. Bell labs put this graph "on-line." They found curious facts such as that the average number of edges needed to join a word to its antonym is 5 or 6! A quite curious graph is formed in Dixon's analysis of the five word classes in the Australian aboriginal language Dyirbal. These appear to be clusters gotten by stringing together related concepts in long chains (cf. Lakoff 1987, pp. 92–102). Once we have this graph, we can talk about domain warping in situations involving high-level cognitive data structures and we find that it corresponds to well-known cognitive operation: namely finding *analogies*, finding a mapping between two sets of concepts that bear the same mutual relation to each other. Matching a heavily shadowed face in front of you with your memory of general face structure is the same warping of a template that is accomplished when you match your knowledge of the Pope with the general concept of a bachelor. Our proposal is that what happens neurally when you analyze the sentence "The Pope is a bachelor" (a classic example of philosophers) is that one cortical area with a "bachelor" template, stored with all sorts of typical properties, activities, life histories of bachelors attempts to fit this activity pattern to the specific data conjured up in a second area describing the Pope and his properties, activities, and life history. A partial match can be achieved, after suitable warping of the archetype. This will also highlight the nonmatching qualities (e.g., the Pope does not date), which is what we want to look at next.

Interruptions and Foreground/Background Coloring

We want to consider the fourth type of deformation in pattern theory, interruptions. Recall that this refers to the fact that we are bombarded by signals from many different objects and events at any given instant and all contribute to the activity being received and processed by the brain. We must locate the boundaries between these objects or events, so we can identify them one at a time. To do this, we have to label or "color" explicitly the parts of the present activity pattern that result from this foreground object or event, suppressing for the time being the rest as background. From the point of view of single cell activity, this is very mysterious: each cell is population coded and, via its collaterals, there is a tendency for a spread of activation. What we need is a mechanism to say a and b are linked but NOT c. Much has been said about this issue, under the names of dynamic linking, compositionality of concepts, etc. In particular, Singer has argued forcefully for synchrony of pulse timing as a possible mechanism (see chapter 10). In the context of pattern theory, the key thing is that whatever mechanism is used, *it must involve correlating activity in reciprocally connected areas*. This is because only by separating foreground from background can the features of the foreground be extracted without confusing them with those of the background. Pattern theory proposes that this is done iteratively: a preliminary foreground/background separation leads to a preliminary computation of features, hence to a preliminary identification, then by feedback a refined foreground/background separation, etc.

We would like to discuss a very simple specific case of this problem, which has been extensively studied in computer vision: the segmentation of a two-dimensional visual signal into distinct objects. Our discussion of the "Old Man" example (earlier) shows that many processes contribute to segmentation. (That example dealt with a photograph, hence it omitted stereo and motion that, in real life, are extremely effective additional processes in segmentation.) One of the processes we discussed was the linking of interrupted edges and the clustering of similarly textured blobs, with the preliminary goal of segmenting the image into homogeneously textured areas. Our hypothesis is that this segmentation is the main goal of one or both of the V1 $\rightleftharpoons$ V2 ($\rightleftharpoons$ V4) feedback loops. Note that in the theory of Anderson and Van Essen, these are the areas holding a pyramid-based description of the image; in their terms, our hypothesis is that segmentation is the main internal computational goal of this pyramid (in its loops with higher areas, V4 may participate in other things, like the computation of shape features for identification of objects). Two quite different mathematical discussions of the segmentation problem can be found in Hong and Rosenfeld (1984), which uses a pyramid-based dynamic linking algorithm, and Lee et al. (1992), which uses Bayesian methods of combining edge and region data.

There are two very specific things to look for if this computation is going on. The first is the need to trace extended edges, that surround the objects in the scene. *Simple Gabor-filter-like cells do not do this*: they are misled by gaps in edges, small texture responses, blur, and local shadows. Lateral inhibition, which is known to occur for a subpopulation of complex cells, is the first step in finding the important edges, as this will often distinguish region boundaries from texture edges. Filling gaps and finding alignments of edge terminators, as von der Heydt has shown is done in V2 (von der Heydt and Peterhans 1989), is another step. But all this information must be put together. A strong suggestion that the the V2 → V1 feedback may be involved was found recently by Mignard and Malpeli (1991): they found that vigorous upper layer activity in V1 can be sustained by feedback from V2 in the absence of direct stimulation from the LGN → V1, layer 4 pathway. It is possible that V2 → V1 paths carry a reconstruction of the extended edges in an image which are then compared with the detailed local signal by the pyramidal cells in layers 2 and 3 of V1, resulting in a new refined signal of edge strength going back to V2, where it is linked up further into larger edges, etc. Algorithms to do this in a computer have been extensively studied both by our students (cf. Nitzberg et al. 1993) and those of Zucker (cf. David and Zucker 1990).

However, correctly tracing extended edges is only one part of the problem. The other is to "color" a region that is surrounded by such a contour, that is, marking explicitly homogeneous areas not interrupted by strong edges. Until a region is so colored, there is no way to compute the features of its shape, such as its center, its area and orientation, etc., hence to begin an identification procedure. The most dramatic evidence that such an active "coloring" process does take place in the cortex is the experiments on masking of Nakayama and Paradiso (1991). Masking seems to freeze the processing at an intermediate stage and they find partial stages at which the homogeneity of part of a region has been made explicit, but not the whole. The underlying neural activity expressed in this coloring process might take

Neuronal Architectures for Pattern-theoretic Problems

place in the cytochrome-oxidase blobs, especially if some mechanism for dynamically linking those blob cells that are responding to two parts of the same object were found. Coloring might mean, for instance, progressive entrainment of larger and larger populations of cells in synchronized firing.

From a computational point of view, it is very important to realize that coloring is not a simple mechanical step (as it seems in artificially simplified stimuli) but requires in real images adaptively determining what homogeneous means, that is, what matters is that the stimulus within the cells receptive field is *relatively* homogeneous compared to variations in a larger surround, and therefore cannot be done by purely local computations. Figure 7.6, for example, shows two images on top of which we have drawn a dotted circle to represent the classical receptive field of a V1 neuron. In these images, the interiors of the dotted circles are identical, hence the V1 neuron "sees" the same blurry contour. But in one, however, the blurry central contour is the perceptual boundary of a foreground object in front of a background; in the other, the blurry contour is merely a shading effect on the surface of a different object. In the first case, the central region is not homogeneous; in the second it is. Thus we predict that at least some V1 neurons with this receptive field would exhibit modulation from outside their classical receptive field that reflects this difference. Whether this modulation was excitatory or inhibitory would depend on whether the local evidence for an edge was strengthened or weakened by the global evidence (as in figure 7.6). It might also have a longer latency than the local response (e.g., this modulation might take effect 50 msec after the initial response). We expect that this modulation is a typical effect of feedback from V2, where the larger receptive fields allow more global integration of the percept.

Another hypothesis for the marking of object boundaries and inhibiting the sideways spread of activity was made by Somogyi and Cowey (1981). They hypothesized that "curtains" of inhibitory double-bouquet cells may activate, cutting off activity in vertical columns from neighboring columns on the other side of this curtain, thus allowing integration of activity within the population of cells responding to one portion of the visible field, but preventing this from interfering with activity related to other parts of the field. This could have a similar effect in dynamically linking cell populations as pulse synchrony.

SPATIOTEMPORAL PATTERNS AND TEMPORAL BUFFERING

There is a strong tendency in analyzing cognition to regard space and time as two quite different things. From the point of view of Pattern Theory, however, the signals received by the brain are functions of both space and time and they exhibit patterns in both dimensions. All of the characteristic deformations present in spatially distributed patterns are present in temporally distributed ones and in signals depending on both space and time. The input to the eyes is a function $I(x, y, t)$ of two spatial and one temporal

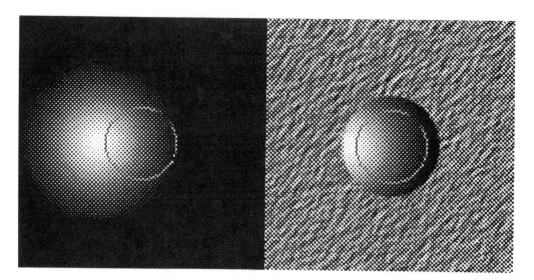

Figure 7.6 Two stimuli, identical locally within a receptive field (indicated by dotted line), differing globally. On the left, a blurry figure on a black ground, its edge within the receptive field; on the right, a single shaded figure on a textured ground, its edge outside the receptive field.

variable; the output of the cochlea is a filtered function $s(\omega, t)$ of frequency and time; the signal from the proprioceptive system is a function $m(k, t)$ indicating the stretch and tension of the kth muscle at time t. In this section, we want to examine the specific problems of computing temporal patterns in signals.

In vision, we often make the assumption that after initial temporal filtering by the ganglion and amacrine cells in the retina, the remainder of the visual system is presented with an instantaneous representation of the image and its optical flow, which can be analyzed as a fixed signal. Observing experimentally the modulation of a response to time varying patterns is difficult because of the apparently stochastic nature of spiking, which requires averaging the cell's response for as long as possible. The standard experimental approach has been finding a way of keeping up a fixed optimal stimulus for as long as possible, "tickling" the cell as it were. In the visual system, this leads the experimenter to prefer repetitively and constantly moving stimuli and this prevents one from analyzing the dependence of the response on subtler temporal variations of the signal.

None of this addresses an obvious aspect of natural stimuli: in general, these are neither still nor moving regularly. Natural stimuli often move and change in complex ways that are essential for the proper identification of their source. A simple example is the identification of people through characteristic gestures and fleeting expressions: it is as though we preserve movie clips of typical things our friends do, and can match this memory against the fleeting temporal signal that we receive. Likewise, it is well known that the recognition of phonemes cannot be done successfully from

Neuronal Architectures for Pattern-theoretic Problems

the analysis of speech at any single instant, but requires the integration of clues hidden in the preceding and succeeding phonemes as well (Liberman 1982). All of these tasks require *temporal buffering*: the temporary storage of the sensory signal or its features while the remainder of the signal continues to unfold. To model this will require neural mechanisms that, as far as we know, have not yet been described and to find these mechanisms will require the presentation and analysis of responses to more complex time varying signals than have been studied as yet.

A specific case is the LGN and the motion pathway (called magnocellular or M in the monkey, and Y pathway in the cat). During a single fixation of the eyes, a small moving object may stimulate many ganglion cells in the M pathway as its image crosses the retina. Often, we may want our eyes to make a transition to pursuit mode, following the object to "freeze" its image on the retina. To do so requires that we predict where the object will be *after* the next 100 msec or so, hence that we have an accurate record of where the object was in the last 100 msec. Since M cells are very transient, some mechanism is needed to sustain activity until the end of the fixation, while its velocity is being calculated. We would like to make the hypothesis that, at least in the cat, *the LGN Y pathway cells are used for this temporal buffering, their activity being sustained by corticothalamic feedback after the moving object passes their receptive field*. This possibility is suggested by the cell counts in the cat LGN that show that there are about 12 times as many LGN cells in its Y pathway as there are retinal ganglion cells (Sherman and Koch 1986). Such a population could encode the time history of the stimulus in many ways. It could store a sequence of activity states in different cells; more likely, the cells might population code this history, or features of this history, like acceleration, stops, and starts.

Other prime candidates for detecting temporal buffering are A1 and M1. In both areas, it seems essential to buffer the temporal activity pattern (i.e., the auditory signal over something like the last 200 msec) or the motor commands over the next 200 msec). In A1, this should be especially simple to check: one needs to record and analyze responses to pairs of sounds, presented sequentially. The null hypothesis, that there is no buffering, would imply that the response to the second part of the stimulus is independent of the first part of the stimulus. Temporal buffering would predict some kind of modulation of the second response. As far as we know, a neurophysiological experiment to look for this kind for buffering has not been done.

LEARNING THE HIDDEN VARIABLES AND THEIR PRIORS VIA MINIMUM DESCRIPTION LENGTH

Bayesian statistics was one of the main inspirations for Pattern Theory. It goes like this: assume that X is a set of variables describing the world—called the hidden variables—and that Y is the data we observe. We assume, moreover, that from experience we know the "prior" probability $pr(X = x)$

[or pr(x) for short] that the variables X take on every possible set of values x (e.g., you know it is very unlikely that your grandmother is wearing a bikini), and that we also know the conditional probability of every possible observation y given the state of the world x, written pr($Y = y|X = x$) [or pr($x|y$) for short]. Then if we have observations y, we will want to estimate the most likely a posteriori values x of the hidden variables describing the world. Bayes says to do this by finding the x that makes the conditional probability pr($x|y$) the largest, which by Bayes's theorem is the x that maximizes $[\text{pr}(y|x) \cdot \text{pr}(x)]$. (So if we think we see Granny in a bikini at a great distance, we reject the conclusion, but if we see her so attired close up, we have to accept it as fact.) The optimal value x so calculated is called the maximal a posteriori or "MAP" estimate of the world variables. This is standard stuff.

To use this rule, one needs to learn, store, and apply via Bayes's rule both the prior probability distribution on the world variables X and the conditional probability on the observations Y given X. In a biological setting, it is possible to imagine that these probability distributions were somehow learned by natural selection and have become encoded into the genes. Perhaps this happens with some animals—for instance the overall structure of a bird's song seems to be genetically encoded—but this does not seem to account for the flexibility of mammalian responses. For instance, a human infant born into a complex technological culture has no trouble learning how to use TV sets. There are various approaches to learning these probability distributions "on the fly," but one that fits in cleanly with both Bayesian statistics and Pattern Theory is to use the Minimum Description Length Principle. This approach is particularly attractive in that it suggests how the world variables X themselves might be learned, not merely their distribution.

The Minimum Description Length (or MDL) Principle says that, starting with many observations $Y = y_n$, you may take advantage of the patterns and repetitions in this string of observations to reencode Y so that, with high probability, if every new observation is reencoded in this way, it will have much shorter length (in bits). For example, suppose five different bird songs are heard regularly in your back yard. You can assign a short distinctive code to each such song, so that instead of having to remember the whole song from scratch each time, you just say to yourself something like "Aha, song #3 again." Note that in doing so, you have automatically learned a world variable at the same time: the number or code you use for each song is, in effect, a name for a species, and you have rediscovered a bit of Linnaean biology. Moreover, if one bird is the most frequent singer, you will probably use the shortest code (e.g., "song #1") for that bird. In this way, you are also learning the probability of different values for the variable "song #x." This is nothing more than the fundamental theorem of Shannon's information theory that provides the link between coding length and probabilities. His theorem states that if you want to encode the different values x of variables X so that the average length of the code is smallest, then the length of the code $c(x)$ in bits will be

$$c(x) = -\log_2 \left[\text{pr}(X = x)\right].$$

(A technical point: in this formula, the log is a positive real number that need not be an integer. But the number of bits in a code is always an integer. So what Shannon did, to get this elegant relationship, was to consider "block coding," codes where several signals were encoded at once. If k signals were encoded, the code length for each signal is $1/k$ times the length of the block code. Then the exact theorem states that by considering longer and longer block codes, the left hand side gets as close as you want to the right.)

How could finding the MAP estimate be implemented cortically? The natural hypothesis is that the probabilities of each set of values x of the hidden world variables and of the probabilities of making an observation $\text{pr}(y|x)$ are stored in the mechanism for pattern synthesis, so that there is a tendency to synthesize the most likely patterns first, the less likely coming to the fore only if the more likely ones are inhibited by mismatch with the input (as in Carpenter and Grossberg 1987). For instance, when a pattern is synthesized to imitate a new signal, the most likely values might be chosen by some summation of activation proportional to $\log[\text{pr}(x, y)]$ (see Lee 1992). In terms of MDL, we can say that the higher level cortical area somehow seeks the most economical way, the simplest pattern of firing, that will generate a top-down synthesized signal close to the true sensory signal.

I would like to give a more elaborate example to show how MDL can lead you to the correct variables with which to describe the world using an old and familiar vision problem: the stereo correspondence problem. The usual approach to stereo vision is apply our knowledge of the three-dimensional structure of the world to show how matching the images I_L and I_R from the left and right eyes leads us to a reconstruction of depth through the "disparity function" $d(x, y)$ such that $I_L(x + d(x, y), y)$ is approximately equal to $I_R(x, y)$. In doing so, most algorithms take into account the "constraint" that most surfaces in the world are smooth, so that depth and disparity vary slowly as we scan across an image. The MDL approach is quite different. First, the raw perceptual signal comes as two sets of N pixel values $I_L(x, y)$ and $I_R(x, y)$ each encoded up to some fixed accuracy by d bits, totaling $2 \cdot d \cdot N$ bits. But the attentive encoder notices how often pieces of the left image code nearly duplicate pieces of the right code: this is a common pattern that cries out for use in shrinking the code length. So we are led to code the signal in three pieces: first the raw left image $I_L(x, y)$, then the disparity $d(x, y)$, and finally the residual $I_R(x, y) - I_L(x + d(x, y), y)$. The disparity and the residual are both quite small, so instead of d bits, these may need only a small number e and f bits, respectively. Provided $d > e + f$, we have saved bits. In fact, if we use the constraint that surfaces are mostly smooth, so that $d(x, y)$ varies slowly, we can further encode $d(x, y)$ by its average value $d_0(y)$ on each horizontal line and its x-derivative $d_x(x, y)$, which is mostly much smaller. The important point is that MDL coding leads you introduce the third coordinate of space, that is, to discover three-dimensional space! A further study of the discontinuities in *d*, and the "nonmatching" pixels visible to one eye only goes further and leads you to *invent a description* of the image containing labels for distinct objects, that is, to *discover that the world is usually made up of discrete objects.* For a more complete discussion, see Mumford (1993, §5d).

Can the learning phase of MDL be implemented in a natural way in cortex? We think this is one of the most interesting challenges to Pattern Theory. We have no proposal except to say that recent work (Intrator 1992; Jordan and Jacobs 1993; Hinton, unpublished observations) shows that many learning rules, more complex than simple Hebbian learning, are possible and suggestive. Hinton's, especially, looks like it might solve the stereo problem along the lines proposed above.

SUMMARY

Starting from the theoretical perspective of Pattern Theory, this chapter has made some specific proposals for the data structures and computational mechanisms to be expected in the cortex. These include (1) the need for feedback loops activating template-like patterns in lower corical areas, (2) a mechanism for shifting or warping patterns of cortical activity, (3) marking both boundaries between unrelated features and the complexes of related activity with a common source, (4) the need for temporal buffering, (5) multiscale population coded representations, and (6) the possibility that the Minimum Description Length Principle can be used as a basis of learning world structures.

A common thread in all the specific proposals above is the need for more sophisticated experimental stimuli, motivated by computational or psychological theory. A well-known experimenter laughed at me 10 years ago when we suggested that one should look for cell responses in higher visual areas correlated to global features of the image outside its "classical" receptive field. Shortly thereafter, von der Heydt's experiments provided the first dramatic proof that this occurs (von der Heydt et al. 1984). Real world stimuli have a huge number of complexities and subtleties not even remotely present in typical laboratory stimuli and these should be studied, one at a time, to see how the cortex handles them. For example, one can present edges that are blurred or noisy, curved or interrupted, embedded in textures or with contrast reversals. One can use complex temporal organization, comparing an extended continuous movement with many small movements that flicker off. Two general paradigms suggest themselves: one is to use pairs of stimuli that are locally identical, but globally quite different. In this case, the higher cortical area can respond to the larger features and so modulate the responses of a cell in the lower area to two stimuli identical within its receptive field. The second is really a special case of this: to present stimuli that are neutral locally, not stimulating a particular cell, but that have major global organization that may imply local structure, and see if it affects the original cell.

A second thread is the need to consider feedback effects when modeling cortical responses. Our observation is that there is a strong bias toward seeking simple feedforward explanations of what the cortex is doing. For instance, Marr's book (Marr 1982) is essentially a purely feedforward theory of vision. If any of the above theorizing is half right, feedback plays a major role in both low- and high-level processing and cannot be ignored, even in primary sensory and motor areas.

8 Observations on Cortical Mechanisms for Object Recognition and Learning

Tomaso A. Poggio and Anya Hurlbert

INTRODUCTION

One of the main goals of vision is object recognition. But there may be many distinct routes to this goal and the goal itself may come in several forms. Anyone who has struggled to identify a particular amoeba swimming on a microscope slide or to distinguish between novel visual stimuli in a psychophysics laboratory might admit that recognizing a familiar face seems an altogether different and simpler task. Recent evidence from several lines of research strongly suggests that not all recognition tasks are the same. Psychophysical results and computational analyses suggest that recognition strategies may depend on the type of both object and visual task. Symmetric objects are better recognized from novel viewpoints than asymmetric objects (Poggio and Vetter 1992); when moved to novel locations in the visual field, objects with translation-invariant features are better recognized than those without (Nazir and O'Regan 1990; Bricolo and Bülthoff 1992). A typical agnosic patient can distinguish between a face and a car, a classification task at the *basic* level of recognition, but cannot recognize the face of Marilyn Monroe, an identification task at the *subordinate* level (Damasio et al. 1990b). A recently reported stroke patient cannot identify the orientation of a line but can align her hand with it if she imagines posting a letter through it, suggesting strongly that there are also multiple outputs from visual recognition (Goodale et al. 1991).

Yet although recognition strategies diverge, recent theories of object recognition converge on one mechanism that might underlie several of the distinct stages, as we will argue in this chapter. This mechanism is a simple one, closely related to template matching and nearest neighbor techniques. It belongs to a class of explanations that we call memory-based models (MBMs), which includes memory-based recognition, sparse population coding, generalized radial basis functions networks, and their extension, hyper basis functions (HBF) networks (Poggio and Girosi 1990b) (figure 8.1). In MBMs, classification or identification of a visual stimulus is accomplished by a network of units. Each unit is broadly tuned to a particular template, so that it is maximally excited when the stimulus exactly matches its template but also responds proportionately less to similar

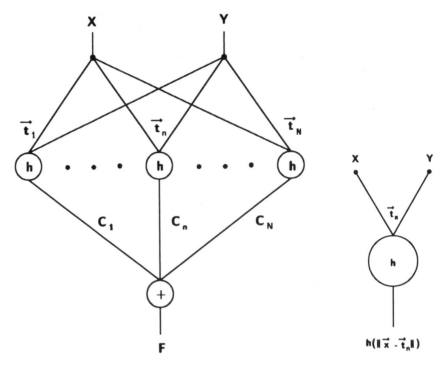

Figure 8.1 An RBF network for the approximation of two-dimensional functions (left) and its basic "hidden" unit (right). x and y are components of the input vector which is compared via the RBF h at each center t. Outputs of the RBFs are weighted by the c_i and summed to yield the function **F** evaluated at the input vector. N is the total number of centers.

stimuli. The weighted sum of activities of all the units uniquely labels a novel stimulus. Several recent and successful face recognition schemes for machine vision share aspects of this framework (Baron 1981; Bichsel 1991; Brunelli and Poggio 1992; Turk and Pentland 1991; Stringa 1992a,b).

We will consider how the basic features of this class of models might be implemented by the human visual system. Our aim is to demonstrate that such models conform to existing physiological data and to make further physiological predictions. We will use as a specific example of the class the HBF network. HBF networks have been used successfully to solve isolated visual tasks, such as learning to detect displacements at hyper-acuity resolution (Poggio et al. 1992) or learning to identify the gender of a face (Brunelli and Poggio 1992). We will discuss how the units of a HBF network might be realized as neurons and how a similar network might be implemented by cortical circuitry and replicated at many levels to perform the multicomponent task of visual recognition. We hope to show that MBMs are not merely toy replicas of neural systems, but viable models that make testable biological predictions.

The main predictions of memory-based models are as follows:

- The existence of broadly tuned neurons at all levels of the visual pathway,

Poggio and Hurlbert

tuned to single features or to configurations in a multidimensional feature space.

- At least two types of plasticity in the adult brain, corresponding to two stages of learning in perceptual skills and tasks. One stage probably involves changes in the tuning of individual neuron responses; this resembles adaptation. The other probably requires changes in cortical circuitry specific to the task being learned, connecting many neurons across possibly many areas.

OBJECT RECOGNITION: MULTIPLE TASKS, MULTIPLE PATHWAYS

Recognizing an object should be difficult because it rarely looks the same on each sighting. Consider the prototypical problem of recognizing a specific face. (We believe that processing of faces is not qualitatively different from processing of other 3D objects, although the former might be streamlined by practice, and biological evidence supports this view [Gross, 1992].) The 2D retinal image formed by the face changes with the observer's viewpoint, and with the many transformations that the face can undergo: changes in its location, pose, and illumination, as well as nonrigid deformations such as the transition from a smile to a frown. A successful recognition system must be robust under all such transformations.

Here we outline an architecture for a recognition system that contains what we believe are the rudimentary elements of a robust system. It is best considered as a protocol for and summary of existing programs in machine vision, but it also represents an attempt to delineate the stages probably involved in visual recognition by humans. The scheme (diagrammed in figure 8.2) has dual routes to recognition. The first is a streamlined route to recognition in which the features extracted in the early stages of image analysis are matched directly to samples in the database. The second potential route to recognition diverges from the first to allow for the possibility that both the database models and the extracted image features might need further processing before a match can be found.

Our task in recognizing a face—or any other 3D object—consists of multiple tasks, which fall into three broad categories that characterize both routes:

- *Segmentation*: Marking the boundaries of the face in the image. This stage typically involves segmenting the entire image into regions *likely* to correspond to different materials or surfaces (and thereby subsumes figure-ground segregation) and is a prerequisite for further analysis of a marked region. *Image measurements* are used to convert the retinal array of light intensities into a *primal image representation*, by computing sparse measurements on the array, such as intensity gradients, or center-surround outputs. The result is a set of vector measurements at each of a sparse or dense set of locations in the image. *Features* may then be found, and used to partition the image.

Cortical Mechanisms for Object Recognition and Learning

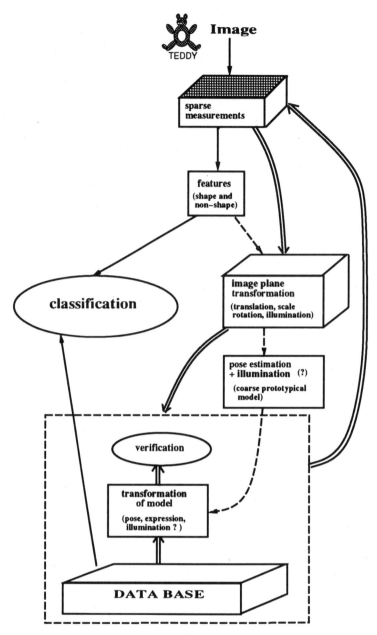

Figure 8.2 A sketch of an architecture for recognition with two hypothetical routes to recognition. Single arrows represent the classification and indexing route described in appendix A. Double arrows represent the main visualization route, and dashed arrows represent alternative pathways within it.

- *Classification*, or basic-level recognition: Distinguishing objects that are faces from those that are not. Parameter values estimated in the preceding stage—(e.g., the distance between eyes and mouth)—are used in this stage for classification of a set of features—(e.g., as a potential face, animal, or man-made tool). This stage requires that the boundaries or the location of at least potential faces be demarcated, and hence generally depends on the preceding step of image segmentation, although it may work without it at an added computational cost.

- *Identification*, or subordinate-level recognition: Matching the face to a stored memory, and thereby labeling it. This stage requires some form of indexing of the database samples. Because it is computationally implausible that the recognition system contains a stored sample of the face in each of its possible views or expressions, or under all possible illumination conditions at all possible viewing distances, this step in general also requires that the face be transformed into a standard form for matching against its stored template. Thus in parallel with the direct route from classification to identification there exists a second route that we call the *visualization* route, which may include an iterative sequence of transformations of both the image-plane and the database models until it converges on a match.

These stages, and some open questions on the overall architecture, are discussed further in appendix A.

As outlined here, the stages are distinct and could be implemented in series within each route to recognition. Most artificial face recognition systems tackle the stages separately, being designed either to detect and localize a face in an image cluttered with other objects (segmentation and classification), or to identify individual faces presented in an expected format (database indexing and identification). Some artificial recognition systems have been constructed to achieve invariant recognition under isolated transformations (visualization). Examples are systems that recognize frontal views of faces under varying illuminations (Brunelli and Poggio 1992); recognize simple paper-clip-like objects independently of viewpoint (Poggio and Edelman 1990); or identify simple objects solely by color under spatially varying illumination (Swain and Ballard 1990).

Yet in biological systems, and in some artificial systems, the stages may act in parallel or even merge. For example, there may be many short-cuts to recognizing a frequently encountered object such as a face.

Finding the face might be streamlined by a quick search at low resolution over the whole image for face-like patterns. The search might employ simplified templates of a face containing anthropometric information (for example, a two-eyes-and-mouth mask). Once located, salient features such as eyes can be used to demarcate the entire object to which they belong, eliminating the need to segment other parts of the image. These detectors would scan the image for the presence of these face-specific features, and using them, locate the face for further processing (translation, scaling, etc.). (Some machine vision systems already implement this idea, using translation-invariant face-feature-detectors such as eye detectors [Bichsel 1991] or symmetry detectors.) Thus segmentation may occur simultaneously with classification. The existence of these face detectors in the human visual system might explain why we so readily perceive faces in the simplest drawings of dots and lines, or in symmetric patterns formed in nature (Hurlbert and Poggio 1986), and why we detect properly configured faces more readily than arbitrary or inverted arrangements of facial features (Purcell and Stewart 1988). Indeed, we wonder whether face recognition may have become so inveterate that the human brain might first classify image regions into face or nonface.

Recognizing an expected object might also be more speedy and efficient than identifying an unexpected one. In the classification stage, only those features specific for the expected object class need be measured, and correct classification would not require that all features be simultaneously available. This step might therefore be itself a form of template matching, where part-templates may serve as well as whole-templates to locate and classify the object. In many cases the classification stage may lead by itself to unique recognition, especially when situational information, such as the expectedness of the object, restricts the relevant database.

Yet many questions are left hanging by this sketch of a recognition system. In biological systems, is matching done between primal image representations, like center-surround outputs at sparse locations, or between sets of higher level features? Computational experiments on face recognition suggest that the former strategy performs much better. What exactly are the key features used for identifying, localizing and normalizing an object of a specific class? Is there an automatic way to learn them (Huber 1985)? Do biological visual systems acquire recognition features through experience (Edelman 1991)? Do humans use expectation to restrict the database for categorization? Some psychophysical experiments suggest that we do not need higher-level expectations to recognize objects quickly in a random series of images, but these experiments have used familiar objects such as the Eiffel Tower (M. Potter, personal communication).

A Sketch of a Memory-Based Cortical Architecture for Recognition

We suggest that most stages in face recognition, and more generally, in object recognition, may be implemented by modules with the same intrinsic structure—a memory based module (MBM). At the heart of this structure is a set of neurons each tuned to a particular value or configuration along one or many feature dimensions. Let us take as an example of such a structure the hyper basis functions (HBF) network. This is a convenient choice because HBFs have been successfully applied already to several problems in object recognition as well as an unrestrictive, easily modifiable choice because HBFs are closely related to other approximation and learning techniques such as multilayer perceptrons.

RBF Networks HBF networks are approximation schemes based on, but more flexible than, radial basis functions (RBF) networks (see figure 8.1; Poggio and Girosi 1990b; Poggio 1990). The fundamental equation underlying RBF networks states that any function $f(x)$ may be approximated by a weighted sum of RBFs:

$$f(x) = \sum_{i=1}^{N} c_i h(\|x - t_i\|)^2 + p(x). \tag{8.1}$$

The functions h may be any of the class of RBFs, for example, Gaussians. $p(x)$ is a polynomial that is required by certain RBFs for the validity of the equation. (For some RBFs, e.g., Gaussians, the addition of $p(x)$ is not necessary, but improves performance of the network.) In an RBF network, each "unit" computes the distance $\|x - t\|$ of the input vector x from its center t and applies the function h to the distance value, that is, it computes the function $h(\|x-t\|)^2$. The N centers t, corresponding to the N data points, thus behave like *templates*, to which the inputs are compared for similarity.

A typical and illustrative choice of RBF is the Gaussian $[h(\|x - t\|) = \exp(-(\|x - t\|)^2/2\sigma^2)]$. In the limiting case where h is a very narrow Gaussian, the network effectively becomes a *look-up* table, in which a unit gives a nonzero signal only if the input exactly matches its center t.

The simplest recognition scheme based on RBF networks that we consider is that suggested by Poggio and Edelman (1990) (see figure 8.3) to solve the specific problem of recognizing a particular 3D object from novel views, a subordinate-level task. In the RBF version of the network, each center stores a sample view of the object, and acts as a unit with a Gaussian-like recognition field around that view. The unit performs an operation that could be described as "blurred" template matching. At the output of the network the activities of the various units are combined with appropriate weights, found during the learning stage. An example of a recognition field measured psychophysically for an asymmetric object after training with a single view is shown in figure 8.4. As predicted from the model (see Poggio and Edelman 1990), the shape of the surface of the recognition errors is roughly Gaussian and centered on the training view.

In this particular model, the inputs to the network are spatial coordinates or measurements of features (e.g., angles or lengths of segments) computed from the image. In general, though, the inputs to an RBF network are not restricted to spatial coordinates but could include, for example, colors or configurations of segments, binocular disparities of features, or texture descriptions. Certainly in any biological implementation of such a network the inputs may include measurements or descriptions of any attribute that the visual system may represent. We assume that in the primate visual system such a recognition module may use a large number of primitive measurements as inputs, taken by different "filters" that can be regarded as many different "templates" for shape, texture, color, and so forth. The only restriction is that the features must be directly computed from the image. Hence the inputs are viewer-centered, not object-centered, although some,

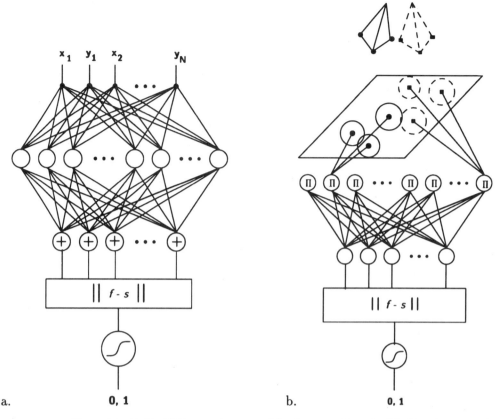

Figure 8.3 (*a*) The HBF network proposed for the recognition of a 3D object from any of its perspective views (Poggio and Edelman 1990). The network attempts to map any view (as defined in the text) into a standard view, arbitrarily chosen. The norm of the difference between the output vector **f** and the standard view **s** is thresholded to yield a 0, 1 answer (instead of the standard view the output of the netwok can be directly a binary classification label). The 2*N* inputs accommodate the input vector **v** representing an arbitrary view. Each of the *n* radial basis functions is initially centered on one of a subset of the *M* views used to synthesize the system ($n \leq M$). During training each of the *M* inputs in the training set is associated with the desired output (i.e., the standard view **s**). (b) A completely equivalent interpretation of (*a*) for the special case of Gaussian radial basis functions. Gaussian functions can be synthesized by multiplying the outputs of two-dimensional Gaussian receptive fields, that "look" at the retinotopic map of the object point features. The solid circles in the image plane represent the 2D Gaussians associated with the first radial basis function, which represents the first view of the object. The dotted circles represent the 2D receptive fields that synthesize the Gaussian radial function associated with another view. The 2D Gaussian receptive fields transduce values of features, represented implicitly as activity in a retinotopic array, and their product "computes" the radial function without the need of calculating norms and exponentials explicitly. (From Poggio and Girosi 1990c)

Poggio and Hurlbert

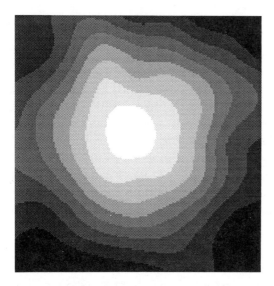

Figure 8.4 The generalization field associated with a single training view. Whereas it is easy to distinguish between, say, tubular and amoeba-like 3D objects, irrespective of their orientation, the recognition error rate for specific objects within each of those two categories increases sharply with misorientation relative to the familiar view. This figure shows that the error rate for amoeba-like objects, previously seen from a single attitude, is viewpoint-dependent. Means of error rates of six subjects and six different objects are plotted versus rotation in depth around two orthogonal axes (Bülthoff et al. 1991; Edelman and Bülthoff 1992). The extent of rotation was ±60° in each direction; the center of the plot corresponds to the training attitude. Shades of gray encode recognition rates, at increments of 5% (white is better than 90%; black is 50%). (From Bülthoff and Edelman 1992). Note that viewpoint independence can be achieved by familiarizing the subject with a sufficient number of training views of the 3D object.

like color, will be viewpoint-independent. The output of the network is, though, object-centered, provided there is a sufficient number of centers. This generality of the network permits a mix of 2D and 3D information in the inputs, and relieves the model from the constraints of either.

This feature of the model also renders irrelevant the question on whether object representations are 2D or 3D. The Poggio–Edelman model makes it clear that 2D-based schemes can provide view invariance as readily as a 3D model can, and compute 3D pose as well (see Poggio and Edelman 1990). So the relevant questions are what is *explicit* in neurons? and what does it mean for information about shape to be explicit in neurons? In a sense, some 2D-based schemes such as the Poggio–Edelman model may be considered as plausible neurophysiological implementations of 3D models.

We do not suggest that the cortical architecture for recognition consists of a collection of such modules, one for each recognizable object. Certainly it is more complex than that cartoon, and not only because viewpoint invariance is not the only problem that the recognition system must solve. For example, the cortex must also learn to recognize objects under varying illumination (photometric invariance) and to recognize objects at the basic as well as subordinate level. (Preliminary results on real objects [faces] suggest that HBF modules can estimate expression and direction of illumination equally as well as pose [Brunelli, personal communication; Beymer, personal communication].) Yet each of these and other distinct tasks in recognition may be implemented by a module broadly similar to the Poggio-Edelman viewpoint-invariance network. We might expect that the system could be decomposed into elementary modules similar in design but different in purpose, some specific for individual objects (and therefore solving a subordinate-level task), some specific to an object class (solving a basic-level task), and others designed to perform transformations or feature extractions, for example, common to several classes. The modules may broadly be categorized as follows:

- *Object-specific.* A module designed to compensate for specific transformations that a specific object might undergo. As in the Poggio-Edelman network, the module would consist of a few units, each maximally tuned to a particular configuration of the object—or the face, say, a particular combination of pose and expression. A more general form of the network may be able to recognize a few different faces: its hidden units would be tuned to different views but of not just one face, and therefore behave more like eigenfaces.

- *Class-specific.* A module that generalizes across objects of a given class. For example, the network may be designed to extract a feature or aspect of *any* of a class of objects, such as pose, color, or distance. For example, there might be a network designed to extract the pose of a face, and a separate network designed to extract the direction of illumination on it. Any face fed as input to network would elicit an estimate of its pose or illumination.

- *Task-specific.* Networks that solve tasks, such as shape-from-shading, *across* classes of objects. An example would be a generic shape-from-shading network that takes as input brightness gradients of image regions. It might act in the early stages of recognition, helping to segment and classify 3D shapes even before they are grouped and classified as objects.

The distinctions between these types of recognition module might be blurred if, for example, the visual system overlearns certain objects or transformations. For example, a shape-from-shading network might develop for a frequently encountered type of material, or for a specific class of object. Indeed, our working assumption is that any apparent differences between recognition strategies for different types of objects arise not from fundamental differences in cortical mechanisms but from imbalances in the distribution of the same basic modules across different objects and different environments. Savanna Man, like us, probably had task-specific modules dedicated to faces, but although we might have shape-from-shading modules specific to familiar pieces of office furniture, he might not be able to recognize a filing cabinet at all, much less under varying illumination. This suggests a decomposition into modules that are both task- and object-specific, which is a rather unconventional but plausible idea.

Transformations specific to a particular object may also be generalized from transformations learned on prototypes of the same class. For example, the deformation caused by a change in pose or, for a face, a change in expression or age, may be learned from a set of examples of the same transformation acting on prototypes of the class. Some transformations may be generalized across all objects sharing the same symmetries (Poggio and Vetter 1992).

The big question is, if the recognition system does consist of similar modules performing interlocking tasks, how are the modules linked, and in what hierarchy (if it makes sense at all to talk of ordered stages)? In constructing a practical system for face recognition, it would make sense first to estimate the pose, expression, and illumination for a generic face and then to use this estimate to "normalize" the face and compare it to single views in the database (additional search to fine tune the match may be necessary). Thus the system would first employ a class-specific module based on invariant properties of faces to recover, say, a generic view—analogous to an object-centered representation—that could feed into face-specific networks for identification. The information that the system extracts in the early stages concerning illumination, expression, context, etc. would not be discarded. Within each stage, modules may be further decomposed and arranged in hierarchies: one may be specific for eyes, and may extract gaze angle, a parameter that may then feed into a module concerned with the pose of the entire face.

For face recognition, the generic view may be recovered by exploiting prior information such as the approximate bilateral symmetry of faces. In general a single monocular view of a 3D object (if shading is neglected) does not contain sufficient 3D information for recognition of novel views. Yet humans are certainly able to recognize faces rotated 20–30° away from frontal after training on just one frontal view. One of us has recently discussed (Poggio 1991) different ways for solving the following problem: *from one 2D view of a 3D object generate other views, exploiting knowledge of views of other, "prototypical" objects of the same class.* It can be shown theoretically (Poggio and Vetter 1992) that prior information on generic shape constraints does reduce the amount of information needed to recognize a 3D object, since additional virtual views can be generated from given model views by the appropriate symmetry transformations. In particular, for bilaterally symmetric objects, a single nonaccidental "model" view is theoretically sufficient for recognition of novel views. Psychophysical experiments (Vetter et al. 1992) confirm that humans are better in recognizing symmetric than nonsymmetric objects.

An interesting question is whether there are indeed multiple routes to recognition. It is obvious that some of the logically distinct steps in recognition of figure 8.2 may be integrated in fewer modules, depending on the specific implementation. Figure 8.5 shows how the same architecture may appear if the classification and the visualization routes are implemented with HBF networks. In this case, the database of face models would es-

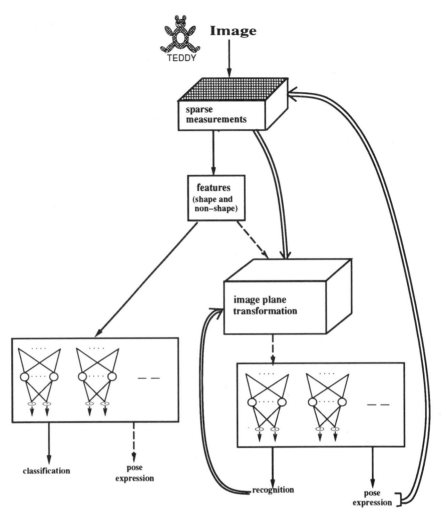

Figure 8.5 A sketch of possibly the most compact (but not the only!) implementation of the proposed recognition architecture in terms of modules of the HBF type.

sentially be embedded in the networks (see Poggio and Edelman 1990).

There are of course several obvious alternatives to this architecture and many possible refinements and extensions. Even if oversimplified, this token architecture is useful to generate meaningful questions. The preceding discussion may in fact be sufficient for performing computational experiments and for developing practical systems. It is also sufficient for suggesting psychophysical experiments. It is of course not enough from the point of view of a physiologist, yet the physiological data in the next section provides broad support for its ingredients.

Physiological Support for a Memory-Based Recognition Architecture
At least superficially, physiological data seems to support the existence of elements of each these modules. Perrett et al. (1985, 1989) report evidence from inferotemporal cortex (IT) not only for cells tuned to individual

faces but also for face cells tuned to intermediate views between frontal and profile, units that one would expect in a class-specific network designed to extract pose of faces. Such cells also support the existence of the view-centered units predicted by the basic Poggio-Edelman recognition module. Young and Yamane (1992) describe cells in anterior IT that respond optimally to particular configurations of facial features, or "physical prototypes." These may conceivably provide input to the cells described by Perrett et al. as "person recognition units," or to the approximately view-independent cells described by Hasselmo et al. (1989) that would in turn correspond almost exactly to the object-centered output of the Poggio-Edelman model. Perrett et al. (1985, 1989) also report cells that respond to a given *pose* of the face regardless of illumination—even when the face is under heavy shadow. Such cells may resemble units in a *task-specific* network. In the superior temporal sulcus, Hasselmo et al. (1989) also find cells sensitive to head movement or facial gesture, independent of the view or identity of the face. Such cells would also appear to be both *class-* and *task-specific*. (See Perrett and Oram [1992] for a more detailed review of relevant physiological data.)

Fujita and Tanaka (1992) have also reported cells in IT that respond optimally to certain configurations of color and shape. These may well represent elements of networks that generalize across objects, classifying them according to their geometric and material constitution. More significantly, Fujita and Tanaka (1992) report that cells in the anterior region of IT (cytoarchitectonic area TE) are arranged in columns, within which cells respond to similar configurations of color, shape, and texture. Each configuration may be thought of as a template, which in turn might encode an entire object (e.g., a face) or a part of an object (e.g., the lips). Within one column, cells may respond to slightly different versions of the template, obtained by rotations in the image-plane, for example. Fujita and Tanaka (1992) conclude that each of the 2000 or so columns in TE may represent one phoneme in the language of objects, and that combinations of activity across the columns are sufficient to encode all recognizable objects.

The existence of such columns supports the notion that the visual system may achieve invariance to image-plane transformations of elementary features by replicating the necessary feature measurements at different positions, at different scales and with different rotations.

In the next section we describe how key aspects of the architecture could be implemented in terms of plausible biophysical mechanisms and neurophysiological circuitries.

NEURAL MODELING OF MEMORY-BASED ARCHITECTURES FOR RECOGNITION

In this section we discuss in more detail the possible neural implementations of a recognition system built from MBMs. The main questions we address are: How are MBMs constructed when a new object or class of objects is learned? and How might MBM units be constructed from known biophysical mechanisms? We propose that there are two stages of learning—supervised and unsupervised—and illustrate to which elements of a memory-based network they correspond. Where could they be localized in terms of cortical structures? What mechanisms could be responsible? We discuss the memory-based module itself and the circuitry that might underlie it.

The Learning-from-Examples Module

The simple RBF version of an MBM, discussed earlier in this chapter, learns to recognize an object in a straightforward way. Its centers are fixed, chosen as a subset of the training examples. The only parameters that can be modified as the network learns to associate each view with the correct response ("yes" or "no" to the target object) are the coefficients c_i, the weights on the connections from each center to the output.

The full HBF network permits learning mechanisms that are more biologically plausible by allowing more parameters to be modified. HBF networks are equivalent to the following scheme for approximating a multivariate function:

$$f^*(\mathbf{x}) = \sum_{\alpha=1}^{n} c_\alpha G(\|(\mathbf{x} - \mathbf{t}_\alpha)\|_W^2) + p(\mathbf{x}) \qquad (8.2)$$

where the centers $\mathbf{t}_\alpha$ and coefficients c_α are unknown, and are in general fewer in number than the data points ($n \leq N$). The norm is a *weighted norm*

$$\|(\mathbf{x} - \mathbf{t}_\alpha)\|_W^2 = (\mathbf{x} - \mathbf{t}_\alpha)^T W^T W (\mathbf{x} - \mathbf{t}_\alpha) \qquad (8.3)$$

where W is an unknown square matrix and the superscript T indicates the transpose. In the simple case of diagonal W the diagonal elements w_i assign a specific weight to each input coordinate, determining in fact the units of measure and the importance of each feature (the matrix W is especially important in cases in which the input features are of a different type and their relative importance is unknown) (Poggio and Girosi 1990a). During learning, not only the coefficients c but also the centers $\mathbf{t}_\alpha$, and the elements of W are updated by instruction on the input–output examples (figure 8.6).

Whereas the RBF technique is similar to and similarly limited as template matching, HBF networks perform a generalization of template matching in an appropriately linearly transformed space, with the appropriate metric. HBF networks are therefore different in both interpretation and capabilities from "vanilla" RBF. An RBF network can recognize an object rotated to novel orientations only if it has centers corresponding to sample rotations of the object. HBFs, though, can perform a variety of more sophisticated recognition tasks. For example, HBFs can

1. Discover the Basri–Ullman result (Basri and Ullman 1989; Brunelli and Poggio, unpublished). (In its strong form, this result states that under

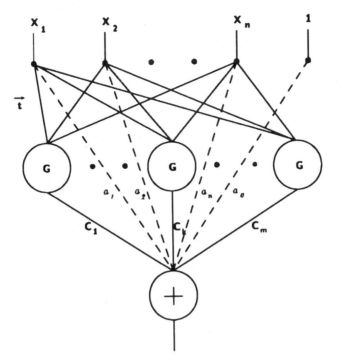

Figure 8.6 A network of the hyper basis functions type. For object recognition the inputs could be image measurements such as values of different filters at each of a number of locations in the image. The network is a natural extension of the template matching scheme and contains it as a special case. The dotted lines correspond to linear and constant terms in the expansion. The output unit may contain a sigmoidal transformation of the sum of its inputs (see Poggio and Girosi 1990b)

orthographic projection any view of the visible features of the 3D object may be generated by a linear combination of two other views.)

2. With a nondiagonal **W**, recognize an object under orthographic projection with only one center.

3. Provide invariance (or near invariance under perspective projection) for scale, rotation, and other uniform deformations in the image plane, without requiring that the features be invariant.

4. Discover symmetry, collinearity, and other "linear-class" properties (see Poggio and Vetter 1992).

Gaussian Radial Basis Functions In the special case where the network basis functions are Gaussian and the matrix **W** diagonal, its elements w_i have an appealingly obvious interpretation. A multidimensional Gaussian basis function is the product of one-dimensional Gaussians and the scale of each is given by the inverse of w_i. For example, a 2D Gaussian radial function centered on **t** can be written as

$$G(\|\mathbf{x} - \mathbf{t}\|_W^2) \equiv e^{-\|\mathbf{x}-\mathbf{t}\|_W^2} = e^{-\frac{(x-t_x)^2}{2\sigma_x^2}} e^{-\frac{(y-t_y)^2}{2\sigma_y^2}}, \tag{8.4}$$

where $\sigma_x = 1/w_1$ and $\sigma_y = 1/w_2$, and w_1 and w_2 are the elements of the diagonal matrix $\mathbf{W}$.

Thus a multidimensional center can be factored in terms of one-dimensional centers. Each one-dimensional center is individually tuned to its input: centers with small w_i, or large σ_i, are less selective and will give appreciable responses to a range of values of the input feature; centers with large w_i, or small σ_i, are more selective for their input and, accordingly, have greater influence on the response of the multidimensional center. The template represented by the multidimensional center can be considered as a conjunction of one-dimensional templates. In this sense, a Gaussian HBF network performs the disjunction of conjunctions: the conjunctions represented by the multidimensional centers are "or"ed in the weighted sum of center activities that forms the output of the network.

Expected Physiological Properties of MBM Units

The Neurophysiological Interpretation of HBF Centers Our key claim is that HBF centers and tuned cortical neurons behave alike.

A Gaussian HBF unit is maximally excited when each component of the input exactly matches each component of the center. Thus the unit is optimally tuned to the stimulus value specified by its center. Units with multidimensional centers are tuned to complex features, formed by the conjunction of simpler features, as described in the previous section.

This description is very like the customary description of cortical cells optimally tuned to a more or less complex stimulus. So-called place coding is the simplest and most universal example of tuning: cells with roughly Gaussian receptive fields have peak sensitivities to given locations in the input space; by overlapping, the cell sensitivities cover all of that space. In V1 the input space may be up to five dimensional, depending on whether the cell is tuned not only to the retinal coordinates x, y but also to stimulus orientation, motion direction, and binocular disparity. In V4 some cells respond optimally to a stimulus combining the appropriate values of speed and color (N. K. Logothetis, personal communication; Logothetis and Charles 1990). Other V4 cells respond optimally to a combination of colour and shape (D. Van Essen, personal communication). In MST cells exist optimally tuned to specific motions in different parts of the receptive field and therefore to different motion "dimensions." Most of these cells are also selective for stimulus contrast. In "later" areas such as IT cells may be tuned to more complex stimuli which can be changed in a number of "dimensions" (Desimone et al. 1984). Gross (1992) concludes that "IT cells tend to respond at different rates to a variety of different stimuli." Thus it seems that multidimensional units with Gaussian-like tuning are not only biologically plausible, but ubiquitous in cortical physiology. This claim is not meant to imply that for every feature dimension of a multidimensionally tuned neuron, neurons feeding into it can be found individually tuned to that dimension. For example, for some motion-selective cells in

MT the selectivities to spatial frequency and temporal frequency cannot be separated. Yet for these, it may be inappropriate to consider time and space as two independent dimensions and more appropriate to consider velocity as the single dimension in which the neuron is tuned. On the other hand, it is well known that at lower levels in the visual system there do exist cells broadly tuned individually to spatial frequency, orientation, and wavelength, for example, and from these dimensions many complex features can be constructed.

We also observe that not all MBMs have the same applicability in describing properties of cortical neurons. In particular, tuned neurons seem to behave more like Gaussian HBF units than like the sigmoidal units typically found in multilayer perceptrons (MLPs): the tuned response function of cortical neurons resembles $exp(-(\|x-t\|)^2/2\sigma^2$, more than it does $\Sigma(x \cdot w)$, where Σ is a sigmoidal "squashing" function and we define w as the vector of connection weights including the bias parameter θ. (The typical sigmoidal response to contrast that most neurons display may be treated as a Gaussian of large variance.) For example, when the stimulus to an orientation-selective cortical neuron is changed from its optimal value in any direction, the neuron's response typically decreases. The activity of a Gaussian HBF unit would also decline with any change in the stimulus away from its optimal value t. But for the sigmoid unit certain changes away from the optimal stimulus will not decrease its activity, for example when the input x is multiplied by a constant $\alpha > 1$.

Lastly, we observe that although the Gaussian is the simplest and most readily interpretable RBF in physiological terms, it might not ultimately provide the best fit to all the physiological data once in. In espousing the general theory of MBMs for cortical mechanisms of object recognition, we do not confine ourselves to Gaussian RBFs as the only model of cortical neurons, but only at present the most plausible.

Centers and a Fundamental Property of Our Sensory World   We can recognize almost any object from any of many small subsets of its features, visual and nonvisual. We can perform many motor actions in several different ways. In most situations, our sensory and motor worlds are *redundant*. In the language of the previous section this means that instead of high-dimensional centers any of *several lower dimensional centers are often sufficient* to perform a given task. This means that the "and" of a high-dimensional conjunction can be replaced by the "or" of its components—a face may be recognized by its eyebrows alone, or a mug by its color. To recognize an object, we may use not only templates comprising all its features, but also subtemplates, comprising subsets of features. This is similar in spirit to the use of several small templates as well as a whole-face template in the Brunelli-Poggio work on frontal face recognition (Brunelli and Poggio 1992).

Splitting the recognizable world into its additive parts may well be preferable to reconstructing it in its full multidimensionality, because a system composed of several independently accessible parts is inherently more robust than a whole, simultaneously dependent on each of its parts. The small loss in uniqueness of recognition is easily offset by the gain against noise and occlusion. This reduction of the recognizable world into its parts may well be what allows us to "understand" the things that we see (see appendix B).

How Many Cells?   The idea of sparse population coding is consistent with much physiological evidence, beginning even at the retinal level where colors are coded by three types of photoreceptors. Young and Yamane (1992) conclude from neurophysiological recordings of IT cells broadly tuned to physical prototypes of faces: "Rather than representing each cell as a vector in the space, the cell could be represented as a surface raised above the feature space. The height of the surface above each point in the feature space would be given by the response magnitude of the cell to the corresponding stimuli and population vectors would be derived by summing the response weighted surfaces for each cell for each stimulus." MBMs also suggest that the importance of the object and the exposure to it may determine how many centers are devoted to its recognition. Thus faces may have a more "punctate" representation than other objects simply because more centers are used. Psychophysical experiments do suggest that an increasing number of centers is created under extended training to recognize a 3D object (Bülthoff and Edelman 1992).

While we would not dare to make a specific prediction on the absolute number of cells used to code for a specific object, computational experiments and our arguments here suggest at least a minimum bound. Simulations by Poggio and Edelman (1990) suggest that in an MBM model a minimum of 10–100 units is needed to represent all possible views of a 3D object. We think that the primate visual system could not achieve the same representation with fewer than on the order of 1000. This number seems physiologically plausible, although we expect that the actual number will depend strongly on the reliability of the neurons, training of the animal, relevance of the represented object and other properties of the implementation. Thus we envisage that training a monkey to one view of a target object may "create" at least on the order of 100 cells tuned to that view in the relevant cortical area, with a generalization field similar to the one shown in figure 8.4. Training to an additional view may *create* or *recruit* cells tuned to that view. Overtraining a monkey on a specific object should result in an overrepresentation in cortex of that object—more cells than normally expected would be tuned to views of the object. Recent results from Kobatake et al. (1993) suggest that up to two orders of magnitude more cells may be "created" in IT (or, rather, the stimulus selectivities of existing cells altered) on overtraining to specific objects.

Note that we do not mean to imply that *only* 10 – 1000 cortical cells would be active on presentation of an object. Many more would be activated than those that are critical for its representation. We suggest only that the activity of approximately 100 cells should be sufficient to discriminate between two distinct objects. This conclusion is broadly supported by the conclusion of Young and Yamane (1992) that the population response of approximately 40 cells in IT is approximately sufficient to encode a particular face, and by the related observation of Britten et al. (1992) that the activity of a small pool of weakly correlated neurons in MT is sufficient to predict a monkey's behavioral response in a motion detection task.

HBF Centers and Biophysical Mechanisms   How might multidimensional Gaussian receptive fields be synthesized from known receptive fields and biophysical mechanisms?

The simplest answer is that cells tuned to complex features are constructed from a hierarchy of simpler cells tuned to incrementally larger conjunctions of elementary features. This idea—a standard explanation— can immediately be formalized in terms of Gaussian radial basis functions, since a multidimensional Gaussian function can be decomposed into the product of lower dimensional Gaussians (Marr and Poggio 1977; Ballard 1986; Mel 1988; Poggio and Girosi 1990b).

The scheme of figure 8.7 is a possible example of an implementation of Gaussian radial basis functions in terms of physiologically plausible mechanisms. The first step applies to situations in which the inputs are place-coded, that is, in which the value of the input is represented by its location in a spatial array of cells—as, for example, the image coordinates x, y are encoded by the spatial pattern of photoreceptor activites. In this case Gaussian radial functions in one, two, and possibly three dimensions can be implemented as *receptive fields* by weighted connections from the sensor arrays (or some retinotopic array of units whose activity encodes the location of features). If the inputs are interval-coded, that is, if the input value is represented by the continuously varying firing rate of a single neuron, then a one-dimensional Gaussian-like tuned cell can be created by passing the input value through multiple sigmoidal functions with different thresholds and taking their difference.

Consider, for example, the problem of encoding color. At the retinal level, color is recorded by the triplet of activities of three types of cell: the cone-opponent red–green (R-G) and blue–yellow (B-Y) cells and the luminance (L) cell. An R-G cell signals increasing amounts of red or decreasing amounts of green by increasing its firing rate. Thus it does not behave like a Gaussian tuned cell. But at higher levels in the visual system, there exist cells that behave very much like units tuned to particular values in 3D color space (Schein and Desimone 1990). How are these multidimensional tuned color cells constructed from one-dimensional rate-coded cells? We suggest that one-dimensional Gaussian tuned cells may be created by the above mechanism, selective to restricted ranges of the three color axes.

Gaussians in higher dimensions can then be synthesized as products of one- and two-dimensional receptive fields. An important feature of this scheme is that the multidimensional radial functions are synthesized

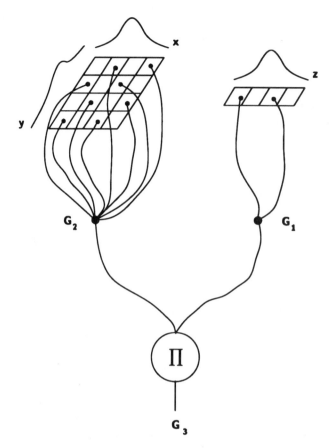

Figure 8.7 A three-dimensional radial Gaussian implemented by multiplying a two-dimensional and a one-dimensional Gaussian receptive field. The latter two functions are synthesized directly by appropriately weighted connections from the sensor arrays, as neural receptive fields are usually thought to arise. Notice that they transduce the implicit position of stimuli in the sensor array into a number (the activity of the unit). They thus serve the dual purpose of providing the required "number" representation from the activity of the sensor array and of computing a Gaussian function. 2D Gaussians acting on a retinotopic map can be regarded as representing 2D "features," while the radial basis function represents the "template" resulting from the conjunction of those lower-dimensional features. (From Poggio and Girosi 1989a)

directly by appropriately weighted connections from the sensor arrays, without any need of an explicit computation of the norm and the exponential. From this perspective the computation is performed by *Gaussian receptive fields* and their combination (through some approximation to multiplication), rather than by threshold functions. The view is in the spirit of the key role that the concept of receptive field has always played in neurophsyiology. It *predicts a sparse population coding* in terms of low-dimensional feature-like cells and multidimensional Gaussian-like receptive fields, somewhat similar to template-like cells, a prediction that could be tested experimentally on cortical cells.

Poggio and Hurlbert

The multiplication operation required by the previous interpretation of Gaussian RBFs to perform the "conjunction" of Gaussian receptive fields is not too implausible from a biophysical point of view. It could be performed by several biophysical mechanisms (see Koch and Poggio 1987; Poggio 1990). Here we mention several possibilities:

1. Inhibition of the silent type and related synaptic and dendritic circuitry (see Poggio and Torre 1978; Torre and Poggio 1978).

2. The AND-like mechanism of NMDA receptors.

3. A logarithmic transformation, followed by summation, followed by exponentiation. The logarithmic and exponential characteristic could be implemented in appropriate ranges by the sigmoid-like pre- to postsynaptic voltage transduction of many synapses.

4. Approximation of the multiplication by summation and thresholding as suggested by Mel (1990).

If the first or second mechanism is used, the product of figure 8.7 can be performed directly on the dendritic tree of the neuron representing the corresponding radial function. In the case of Gaussian receptive fields used to synthesize Gaussian radial basis functions, the center vector is effectively stored in the position of the 2D (or 1D) receptive fields and in their connections to the product unit(s). This is plausible physiologically.

Linear terms (direct connections from the inputs to the output) can be realized directly as inputs to an output neuron that summates linearly its synaptic inputs. An output nonlinearity such as a threshold or a sigmoid or a log transformation may be advantageous for many tasks and will not change the basic form of the model (see Poggio and Girosi 1989).

Circuits There is at least one other way to implement HBFs networks in terms of known properties of neurons. It exploits the equivalence of HBFs with MLP networks for normalized inputs (Maruyama et al. 1991). If the inputs are normalized (as usual for unitary input representations), an HBF network could be implemented as an MLP network by using threshold units. There is the problem, though, in normalizing the inputs in a biologically plausible way. MLP networks have a straightforward implementation in terms of linear excitation and inhibition and of the threshold mechanism of the spike for the sigmoidal nonlinearity. The latter could also be implemented in terms of the pre- to postsynaptic relationship between presynaptic voltage and postsynaptic voltage. In either case this implementation requires one neuron per sigmoidal unit in the network.

Mel (1992) has simulated a specific biophysical hypothesis about the role of cortical pyramidal cells in implementing a learning scheme that is very similar to a HBF network. Marr (1970) had proposed another similar model of how pyramidal cells in neocortex could learn to discriminate different patterns. Marr's model is, in a sense, the look-up table limit of our HBF model.

Mechanisms for Learning

Reasoning from the HBF model, we expect two mechanisms for learning, probably with different localizations, one that could occur unsupervised and thus is similar to adaptation, and one supervised and probably based on Hebb-like mechanisms.

The first stage of learning would occur at the site of the *centers*. Let us remember that a center represents a neuron tuned to a particular visual stimulus, for example, a vertically oriented light bar. The coefficients c_α represent the synaptic weights on the connections that the neuron makes to the output neuron that registers the network's response. In the simple RBF scheme the only parameters updated by learning are these coefficients. But in constructing the network, the centers must be set to values equal to the input examples. Physiologically, then, selecting the centers t_α might correspond to choosing or retuning a subset of neurons selectively responsive to the range of stimulus attributes encountered in the task. This stage would be very much like *adaptation*, an adjustment to the prevailing stimulus conditions. It could occur *unsupervised*, and would strictly depend only on the stimuli, not on the task.

The second stage, updating of the coefficients c_α, could occur only *supervised*, since it depends on the full input and output example pairs, or, in other words, on the task. It could be achieved by a simple Hebb-type rule, since the gradient descent equations for the c are (Poggio and Girosi 1989):

$$\dot{c}_\alpha = \omega \sum_{i=1}^{N} \Delta_i G(\|x_i - t_\alpha\|_W^2) \tag{8.5}$$

with $\alpha = 1, \ldots, n$ and Δ_i the squared error between the correct output for example i and the actual output of the network. Thus equation (8.5) says that the change in the c_α should be proportional to the product of the activity of the unit i and the output error of the network. In other words, the "weights" of the c synapses will change depending on the product of pre- and postsynaptic activity (Poggio and Girosi 1989; Mel 1988, 1990).

In the RBF case, the centers are fixed after they are initially selected to conform to the input examples. In the HBF case, the centers move to optimal locations during learning. This movement may be seen as task-specific or *supervised* fine-tuning of the centers' stimulus selectivities. It is highly unlikely that the biological visual system chooses between distinct RBF-like and HBF-like implementations for given problems. It is possible, though, that tuning of cell selectivities can occur in at least two different ways, corresponding to the *supervised* and *unsupervised* stages outlined here. We might also expect that these two types of learning of "centers" could occur on two different time scales: one fast, corresponding to selecting centers from a preexisting set, and one slow, corresponding to synthesizing new centers or refining their stimulus specificities. The cortical locations of these two mechanisms, one unsupervised, the other supervised, may be different and have interesting implications on how to interpret data on transfer of learning (see Poggio et al. 1992).

For fast, unsupervised learning, there might be a large reservoir of centers already available, most of them with an associated $c = 0$, as suggested by Mel (1990) in a slightly different context. The relevant ones would gain a nonzero weight during the adaptive process. Alternatively, the mechanism could be similar to some of the unsupervised learning models described by Linsker (1990), Intrator and Cooper (1991), Földiak (1991), and others.

Slow, supervised learning may occur by the stabilization of electrically close synapses depending on the degree to which they are co-activated (see, e.g., Mel 1992). In this scheme, the changes will be formation and stabilization of synapses and synapse clusters (each synapse representing a Gaussian field) on a cortical pyramidal cell simply due to correlations of presynaptic activities. We suggest that this synthesis of new centers, as would be needed in learning to recognize unfamiliar objects, should be slower than selecting centers from an existing pool. But some recent data on perceptual learning (e.g., Fiorentini and Berardi 1981; Poggio et al. 1992; Karni and Sagi 1990) indicate otherwise: the fact that human observers rapidly learn entirely novel visual patterns suggests that new centers might be synthesized rapidly.

It seems reasonable to conjecture, though, that updating of the elements of the **W** matrix may take place on a much slower time scale.

Do the update schemes have a physiologically plausible implementation? Methods like the random-step method (Caprile and Girosi 1990), that do not require calculation of derivatives, are biologically the most plausible. (In a typical random-step method, network weight changes are generated randomly under the guidance of simple rules; for example, the rule might be to double the size of the random change if the network performace improves and to halve the size if it does not.) In the Gaussian case, with basis functions synthesized through the product of Gaussian receptive fields, moving the centers means establishing or erasing connections to the product unit. A similar argument can be made also about the learning of the matrix **W**. Notice that in the diagonal Gaussian case the parameters to be changed are exactly the σ of the Gaussians (i.e., the spread of the associated receptive fields). Notice also that the σ for all centers on one particular dimension is the same, suggesting that the learning of w_i may involve the modification of the scale factor in the input arrays rather than a change in the dendritic spread of the postsynaptic neurons. In all these schemes the real problem consists in how to provide the "teacher" input.

PREDICTIONS AND REMARKS

To summarize, we highlight the main predictions made by our interpretation of memory-based models of the brain.

1. *Sparse population coding.* The general issue of how the nervous system represents objects and concepts is of course unresolved. "Sparse" or "punctate" coding theories propose that individual cells are highly specific and encode individual patterns. "Population" theories propose that distributed activity in a large number of cells underlies perception. Models of the HBF type suggest that a small number of cells or groups of cells (the centers), each broadly tuned, may be sufficient to represent a 3D object. Thus our interpretation of MBMs *predicts* a "sparse population coding," partway between fully distributed representations and grandmother neurones. Specifically, we predict that the activity of approximately 100 cells is sufficient to distinguish any particular object, although many more cells may be active at the same time.

2. *Viewer-centered and object-centered cells.* Our model (see the module of figure 8.3) predicts the existence of viewer-centered cells (the centers) and object-centered cells (the output of the network). Evidence pointing in this direction in the case of face cells in IT is already available. We predict a similar situation for other 3D objects. It should be noted that the module of figure 8.3 is only a small part of an overall architecture. We predict the existence of other types of cells, such as pose-tuned, expression-tuned, and illumination-tuned cells. Very recently N. Logothetis (personal communication) has succeeded in training monkeys to recognize the same objects used in human psychophysics, and has reproduced the key results of Bülthoff and Edelman (1992). He also succeeded in measuring generalization fields of the type shown in figure 8.4 after training on a single view. We believe that such a psychophysically measured generalization field corresponds to a group of cells tuned in a Gaussian-like manner to that view. We expect that in trained monkeys, cells exist corresponding to the hidden units of an HBF network, specific for the training view, as well as possibly other cells responding to subparts of the view. We conjecture (although this is not a critical prediction of the theory) that the step of creating the tuned cells (i.e. the centers) is unsupervised: in other words, that to create the centers it would be sufficient to expose monkeys to target views without actually training them to respond in specific ways.

3. *Cells tuned to full views and cells tuned to parts.* Our model implies that both high-dimensional and low-dimensional centers should exist for recognizable objects, corresponding to full templates and template parts. Physiologically this corresponds to cells that require the whole object to respond (say a face) as well as cells that respond also when only a part of the object is present (say, the mouth).

4. *Rapid synaptic plasticity.* We predict that the formation of new centers and the change in synaptic weights may happen over short time scales (possibly minutes) and relatively early in the visual pathway (see Poggio et al. 1992). As we mentioned, it is likely that the formation of new centers is unsupervised while other synaptic changes, corresponding to the c_i coefficients, should be supervised.

HBF-LIKE MODULES AND THEORIES OF THE BRAIN

As theories of the brain (or of parts of it) HBFs networks replace computation with memory. They are equivalent to modules that work as *interpolating look-up tables*. In a previous paper one of us has discussed how theories of this type can be regarded as a modern version of the "grandmother cell" idea (Poggio 1990).

The proposal that much information processing in the brain is performed through modules that are similar to *enhanced look-up tables* is attractive for many reasons. It also promises to bring closer apparently orthogonal views, such as the *immediate perception* of Gibson (1979) and the *representational theory* of Marr (1982), since almost iconic "snapshots" of the world may allow the synthesis of computational mechanisms equivalent to vision algorithms. The idea may change significantly the computational perspective on several vision tasks. As a simple example, consider the different specific tasks of hyperacuity employed by psychophysicists. The proposal would suggest that an appropriate module for the task, somewhat similar to a new "routine," may be synthesized by learning in the brain (see Poggio et al. 1992).

The claim common to several network theories, such as multilayer perceptrons and HBF networks, is that the brain can be explained, at least in part, in terms of approximation modules. In the case of HBF networks these modules can be considered as a powerful extension of look-up tables. MLP networks cannot be interpreted directly as modified look-up tables (they are more similar to an extension of multidimensional Fourier series), but the case of normalized inputs shows that they are similar to using templates.

The HBF theory predicts that population coding (broadly tuned neurons combined linearly) is a consequence of extending a look-up table scheme—corresponding to interval coding—to yield interpolation (or more precisely approximation), that is, generalization. In other words, *sparse population coding* and *neurons tuned to specific optimal stimuli* are direct and strong predictions of HBF schemes. It seems that the hidden units of HBF models bear suggestive similarities with the usual descriptions of cortical neurons as being tuned to optimal multidimensional stimuli. It is of course possible that a hierarchy of different networks—for example, MLP networks—may lead to tuned cells similar to the hidden units of HBF networks.

APPENDIX A: AN ARCHITECTURE FOR RECOGNITION

The Classification and Indexing Route to Recognition

Here we elaborate on the architecture for a recognition system introduced in section 2. Figure 8.2 illustrates the main components of the architecture and its two interlocking routes to recognition. The first route, which we call the classification and indexing route, is essentially equivalent to

Cortical Mechanisms for Object Recognition and Learning

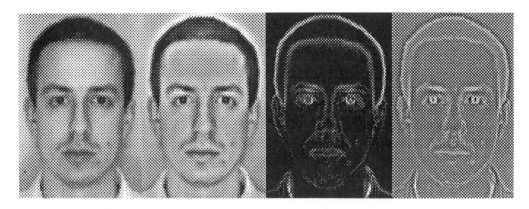

Figure 8.8 Different M-arrays corresponding to different types of measurements (from left to right: I, $I/\langle I \rangle$, $|\Delta I|_{L_1}$ and $\partial_{xx}I + \partial_{yy}I$). The measurements to be used are obtained on a much coarser grid than the original image. (From Brunelli and Poggio 1992)

an earlier proposal Poggio and Edelman 1990) in which an HBF network receives inputs in the form of feature parameters and classifies inputs as *same* or *different* from the target object. This is a streamlined route to recognition that requires that the features extracted in the early stages of image analysis be sufficient to enable matching with samples in the database. Its goal may be primarily basic level recognition, but it is also the route that might suit best the search for and recognition of an expected object. In that case it may be used to identify objects (at the subordinate level) whose class membership is known in advance. It consists of three main stages:

1. *Image measurements* The first step is to compute a *primal image representation*, which is a set of sparse measurements on the image, based on appropriate smoothed derivatives, corresponding to center-surround and directional receptive fields. It can be argued that the (vector) measurements to be considered should be multiple nonlinear functions of differential operators applied to the image at sparse locations (for a discussion of linear and nonlinear measurement or "matching" primitives see Appendix in Nishihara and Poggio 1984). (Similar procedures may involve using Gaussians of different scales and orientations [e.g., Marr and Poggio 1977], Koenderink's "jets" [Koenderink and VanDoorn 1990], Gabor filters, or wavelets. A regularized gradient of the image also works well.) We call this array of measurements an M-array; in general, it is a vector-valued array (figure 8.8). For recognition of frontal images of faces an M-array as small as 30×30 has been found sufficient to encode an image of initial size 512×512 (Brunelli and Poggio 1992).

2. *Feature detection and measurements.* Key features, encoded by the primal measurements, are then found and localized. These features may be specific for a specific object class—for the expected class, if it is known in advance, or for an alternative class considered as a potential match. This

step can be regarded as performing a sort of template matching with several appropriate examples; when a face is the object of the search, templates may include eye pairs of different size, pose, and expression. In the HBF case the templates would effectively correspond to different centers, and matching would proceed in a more sophisticated way than direct comparison. It is clear that this step may by itself accomplish segmentation.

3. *Classification and indexing.* Parameter values estimated by the preceding stage for the features of interest (e.g., the distance between eyes and mouth) are used in this stage for classification and indexing in a database of known examples. In many cases this may lead by itself to unique recognition, especially when situational information, such as the expectedness of a particular object, restricts the relevant data base. Classification could be done via a number of classical schemes such as nearest neighbor or with modules that are more biologically plausible such as HBF networks.

Some open questions remain:

• What are the features used by the human visual system in the feature detection stage? A plausible idea is that there is a large set of filters tuned to different 2D shape features and efficiently doing a kind of template matching on the input. Some functional of the correlation function is then evaluated (such as the max of the correlation or some robust statistics on the correlation values). The results represent some of the components (for that particular filter, i.e., template) of the input vector to object-specific networks consisting of hidden units each tuned to a view and an output unit which is view-invariant. Networks of this type may also exist not only for specific objects but also for general object components, perhaps similar to more precise versions of some of Biederman's geons (Biederman 1987). They would be synthesized by familiarity and their output may have a varying degree of view invariance depending on the type and number of the tuned cells in the hidden layer. Networks of this type, tuned to a particular shape, could easily be combined conjunctively to represent more complex shapes (but still exploiting the fundamental property of additivity). This general scheme avoids the correspondence problem since the components of the input vectors are statistics taken over the whole image, rather than individual pixel values or feature locations. It may well be that in the absence of a serial mechanism such as eye motions and attentional shifts the visual system does not have a way to keep and use spatial relations between different components or feaures in an image and that it can only detect the likely "presence" of, say, a few hundred features of various complexity.

• The architecture described here consists of a hierarchy of HBF-like networks. Does the human visual system operate with a similar hierarchy? For instance, an eye-recognizing MBM network may provide some of the inputs to a face recognition network that will combine the presence (and possibly relative position) of eyes with other face features (remember that

Cortical Mechanisms for Object Recognition and Learning

an MBM network can be regarded as a disjunction of conjunctions). The inputs to the eye-recognizing networks may be themselves provided by other RBF-like networks; this is similar to the use in the eye-recognizing networks of inputs that are the result of filtering the image through a few basic filters out of a large vocabulary consisting of hundreds of "elementary" templates, representing a vocabulary of shapes of the type described by Fujita and Tanaka (1992). The description of Perrett and Oram (1992) is consistent with this scenario. At various stages in this hierarchy more invariances may be achieved for position, rotation, scaling, etc., in a manner similar to how complex cells are built from simple ones.

The Visualization Route to Recognition

The second potential route to recognition takes a necessary detour from the first route to fine-tune the matching mechanisms. Like the classification pathway it begins with the two stages of *image measurement* and *feature detection*, but diverges because it allows for the possibility that a match between the database and measured image features might not directly be found. Further processing may take place on the image or on the stored examples to bring the two into registration or to narrow the range of the latter. The main purpose of this loop is to correct for deformations before comparing image to data base.

Computational arguments (Breuel 1992) suggest that this route should separate transformations to be applied to the image (to redress image-plane deformations such as image-plane translations, scaling, and rotations) from those to be applied to the database model (which may include rotations-in-depth, illumination changes, and alterations in facial expression, for example). The system may try a number of transformations in parallel and on multiple scales of spatial resolution (see chapter 13 by Van Essen, Anderson, and Olshausen) until it finds the one that succeeds. In general the whole process may be iterated several times before it achieves a satisfactory level of confidence (see chapter 7 by Mumford and chapter 12 by Ullman for similar proposals). In the primate visual system, the likely site for the latter transformations is cortical area IT, whereas the former would probably take place earlier, as available results on properties of IT seems to suggest (Gross 1992; Perrett et al. 1982; Perrett and Harries 1988; Perrett et al. 1989). The main steps of this hypothetical second route to recognition are as follows:

1. *Image measurement.*

2. *Feature detection.*

3. *Image rectification.* The feature detection stage provides information about the location of key features that is used in this stage to normalize for image-plane translation, scaling and image-plane rotation of the input M-array.

4. *Pose estimation.* 3D pose (two parameters), illumination, and other parameters (such as facial expression) are estimated from the M-array. This computation could be performed by an MBM module that has "learned" the appropriate estimation function from examples of objects of the same class.

5. *Visualization.* The models (M-arrays in the database corresponding to known objects) are warped in the dimensions of pose and expression and illumination, to bring them into register with the estimate obtained from the input image. The transformation of the models is performed by exploiting information specific to the given object (several views per object may have been stored in memory) or by applying a generic transformation (e.g., for a face, from "serious" to "smiling") learned from objects of the same class. Several transformations may be attempted at this stage before a good match is found in the next step.

6. *Verification and indexing.* The rectified "image" is compared with the warped database of standard representations. Open questions remain on how the database may be organized and what are the most efficient means of indexing it.

APPENDIX B: ON THE DECOMPOSITION OF MULTIDIMENSIONAL INPUTS

It is well known (see earlier, and Poggio and Girosi 1989) that the simplest version of a regularization network approximates a vector field $y(x)$ as

$$y(x) = \sum_{i=1}^{N} c_i G(x - x_i) \tag{8.6}$$

with G being the chosen Green function and

$$(G)c_m = y_m. \tag{8.7}$$

It follows that the vector field is approximated as the linear combination of example fields, that is

$$y(x) = \sum_{l=1}^{N} b_l(x)y_l \tag{8.8}$$

where the b_l depend on the chosen G.

Thus *for any choice of the regularization network—even HBF—and any choice of the Green function—including Green functions corresponding to additive splines and tensor product splines—the estimated output (vector) image is always a linear combination of example (vector) images with coefficients that depend (nonlinearly) on the desired input value.* This observation together with the fundamental hypothesis suggests that an output vector (say a vectorized image) can be represented as a linear combination of examples. This is similar to decomposition in parts.

ACKNOWLEDGMENTS

This chapter describes research done within the Center for Biological Information Processing in the Department of Brain and Cognitive Sciences, and at the Artificial Intelligence Laboratory at MIT. This research is sponsored by a grant from NATO to Dr. Hurlbert and Dr. Poggio; by a grant from the Office of Naval Research, Cognitive and Neural Sciences Division; by the Office of Naval Research contract N00014–89–J–3139 under the DARPA Artificial Neural Network Technology Program, and by a grant from the National Science Foundation under contract ASC-9217041 (including funds from DARPA provided under the HPCC program). Support for the A.I. Laboratory's artificial intelligence research is provided by ONR contract N00014–91–J–4038. Tomaso Poggio is supported by the Uncas and Helen Whitaker Chair at the Whitaker College, Massachusetts Institute of Technology. Anya Hurlbert's research is supported by the JCI (ESRC, SERC, and MRC), SERC III, and the Royal Society, England.

9   Constructing Neuronal Theories of Mind

Michael I. Posner and Mary K. Rothbart

What should a theory of higher brain function be about? If it is to be a theory of the brain, it is clear that it should be about neurons and neuroanatomy. Yet it is widely agreed that even a complete description of the functions of neurons is unlikely to be adequate to provide a theory of higher brain function. Nor do studies of the brain's structure provided by the new neuroimaging methods, such as positron emission tomography or magnetic resonance imagery, by themselves provide a clear answer to the question of brain functions. In our view, the term "higher" indicates that such a brain theory should also be concerned with mind. If so, it becomes important to know what "mind" is like, how to measure it, and what a theory of mind must explain.

Many descriptions of mind begin with subjective experience and measure "mind" by self reports given verbally or nonverbally about that experience. We believe the task of connecting brain to mind requires as fine an analysis of mind as we have been able to make of neuronal activity. In our view, the analysis of mind necessary to make connections with brain systems involves its specification into the elementary operations that provide a basis for localization of function within neural systems. Over the past 25 years researchers in cognitive science have developed ways of defining and measuring mental operations.

FRAMEWORK

Cognitive Systems

We find it useful to view the connection between cognitive systems and neurosystems in terms of a very general framework shown in figure 9.1. This framework involves five levels of analysis. At the highest level is performance of the tasks of every day life. These include activities like reading, recognizing faces, daydreaming, moving from place to place, playing music, writing, and planning a trip. Verbal report is often informative at this level; we can report what faces look like or our intention to go someplace.

A great deal of evidence from studies of lesioned patients indicates that these tasks may be grouped together into a somewhat lesser number of

Level	Example	Methods
COGNITIVE SYSTEM	LANGUAGE, ATTENTION, MOTOR CONTROL	VERBAL REPORT
ELEMENTARY OPERATION	NEXT, ROTATE, ZOOM	COMPUTER SIMULATIONS (AI)
PATHWAY ACTIVATION	FACILITATE, iNHIBIT	COGNITIVE STUDIES NEURAL NETWORKS
NEURO SYSTEM	PARIETAL CORTEX BASAL GANGLIA	PET, ERP, MRI
MICROCIRCUITS	VISUAL LAMINA AREA V4	CELLULAR RECORDING

Figure 9.1 Framework for linking cognitive and neural systems.

"cognitive systems," with many tasks of daily life draw on the same cognitive system. Thus reading, writing, speaking, and conversing are all tasks of daily life and they all involve the cognitive system we call "language." Brain damage (lesions) at various locations of the left hemisphere impairs aspects of this language system. The idea of a cognitive system has some analogies with an organ system in that it is a set of structures functioning together to allow the performance of a general function. Sometimes "planning ahead" is thought to involve a common cognitive system. Lesions of the frontal lobes often impair planning and thus there may be a brain system that underlies it. Similarly, selective attention appears to involve particular brain systems.

Elementary Operations

Complex cognitive tasks such as playing chess, reading, or manipulating visual images have also been subjected to detailed analysis. These analyses have divided a task into logical operations that might form the basis for programming a computer to simulate human performance. Consider the task of imagining you are walking along a familiar route. One analysis of the imagery process identifies 12 elementary operations (Kosslyn 1980). Each operation has an input, a computation, and an output. When organized into an appropriate sequence, one has a computational model capable of performing an imagery task.

This approach to mental imagery is a form of artificial intelligence (AI) called symbol processing. Its emphasis is on the logic of a set of operations that would be sufficient to program or simulate the task being studied. Although AI models do not directly model the brain, they sometimes help in the construction of neural models by making clear what the logical operations are. The logical analysis of the model may or may not relate to what happens when people actually perform the task. The model provides a kind of sufficiency analysis, in that it shows the task can be analyzed into a set of subroutines sufficient to perform it.

Psychological Pathways

The next step in our effort to link cognitive processes to neural systems is to ask how a human mind performs the postulated operation. To achieve that goal, we need to design a model task incorporating the operations under study. How, for example, does one generate a visual image? Suppose you are presented with either a visual or an auditory letter. Your task is to determine as quickly as possible if a second (probe) letter is the same letter as the first (Posner 1978). If the first letter is a visual upper case *A* and a probe letter in the same case is presented immediately (e.g., *AA*), it takes about 80 msec less to process than if the probe is in the opposite case (e.g., *Aa*). Following a spoken letter, both upper- and lowercase visual probes also take about 80 msec longer than would a direct visual physical match. After a half second delay between the first letter and the probe, uppercase probes are handled just as fast, whether the first letter is visual or auditory, but lowercase probes still take longer. It takes about half a second to generate an optimal visual representation of the auditory stimulus. By this objective test, you have now generated an image.

How in detail is an image generated? If you are presented with the letter name *F* (Kosslyn 1988) and asked to form a visual image, how can we tell if you are doing so? We can, of course, ask you. But even if we take your answer as definitive, we would have no way of going further to ask in detail how the image is constructed, because you really have little insight into that level of your processing system. To find out if an image has been formed, a slightly different probe can be used. This time you are asked whether a probe *X* that appears is located on or off the image you have created. Immediately after the presentation of the first letter, you will be slow to verify whether the *X* lies on the image or not because the image is not yet created. After a short delay you are fast to verify probes that lie on the upright of the letter but slow for those on the cross bar. It is as though you have generated the left part of the image but not yet the rest of it. In fact, these images appear to be generated stroke by stroke. What is most remarkable is that as you generate them, verification of probes lying on the stroke that has already been generated is facilitated.

Suppose you are presented with an upright letter, and asked to rotate clockwise in your mind's eye (Cooper and Shepard 1975). Are you actually performing the rotation? To test whether you are, we probe with letters at varying angles from the upright. You are asked to report by pressing

Constructing Neuronal Theories of Mind

one key if the probe is a correct letter and another key if it is a mirror-image letter. Prior to starting the experiment we calculate your rotation speed from the reaction time to respond to letters presented at different angles of orientation. Suppose your calculated rotation speed is 100°/sec. After 0.3 sec you should be faster to 30° probes and slower to upright or 60° probes. That is exactly what is found; you are actually faster in responding to a probe letter at 30° orientation than to one that is upright at the usual angle at that we experience letters. Your mental rotation has objective consequences that can be measured precisely in terms of the time to verify the probe, and they are strong enough to overcome the usual preference for upright letters based on past experience.

It is also possible to study inhibition of processing performance. Suppose you are shown a red S on top of a green K (Allport 1989). You are asked to name the red letter and ignore the green letter that is under it. On one trial you name S and the rejected letter is K. If on the next trial the red letter is a K you will be slow in naming it. When you select the red item you also inhibit the green one. This inhibition remains present for 1–2 sec and retards your performance on the subsequent trial.

These experiments have shown that the performance of a cognitive operation can be observed in terms of exquisitely time-locked facilitations in the speed of processing probe items. We call this level of analysis the "performance domain" because we are looking at facilitation or inhibition in performance measures such as reaction time or threshold detection. The use of the words facilitation and inhibition is biased to make one inquire whether such patterns are related to the activity of the populations of neural cells that might perform the computation. To answer the question of how facilitation and inhibition measured in performance relate to neural activity requires methods to link mental operation to underlying neural systems (see below).

There is a second objective method for studying mental operations. In addition to requiring time, they also tend to show specific interference when they compete for the same computation. To explain this feature, consider a task involving timing (Keele et al. 1985). In this task, subjects listen to a tone occurring every 0.5 sec and tap a key in synchrony with it. After the tone is turned off they continue to tap at the same interval. Performance in this task is measured by the variability of the key presses. In the focal condition, subjects perform the task by itself, but in a second condition they do it at the same time as a secondary task. Two different secondary tasks are used, both designed so that they do not involve the same input or output mode as the primary task. Both secondary tasks are designed to be equivalent in difficulty and to demand the same amount of attention. One task is to determine if the interval between one pair of tones is the same or different as the interval between a second pair. The other task involves the same tones, but now the person judges if the difference in loudness between one pair is the same or different than the second pair. The interval judging secondary task involves an internal operation of timing to determine the tone time differences. The loudness secondary task involves judgments of intensity that, unlike the interval task, do not overlap with the timing required by the primary tapping task. The tones occur while the subject is steadily pressing the key at an interval that must be timed internally. The finding is that the interval judging secondary task interferes much more with the primary task than does the loudness judging task. When the internal mental operation of timing is shared between the two tasks, performance is affected, thus revealing an underlying hidden operation thought to involve a clock that is common to both perception and action. The use of dual-task methods has been an important one in the objective measurement of mental operations (Posner 1978).

TOOLS FOR THEORY CONSTRUCTION

Methods play a particularly important role in every scientific endeavor and this is certainly true of the effort to relate the facilitations and inhibitions in the performance of mental operations to their underlying neural systems. To move from elementary operations and their effects upon performance to the level of neural systems (see figure 9.1) it is important to have methods of localization. Although theories of localization of mental function have been present for decades, methods for examining the relation between cognitive function and brain activity have been indirect. Much of the classic work has involved examinations of brain sections following death, with damaged areas of the brain related to the prior behavior of the organism. In vivo examinations of the human brain by measurement of electrical activity have been possible for 50 years, but the use of imaging techniques based on X ray, radionuclides, and magnetic resonance is more recent. We are only now developing appropriate strategies to combine these various methods.

The heart of the problem of constructing neural models of cognition is to move from the level of performance to underlying neural systems or micro-circuits. Neuroscience approaches have placed somewhat greater emphasis on spatial methods that give hope for studying localization. Cognitive approaches have tended to place emphasis on the temporal organization of information flow in the nervous system.

Cognitive neuroscience requires the integration of methods that trace the time dynamics of information processing with those that provide information on the location of neural systems activated. Fortunately new methods and adaptations of older methods have become available in the last dozen years. Two methods prominent in this chapter emphasize the advantage of combining spatial and temporal precision.

Spatial Localization

Positron emission tomography (PET) is a radioactive tracer method of measuring cerebral blood flow or metabolic activity that provides a means of tracing cerebral activity during sustained cognitive tasks. As used in the studies discussed here, oxygen 15–labeled water is injected (Raichle 1987) and the water is carried along with the blood to various parts of the brain. The distribution of labeled substance is monitored by radiation generated when positrons are absorbed. This radiation is sensed by an array of detectors. While the spatial resolution of this method is limited, the method becomes quite accurate when successive scans are compared. It is possible to compare scans because the data acquired for blood flow images can be obtained within 40 sec. Differences between the central tendencies of blood flow changes in the two conditions can be measured to within a few millimeters. Although current PET methods are not chronometric in the sense of being sensitive to changes in milliseconds, their probable physical limit involves the rate at which blood vessels reflect neural changes. At present, it is possible to obtain localization of blood flow activity in the range of a few millimeters, and to look at changes during tasks that last for less than a minute.

Temporal Dynamics

So far, the spatial imaging methods used with humans (e.g., PET) have not provided the kind of temporal precision of information required for the analysis of many cognitive tasks, where differences of 20–100 msec are frequently of theoretical importance (Posner 1978). Event-related electrical activity recorded from the scalp of humans provides one method for achieving high temporal information and limited spatial resolution (Mangun et al. 1992). The use of event-related potentials (ERP) has been quite helpful in linking mental operations studied by chronometric methods to brain systems in general. By combining PET and ERP studies we have been working on using the former to compensate for the more limited localization possible with measurements on the scalp (Compton et al. 1991).

BUILDING A NEURAL THEORY OF MIND

In 1992 (Posner and Rothbart 1992) we summarized evidence arising from PET studies suggesting that the anterior cingulate gyrus plays a critical role in one aspect of attention. The anterior cingulate lies on the midline of the frontal lobe and has strong connections with a variety of other neural areas. The aspect of attention related to the anterior cingulate is close to what is often meant by consciousness and relates to both awareness and to voluntary control (figure 9.2).

Below we summarize the major pieces of evidence that formed the core of the paper by Posner and Rothart (1992).

The anterior attention network seems to be much more directly related to awareness than the posterior network, as has been indicated by the PET studies cited previously. The use of subjective experience as evidence for a brain process related to consciousness has been criticized by many authors. However, we note that the evidence for the activation of the an-

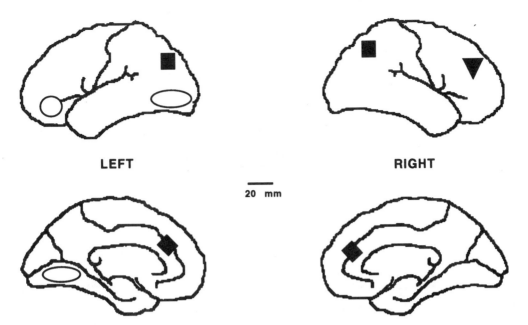

Figure 9.2 The cortical projections of the attention networks. The data are mainly from PET studies. The attentional networks are shown by solid shapes on the lateral and medial surfaces of the right and left hemisphere. Squares are the posterior attention network (parietal lobes), triangles, the vigilance network (right frontal), and diamonds, the anterior attention network (anterior cingulate, supplementary motor area). The open shapes refer to word-processing systems (ellipse, visual word form; circle, semantic associates) that have been shown to relate to the posterior and anterior attention systems, respectively.

terior cingulate is entirely objective; it does not rest upon any subjective report. Nevertheless, if one defines consciousness in terms of awareness, it is necessary to show evidence that the anterior attention network is related to phenomenal reports in a systematic way. In this section, we note *five points*, each of which appears to relate *subjective experience* to activation of the anterior attention system. First, the degree of activation of this network increases with the number of targets presented in a semantic monitoring task and decreases with the amount of practice in the task. At first one might suppose that target detection is confounded with task difficulty. But in our semantic monitoring task the same semantic decision must be made irrespective of the number of actual targets. In our tasks no storage or counting of targets was needed. Thus we effectively dissociated target detection from task difficulty. Nonetheless, anterior cingulate activation was related to number of targets present. The increase in activation with number of targets and reduction in such activation with practice corresponds to the common finding in cognitive studies that conscious attention is involved in target detection and is required to a greater degree early in practice (Fitts and Posner 1967). As practice proceeds, feelings of effort and continuous attention diminish and details of performance drop out of subjective experience.

Second, the anterior system appears to be active during tasks requiring the subject to detect visual stimuli, when the targets involve color form, motion, or word semantics (Petersen et al. 1989; Corbetta et al. 1990).

Third, the anterior attention system is activated when listening passively words, but not when watching those words. This finding appears to correspond subjectively to the intrusive nature of auditory words to consciousness when they are presented in a quiet background. They seem to capture awareness. Reading does not have this intrusive character. For a visual word to dominate awareness, an act of visual orienting is needed to boost its signal strength.

Fourth, the anterior attention system is more active during conflict blocks of the Stroop test than during nonconflict blocks (Pardo et al. 1990). This is consistent with the commonly held idea that conflict between word name and ink color produces a strong conscious effort to inhibit saying the written word (Posner 1978). Finally, there is a relation between the vigilance system and awareness. When one attends to a source of sensory input in order to detect an infrequent target, the subjective feeling is of emptying the head of thoughts or feelings. This subjective "clearing of consciousness" appears to be accompanied by an increase in activation of the right frontal lobe vigilance network and a reduction in the anterior cingulate. Just as feelings of effort associated with target detection or inhibiting prepotent responses are accompanied by evidence of cingulate activation, so the clearing of thought is accompanied by evidence of cingulate inhibition. (pp. 97–99)

Both those who read the paper and the authors were somewhat uneasy about one aspect of its content. It is not an easy thing to be writing something that might be interpreted as meaning there is a spot inside the nervous system that represents the neural correlate of consciousness. This is probably even more true following Dennett's (1992) philosophical critique of those who implicitly cling to a view that one area is the arena of consciousness or what he calls the Cartesian Theater of the Mind. Nonetheless, the specific points made above seemed to us to identify cingulate activation with aspects of awareness in so much tighter a way than previous efforts that it was reasonable to set them down with as much clarity as possible.

Since writing that paper, much has happened both to increase our uneasiness with identifying consciousness with precise brain coordinates and to allow us to build out from that rather uncomfortable position by specifying in somewhat more detail what role the anterior cingulate might actually play.

Reentry

One new development was our increased understanding of how the brain actually executes a voluntary instruction to attend to something (Grossenbacher et al. 1991). Studies employing PET have revealed important anatomical aspects of word reading (Petersen et al. 1989, 1990). Two major areas of activation appear within the visual system. A right posterior temporal parietal area is activated passively by both consonant strings and words. This activation appears to be enhanced when subjects are required to detect a feature. This area is thought to be a visual representation that is prelexical. A left ventral occipital area is activated by both words and pseudowords (e.g., tweal), but not by consonant letter strings. The location and properties of this left posterior activation suggest it is involved in what is called the visual word form. The visual word form is a representation of the orthography of the letter string in which individual letters are combined into a single chunk.

These recent PET findings, together with many cognitive studies, imply that letter strings are represented within the visual system both as unorganized features or letters and within a unified visual word form. However, the PET images used in these studies require an average of 40 sec and could also involve feedback from more anterior to posterior areas. For example, the occipital visual word form may be set up only after the subject accesses the word meaning.

To study the time course of word processing, we developed tasks that take advantage of another recent PET result showing that when subjects attend to color, motion, or form, appropriate posterior prestriate areas are increased in activation. Attention appears to amplify the activity of anatomical areas in which the related computations occur (Corbetta et al. 1990). We considered a task that requires the subject to deal with the individual features of a letter (Compton et al. 1991). To study attention to visual features, subjects are required to detect a line thickening within one letter of a four- or six-letter word or consonant string. To link cognitive results with anatomy we studied these tasks along with passive perception while recording electrical activity from 32–64 electrodes positioned over occipital, temporal, parietal, and frontal areas.

The results of this study provide encouragement for the effort to relate PET anatomical data to time dynamic ERP data. We found a very strong posterior tempoparietal asymmetry in which electrical activity at about 100 msec is larger from the right hemisphere than from the left. This effect is quite strong, but only at temporal and inferior parietal sites. The effect fits with the idea of a right posterior generator related to visual features because it occurs within the first 100 msec and similarly for word and nonword strings in all task blocks.

We asked subjects to respond to the presence of a thick feature by pressing one key if it was present and another if it was not. There was no difference between the strings that made words and those that did not. On trials when a target was present, reaction times appeared to reflect the distance of the target from the center of vision. Moreover, the differences in reaction time between four and six letters (the slope of the search function) was about the same whether the target was present or absent. If subjects had been searching serially and stopped when they found a target, the slopes for the target absent trials would be twice that for the target present trials, since when a target was present subjects could respond as soon as they detected a target, which on the average would require searching only half the list. These results make it seem reasonable that when attending to features, subjects search a representation located in the right posterior temporal lobe.

The second anatomical area relates to the visual word form system of the left ventral occipital lobe. To study the difference between words and consonant strings we superimposed their event-related potentials. While differences between the two stimulus types are found in a number of electrodes, these appear to occur first along a posterior band of areas extending from the right posterior temporal lobe to the left anterior temporal lobe. These findings are generally consistent with the location of the PET generator in the anterior left occipital lobe near the midline, although the degree of anatomical localization from the scalp voltage data is low and the evidence for lateral asymmetry in the voltage data is not strong. We are seeing differences over a large part of the posterior scalp of both hemispheres. It is possible to gain a somewhat better notion of the localization of these effects if the data are transformed by use of an average reference based on all of the electrodes other than the one being considered, weighted by their distance from the active electrode site. This transformation suggests a left posterior generator in the neighborhood of the posterior temporal lobe.

If the ERP effect is coming from the visual word form area, our findings suggest this area is making the initial discrimination between words and nonwords starting at about 200 msec after input. Since the PET result is averaged over 40 sec of activity, activation in posterior locations could have been fed back from some more anterior area. However, the ERP data clearly suggest that in the passive conditions, the posterior discrimination is being made first, because no other electrodes show this difference prior to the posterior ones.

In PET studies an area of the left frontal area is active when subjects deal with the meaning of a word. When the process is extended by requiring association of several words to a given input or by slowing the rate of presentation, this left frontal area is joined by activation in Wernicke's area (Fiez and Petersen, 1993). To study semantic and feature activation together we used tasks that clearly involved both. The feature task was again looking for a thick letter; the semantic task required the subject to determine if the word referred to a natural or a manufactured item. Our reasoning was that in the first task, subjects would be attending to the feature level and in the semantic task to the meaning of the word. If, as has been described in the PET work, attention serves to amplify computations, it should be possible to see amplifications of the voltages in the waveforms in the right posterior area in feature analysis and in the left frontal area in semantics, depending on the task used. We used exactly the same stimuli in the two tasks.

Results showed that the left frontal area was more positive at about 200–300 msec when the task was semantic, while the right posterior area showed more positivity when the task was feature search. These effects were not confined to single electrode sites. For the posterior area, it was possible to compare the electrode sites first showing the greater right hemisphere activation at 100 msec associated with the visual attribute, with those showing the amplification due to attribute search at about 250 msec. Our comparison generally supported the idea that roughly the same areas that first carried out the visual attribute computations on the letter string were reactivated 150 msec later when subjects were looking for the thick letter. This fit with the idea that subjects can voluntarily reactivate areas of the brain that performed the task automatically when they are instructed to deal with that computation voluntarily. The semantic effect was found at several frontal sites bilaterally. This differs from the PET data, which were strictly left lateralized, although there was some evidence that the left frontal area showed the effect more strongly than the right.

A popular idea in modern physiology is called reentrant processing (Edelman and Mountcastle 1978). Basically, this is the idea that higher level associations are made by fibers that reenter the brain areas that processed the initial input. Mountcastle has written about the basic organization of cortical anatomy as follows:

It is well known from classical neuroanatomy that many of the large entities of the brain are interconnected by extrinsic pathways into complex systems, including massive reentrant circuits.

Simulations based on the coordination of wide spread neural systems also rely upon this principle, as described by Sporns et al. (1989):

Signaling between neuronal groups occurs via excitatory connections that link cortical areas, usually in a reciprocal fashion. According to the theory of neural group selection, selective dynamical links are formed between distant neural groups via reciprocal connections in a process called reentry. Reentrant signaling establishes correlations between cortical maps, within or between different levels of the nervous systems.

Reentrant processing may be contrasted with more traditional notions that higher functions are confined to higher associational areas of the brain. A similar viewpoint to reentrant processing is expressed by Damasio and Damasio (chapter 3). In our studies, the visual computation occurs at 100 msec, followed by a semantic computation which might be complete by 200–300 msec. When the instruction is to search the string for a feature, the electrodes around the area originally performing the visual computation are reactivated. Similarly, when asked to make a semantic computation, the area thought to perform such computations was amplified in electrical activity about 100 msec after its initial computation.

If the brain operates in this way we might then be able to instruct the subject to compute the same functions in different orders and thus repro-gram the order of the underlying computations. To investigate this, we (Grossenbacher et al. 1991) defined what we call a conjunction task. We ask the subjects to respond with a key if a word refers to an object that is manufactured (e.g., paper) and has a thick letter and otherwise to respond with a second nontarget key. On one day we have the subjects perform the thick letter task and then ask them to respond with the target key if the stimulus has a thick letter and refers to a manufactured object. On an-other day we have subjects perform the semantic task and then ask them to respond with the target key if the word is manufactured and has a thick letter. The function to be computed is thus exactly the same. The inputs are identical and the responses (if correct) are identical, but the order of the underlying computations is reversed.

We did not expect the subjects to actually compute the functions in a serial fashion. Our hope was only that they would emphasize the priority computation and perhaps complete it somewhat earlier than the second nonpriority computation. To see if this happens one can look at the reaction time data from this task. We examined the nontarget reaction times where subjects can quit when they find either computation inappropriate for a target. In general the thick letter task is somewhat faster; subjects quit sooner when there is no thick letter present. They are also relatively faster if the thick letter task has been given priority by training. On the other hand, if a thick letter is present and responses must be based on semantic analysis, subjects are faster if the semantics was given priority by training. Thus the reaction time data suggest that we have been successful in reordering the computation times.

We can now look at images of the underlying brain activity recorded from above the left frontal or right posterior areas. The two forms of the conjunction task differ at about 300 msec following input. In the left frontal area, the semantic priority task returns to baseline first; later, the physical priority task returns to baseline. A reversed effect is found in the posterior area. This time the physical priority task returns first to baseline followed by the semantic priority task. The two brain areas seem to reflect the relative priority emphasized in the directions. We believe that the subject has used attention to program the relative order of the two computations represented by the two anatomical areas.

These results suggest that the person is able to reorder the priority of the underlying computations in the conjunction task. They also provide us with a basis for understanding how the brain can carry out so many different tasks on visual input. Aspects of the underlying computations do not seem affected by the instructions. The visual attribute area of the right posterior brain seems to carry out the computation on the input string at 100 msec irrespective of whether the person is concerned with visual features as a part of the task or not. However, when the task is identified as looking for a thick letter these same brain areas are reactivated and presumably carry out the additional computations necessary to make sure that one of the letters has just enough thickening to constitute a target. Attention thus can amplify computations within particular areas, but often does so by reentering the area, not by amplifying its initial activation.

Models of Control

The data on reentrant processing imply two important ideas related to the attentional control of information processing. First, the results of attentional control are widely distributed, resulting in amplification of activity in the anatomical areas that originally computed that information. Second, the source of this attentional control need not involve a system that has access to the information being amplified, but can be a system that has connections to places where the computations occur. This sense of control by a separate attentional system over widely distributed computations is suggested physiologically by Van Essen, Anderson, and Olshausen (chapter 13) and computationally by Ullman (chapter 12).

As the result of activity within the attention network, the relevant brain areas will be amplified and/or irrelevant ones inhibited, leaving the brain to be dominated by the selected computations. If this were the correct theory of attentional control, one would expect to find the source of attention to lie in systems widely connected to other brain areas, but not otherwise unique in structure. As pointed out by Goldman-Rakic (1988), this appears to be the basic organization of frontal networks. Anterior cingulate connections to limbic, thalamic, and basal ganglia pathways would distribute its activity to the widely dispersed connections we have seen to be involved in cognitive computations.

To illustrate this framework for attention, we use a recent model of control of covert visual attention developed at our center by Jackson and Houghton (1992). The model involves location expectations held by the anterior attention network interacting with location cues that influence the posterior attention network. The basic architecture of the model is shown in figure 9.3.

The posterior attention network (including the parietal lobe and associated thalamic and midbrain areas) and the anterior attention network (including the anterior cingulate) influence each other via direct cortical projections, but also indirectly through a comparator operation involving the basal ganglia. The direct loops have the effect of allowing activations at common locations in the two systems to support one another. A sensory event facilitates processing at a location due to activation of the posterior network, but an expectation also operates via the anterior network to facilitate the expected location.

The role of the basal ganglia loops are more complex. A direct pathway between the anterior cingulate and striatum serves as a reverberating circuit to maintain expected locations and to amplify them when their locations match. The indirect pathway operates when there is a mismatch between the two attention systems to dampen down activation within the anterior system at any locations for which there is no input from the posterior attention system. This allows expectations to be overcome.

The resulting system is rather complicated but it is constrained by the anatomical structures of the relevant components. It make predictions of performance in cognitive experiments using cues and targets. For example, the model can predict some reaction time results accumulated from cognitive experiments involving manipulations such as lesions of the posterior system, blocking of NE or DA into the system, competition from dual tasks, etc. We do not wish to present the model as a final answer to the coordination of attentional systems, but merely to show that the logic of the operations we have discussed can be embodied in a functioning computer model.

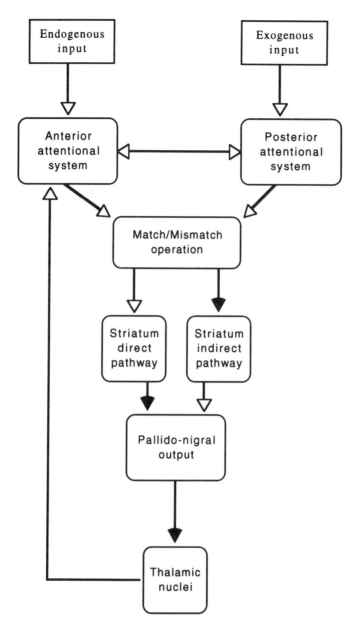

Figure 9.3 Architecture indicating the role of attention networks in covert orienting to visual locations. The top of the figure indicates the areas of the posterior attention network (e.g., parietal cortex) and anterior attention network (e.g., anterior cingulate). The lower parts of the figure indicate areas of basal ganglia and its connections. The open arrows represent excitatory and the closed arrows inhibitory connections.

The type of attentional control of orienting suggested above can thus lead to specific simulations allowing tests at the cognitive level in considerable detail. Recent PET data (Corbetta et al. 1993) show that involuntary shifts of attention to visual locations induced by cues produce strong superior parietal activation within the posterior system. When a subject must shift attention endogenously to report a target, this activation is accompanied by frontal and anterior cingulate activation, as would be expected if endogenous shifts are controlled from frontal areas.

Some recent developmental findings (Posner and Rothbart 1992) show that the distributed connections by which the attention systems assume control over various functions may develop over a considerable period of life. The control of orienting by the posterior attention system appears to develop mainly between 4 and 6 months. During this period the infant develops the ability to disengage from visual stimuli and to control the areas of the visual field to which they will attend. An important goal of early infant development is the control of pain and distress. PET studies suggest that the highest representation of pain is within the anterior cingulate (Talbot et al. 1991). This finding suggests that attentional manipulations might be important in the control of distress. We have shown the evidence of such control at about three months. Orienting to visual events can be employed to quiet or calm negative vocalizations. However, the distress appears to be maintained and reappears when the infant's attention to the stimulus is reduced. Caregivers also report the use of visual orienting to block overt manifestations of distress at about this age. It appears that the attention system continues to develop later in the first year of life and beyond. One the hallmarks of the higher level attentional system involving the anterior cingulate is in involvement in tasks that involve conflict between signals such as the stroop effect (Pardo et al. 1990). At 9 to 12 months one begins to see how control of reaching behavior allows the infant to reach separately from the line of regard (Diamond 1988). Control of language behavior by the attentional system occurs even later. Studies of development provide another method for observing attention and for testing models.

PRINCIPLES CONNECTING COGNITIVE AND NEURAL SYSTEMS

It is often difficult to grasp the principles that arise out of various experimental demonstrations and modeling efforts, particularly when they are presented in abbreviated form such as in this chapter. Below we attempt to summarize some of the general ideas related to our framework that appear to arise from our review. Each of these principles seeks to connect mental experience to neural areas via the methods we have described above.

1. *Elementary mental operations are localized in discrete neural areas.* Evidence supporting this point rests both upon work in attention and in language. Operations involved in selective attention discussed in this chapter are

carried out by diverse networks. The study of areas active during auditory and visual words processing leads us to a similar conclusion (Petersen et al. 1989). In addition, we view current discussions of motor control, object and face recognition, and memory as providing additional support for this general idea.

2. *Cognitive tasks are performed by a network of widely distributed neural systems.* We have illustrated this idea in this chapter by showing that the attention involves several networks of cortical and subcortical areas. Studies of visual and auditory word processing also suggest networks of anatomical areas are involved even in very simple word-association tasks (Petersen et al. 1989).

3. *Computations in a network interact by means of "reentrant" processes.* Cognitive experiments give good evidence that the successful ordering of computations is necessary for performance. Ordering does not take place by a strict serial organization. Instead, computations appear to pass information back and forth to coordinate their results. While it has been clear that precise connections exist between anatomically distant areas, it appears that a particular anatomical area is active whenever its computation is required. Since computations are often contingent on information from another area, this can take place only if that information is fed back to reenter the critical areas.

4. *Hierarchical control is a property of network operation.* Discovery of a separate network of anatomical areas devoted to attention has been described. This view provides a basis for establishing executive control over widely distributed networks. The requirement for such control systems is clear from cognitive experiments showing interference between simultaneous performance of cognitive tasks irrespective of the nature of their constituent computations. Moreover, the control appeared to be largely inhibitory. That is, attention to one concept reduces the probability that other concepts receive attention. Selection between simultaneously operating representations appears to require inhibition of one of them. These findings supported the idea of executive control by attention systems.

5. *Activation of a computation produces a temporary reduction in the threshold for its reactivation.* This principle underlies the cognitive phenomenon of priming. Whenever a code has been active it becomes easier for a stimulus to reactivate it. For the processing of words, priming exists at the level of attributes, word forms, phonology, and semantics.

6. *When a computation is repeated its reduced threshold is accompanied by reduced effort and less attention.* This principle may seem the inverse of the last, but in fact it is a corollary. The repetition of a computation improves its efficiency; as a result, the overall activity accompanying the computation is reduced. Blood flow is less, electrical activity reduced, and the interference between the repeated computation and other activity is reduced. Thus a habituated activity will produce less orienting of attention, the memory of it having

been performed will be reduced, and other signs that the activity has been automated will be found.

7. *Activating a computation from sensory input (bottom-up) and from attention (top-down) involves many of the same neurons.* Attention to motion, color, or form activates many of the same prestriate areas that were active when passively receiving information of the same type. There is some evidence that the size and/or number of prestriate areas active in attention conditions are greater than during the comparable passive perception condition. The same principle was discussed above from recording scalp electrical activity during word processing. We believe attention can be used to mark the activity underlying a particular computation.

8. *Practice in the performance of any computation will decrease the neural networks necessary to perform it.* The idea that repetition of events leads eventually to their automation, that is, to performance without attention, is well established in psychology (Posner 1978). Recent PET data suggest that repetition of the same performance leads to reduced blood flow in the neural areas that are originally required to generate the response (Fiez and Petersen, 1993). We believe that this principle, like others in this section, will apply to cognitive tasks in general.

ACKNOWLEDGMENTS

This research was supported by the Office of Naval Research contract N 0014-89-J3013 and by the Center for the Cognitive Neuroscience of Attention supported by the James S. McDonnell Foundation and Pew Memorial Trusts. We are grateful to Anne Clohessy, Paul Compton, Peter Grossenbacher, Cathy Harman, and Steve Jackson for the use of unpublished ideas and data from their research projects.

10 Putative Functions of Temporal Correlations in Neocortical Processing

Wolf Singer

INTRODUCTION

The ability to record from individual neurons in the central nervous system of animals while these engage in perceptual tasks, memorize, or perform a motor response has revealed numerous and fascinating correlations between the activity of individual nerve cells and complex behavioral patterns. Cells have been found whose responses distinguish between familiar and unfamiliar objects (Baylis and Rolls 1987; Miyashita and Chang 1988; Fuster 1990; Miller et al. 1991a,b), are selective for particular aspects of faces (for review see Rolls 1991; Gross 1992), reflect precisely the location of a remembered target (Goldman-Rakic et al. 1990), or predict with accuracy the direction of an eye movement (Goldberg et al. 1990; Wurtz et al. 1990). Cells have been described in the visual cortex whose response thresholds correspond precisely to the behavioral thresholds of the animal (Parker and Hawken 1985; Newsome et al. 1990). Finally, activating a local cluster of neurons by microstimulation in a visual area specialized for motion processing biases the perception of motion as if additional moving targets were added to the visual stimulus (Salzman et al. 1992). Results of this kind are strong support that the activation of individual neurons can represent a code for highly complex functions, a notion that is commonly addressed as the "single neuron doctrine" (Barlow 1972). However, there have always also been speculations that additional coding principles might be realized. Most of these begin with Donald Hebb's proposal that representations of sensory or motor patterns should consist of assemblies of cooperatively interacting neurons rather than of individual cells. This coding principle implies that information is contained not only in the activation level of individual neurons but also, and actually to a crucial extent, in the relations between the activities of distributed neurons. If true, the description of a particular neuronal state would have to take into account not only the rate and the specificity of individual neuronal responses but also the relations between discharges of distributed neurons. Over the last decade, these speculations have received some support both from experimental results and theoretical considerations. Search for individual neurons responding with the required selectivity to individual objects was only partly successful and has so far revealed specificity only for faces and for a limited set of objects with which the animal had been familiarized extensively before (see below). And even in these cases it is likely that a particular face or object evokes responses in a very large number of neurons. Recordings from motor centers such as the deep layers of the tectum and areas of the motor cortex provided no evidence for command neurons such as exist in simple nervous systems and code for specific motor patterns. Rather, these studies provided strong support for a population code as the trajectory of a particular movement could be predicted correctly only if the relative contributions of a large number of neurons were considered (Georgopoulos 1990; Mussa-Ivaldi et al. 1990; Sparks et al. 1990). Arguments favoring the possibility of relational codes have also been derived from the growing evidence that cortical processes are highly distributed (Zeki 1973; Ungerleider and Mishkin 1982; Maunsell and Newsome 1987; Desimone and Ungerleider 1989; Newsome et al. 1990; Felleman and Van Essen 1991; Zeki et al. 1991; Goodale and Milner 1992). Further indications for the putative significance of relational codes are provided by theoretical studies that attempted to simulate certain aspects of pattern recognition and motor control in artificial neuronal networks. Single cell codes were found appropriate for the representation of a limited set of well-defined patterns but the number of required representational elements scaled very unfavorably with the number of representable patterns. Moreover, severe difficulties were encountered with functions such as figure-ground distinction because single cell codes turned out to be too rigid and inflexible, again leading to a combinatorial explosion of the required representational units. By implementing population or relational codes some of these problems could be solved or at least alleviated. Exploiting relational codes also opens up the possibility to use time as an additional coding space. By defining a narrow temporal window for the evaluation of coincident firing and by temporal patterning of individual neuronal responses, relations between the activities of spatially distributed neurons can be defined very selectively (Milner 1974; von der Malsburg 1985). If such temporal coding is added to the principle of population coding the number of different patterns or representations that can be generated by a given set of neurons increases substantially. Moreover, it has been demonstrated that perceptual functions like scene segmentation and figure-ground distinction that require flexible association of features can in principle be solved if one relies on relational codes in which the relatedness of distributed neurons is expressed by the temporary synchronization of their respective discharges (Milner 1974; von der Malsburg 1985; von der Malsburg and Schneider 1986; Shimizu et al. 1986).

Arguments emphasizing the importance of temporal relations between the discharges of cortical neurons have also been derived from recent data on connectivity and synaptic efficacy. Cortical cells receive many thousand synaptic inputs but on the average a particular cell contacts any of its target cells only with one synapse (Braitenberg and Schütz 1991). In vitro studies from cortical slices indicate that the efficacy of individual synapses is low and that not every presynaptic action potential triggers the release of transmitter (Stevens 1987; see chapter 11 by Stevens). Thus, many presynaptic afferents need to be activated simultaneously to drive a particular cell above threshold and to ensure reliable transmission. Even more cooperativity is required for the induction of synaptic modifications such as long-term potentiation and long-term depression. These modifications have high thresholds and require substantial and prolonged postsynaptic activation (Artola and Singer 1987, 1990). Temporal coordination of cortical responses appears thus necessary both for successful transmission across successive processing stages and for the induction of use-dependent synaptic modifications.

Despite these numerous arguments supporting the putative importance of temporal relations among distributed neuronal responses in the neocortex, systematic search for temporal relations among the activities of simultaneously recorded cortical neurons is still at an early stage. Initially, cross-correlation analysis of multielectrode recordings has been used primarily as a tool of functional anatomy to reveal excitatory and inhibitory connections among neurons. Hence, analysis was often confined to spontaneous activity. If cells were activated with sensory stimuli this was done to increase activity and to reduce the duration of the measurements, but not with the goal to disclose dynamic, stimulus-related interactions. It is only recently that cross-correlation analysis has been extended to responses evoked by selected stimulus configurations to test whether the responses of spatially distributed cortical neurons exhibit temporal relations that are sufficiently consistent to serve a functional role in cortical processing. Many of these latter experiments were designed to test specific predictions derived from recent theories on population coding. Therefore, the review of these cross-correlation studies will be preceded by a brief description of the conceptual background. As most of the theoretical models have dealt with problems of visual pattern processing and recognition and as most of the experimental studies have been performed in the visual system, the conceptual background will be illustrated mainly on the basis of visual processes.

REPRESENTATIONS AND THE BINDING PROBLEM

Most perceptual objects can be decomposed into components and in general the features of these components are not unique for a particular object. The individuality of objects results from the specific composition of elementary features and their relations rather than from the specificity of the component features. Hence, for a versatile representation of sensory patterns in the nervous system three basic functions have to be accomplished: (1) elementary features need to be represented by neuronal responses, (2) responses to features constituting a particular object have to be distinguished and bound together in a flexible way, and (3) the specific relations among these features have to be encoded and preserved.

One way to achieve the grouping of features and to establish an unambiguous code for their specific relations is to connect the set of neurons that responds to the component features of a particular object to a higher order neuron that will represent the object. If the thresholds of these higher order neurons are adjusted so that each cell responds only to one particular combination of feature detectors, the responses of these higher order neurons would provide an unambiguous description of the relations between the component features and hence would be equivalent to the representation of the pattern. In this scheme the features of the object are bound together by convergence of fixed connections that link neurons representing component features with neurons representing the whole pattern. The relations between features are encoded by the specific architecture of these convergent connections.

However, not all of the predictions following from this latter assumption are supported by experimental evidence. First, while cells occupying higher levels in the processing hierarchy tend to be selective for more complex constellations of features than cells at lower levels, many continue to respond to rather simple patterns such as edges, gratings, and simple geometric shapes (Tanaka et al. 1991; Gallant et al. 1993). Second, apart from cells responding preferentially to aspects of faces and hands (Gross et al. 1972; Desimone et al. 1984, 1985; Baylis et al. 1985; Rolls 1991) it has been notoriously difficult to find other object-specific cells except in cases where animals had been familiarized with a limited set of objects during extensive training (Miyashita 1988; Sakai and Miyashita 1991). Third, no single area in the visual processing stream has yet been identified that could serve as the ultimate site of convergence and that would be large enough to accommodate the vast number of neurons that are required if all distinguishable objects including their many different views were represented by individual neurons. Finally, the point has been made that "binding by convergence" may not be flexible enough to account for the rapid formation of representations of new patterns. To allow for the representation of new, hitherto unknown objects one would have to postulate a large reservoir of uncommitted cells. These neurons would have to maintain latent input connections from all feature-selective neurons at lower processing stages and subsets of these connections would have to be selected and consolidated instantaneously when a new representation is established.

Similar combinatorial problems arise in the case of motor control but they have received less theoretical consideration. Here, the solution equivalent to "binding by convergence" is that individual command neurons at the top of a hierarchically organized motor system each triggers one complex motor act. Their activity would have to become distributed through divergent and highly selective connections to subsets of effector neurons that eventually activate particular muscle groups. This coding concept encounters the same problem as its homologous concept on the sensory side: First, no such command neurons were found in areas that could perhaps be regarded as being on the top of the processing hierarchy such as the supplementary motor field or prefrontal motor areas. Second, given the sparseness of connections between individual cortical cells (see above) it is hard to see how activation of only a few neurons could give rise to the mass action required for the execution of a movement. Third, one would again have to postulate a large reservoir of uncommitted cells to allow for the representation of newly learned motor patterns. These uncommitted command cells would have to maintain latent connections to virtually all effector muscles and the appropriate subsets of these connections would have to become functional only, but then would have to be recruited permanently, when the particular motor skill is established for which these connections are required. Finally, there is the problem of temporal patterning. This problem needs to be solved also for the processing of sensory patterns if these are spread out in time, but it is particularly obvious in motor programming. For the execution of a motor act it is necessary to generate complex and precisely coordinated temporal sequences according to which the distributed muscle groups are activated. One solution would be sets of delay lines to distribute the activity of the command neuron in the appropriate temporal order to the effector neurons at more peripheral levels. But this would further increase the number of command neurons because each motion executed at different speeds would require a command cell connected to a different set of delay lines. The inverse problem exists for the representation of sensory patterns that have not only a spatial but also a temporal structure.

Because these difficulties cannot be overcome easily in architectures that solve the binding problem by serial recombination of converging (in the motor path "diverging") feedforward connections alternative proposals have been developed.

Before discussing these alternative concepts it is necessary to emphasize that "binding by convergence" may be a viable solution for specialized representational systems. However, because of the limitations discussed above this coding strategy can be used only for the representation of a limited set of stereotyped patterns.

Alternative proposals for the solution to the binding problem are based on the assumption that representations consist of assemblies of a large number of simultaneously active neurons that may be contained in a single cortical area but that may also be distributed over many cortical areas (Hebb 1949; Braitenberg 1978; Edelman and Mountcastle 1978; Crick 1984; Grossberg 1980; Palm 1982, 1990; Singer 1985, 1990; von der Malsburg 1985; Edelman 1987, 1989; Abeles 1991). The essential feature of assembly coding is that individual cells can participate at different times in the representation of different objects. The assumption is that just as a particular feature can be present in many different patterns, a neuron coding for this feature can be shared by many different representations. This reduces substantially the number of cells required for the representation of different objects and allows for considerably more flexibility in the generation of new representations.

Basic requirements for representing objects by such assemblies are as follows: First, the responses of the cells responding to a visual scene need to be compared with one another and examined for possible, "meaningful" relations. Second, cells coding for features that can be related need to become organized into an "assembly." This should be the case for the cells that are, for example, activated by the constituent features of a particular object. Third, if patterns change, neurons must be able to rapidly change partners and to form new assemblies. Fourth, neurons that have joined a particular assembly must become identifiable as members of this very assembly. Their responses must be tagged so that they can be recognized as being related (i.e., the distributed responses of the assembly must be recognizable as representing a "whole"). It is commonly assumed that these organizing steps, the probing of possible relations, the formation of an assembly, and the labeling of responses are achieved in a single self-organizing process by selective reciprocal connections between the distributed neuronal elements. The idea is that the probabilities with which neurons become organized into particular assemblies are determined, first, by the respective constellation of features in the pattern and, second, by the functional architecture of the assembly forming coupling connections. Several proposals have been made concerning the mechanisms by which these connections could serve to "label" the responses of neurons that have joined into the same assembly. Most of them assume that the assembly-generating connections are excitatory and reciprocal and serve to enhance and to prolong the responses of neurons that were organized in an assembly (Hebb 1949; Singer 1979, 1985; Grossberg 1980; Palm 1982).

Another proposal is that assemblies should be distinguished in addition by a temporal code (von der Malsburg 1985; von der Malsburg and Schneider 1986). A similar suggestion, although formulated less explicitly, had been made previously by Milner (1974). This hypothesis assumes that the assembly-forming connections should establish temporal coherence on a millisecond time scale between the responses of the coupled cells. Thus, neurons having joined into an assembly would be identifiable as members of the assembly because of the synchronization of their discharges. Expressing relations between members of an assembly by the temporal coherence rather than the amplitude of their responses has several advantages: First, it reduces the ambiguities that result from the fact that discharge rates depend strongly on variables such as stimulus intensity and quality of fit between stimulus features and receptive field properties. If assemblies were solely defined by a rate code it would be impossible to decide whether a strongly active cell is discharging at a high rate because it joined an assembly or because it was activated by a particularly effective stimulus. Relying on temporal relations preserves the important option to use discharge rates as a code for stimulus parameters. This is essential in systems using coarse codes because the information about the presence of particular features and about their precise location is contained in the graded responses of populations of cells. Second, exploiting temporal relations increases the number of assemblies that can be active simultaneously without becoming confounded. In most cases simultaneously active assemblies will be distinguished by spatial segregation due to retinotopy and compartmentalization of cortical areas. But there may be conditions in which additional distinctions are required to avoid fusion of unrelated assemblies. Responses of neurons could overlap on a coarse time scale but still remain distinguishable as coming from a particular assembly if they are correlated at a fine time scale. Third, cells that succeeded in synchronizing their discharges have a stronger impact on target cells. This follows from the plausible assumption that afferents to cortical neurons will be more efficient in driving a postsynaptic cell if they discharge in synchrony. This effect will be particularly strong when the activation levels of the afferent fibers are low and when the postsynaptic potentials evoked by the individual fibers are small. Both conditions seem to be fulfilled for cortical networks (see above). Thus, formation of coherently active assemblies can serve to enhance the saliency of responses to features that can be associated in a "meaningful" way. This may contribute to the segregation of object-related features from unrelated features of the background. This concept of "binding by synchrony" has also been applied to intermodal integration (Damasio 1990) and even to high level processes underlying phenomena such as attention (Crick 1984) and consciousness (Crick and Koch 1990a).

PREDICTIONS

A network that allows for the self-organization of pattern specific assemblies must meet the following constraints:

1. Neurons within the same cortical area as well as neurons distributed across different areas must be coupled reciprocally by connections ensuring the selection and dynamic stabilization of specific assemblies.

2. These connections must be exceedingly numerous because their number, together with the number of cells, limits the number of possible constellations.

3. The assembly forming connections must be highly specific as the grouping criteria according to which features are bound together into object representations reside in the functional architecture of these connections.

4. The network must allow for highly dynamic interactions to enable individual cells to link at different times with different partners.

5. The coupling connections must have adaptive synapses allowing for use-dependent long-term modifications of synaptic gain to permit the formation and stabilization of new grouping criteria when new object representations are to be installed during learning.

6. These use-dependent synaptic modifications should follow a correlation rule whereby synaptic connection should strengthen if pre- and postsynaptic activity is often correlated, and they should weaken in case there is no correlation. This is required to enhance grouping of features that often occur in consistent relations as is the case for features constituting a particular object.

7. These grouping operations should occur over multiple processing stages because search for "meaningful" groupings has to be performed at different spatial scales and according to different feature domains. This could be achieved by distributing the grouping operations over different cortical areas in which different neighborhood relations are realized with respect to the representation of retinal location and of feature domains by remapping of inputs.

These seven predictions need to be fulfilled irrespective of whether assemblies are defined by rate or temporal codes. If cells having joined an assembly are distinguished by a rate code the prediction is that cells activated by features of a particular object engage in stronger and perhaps also more sustained responses than cells responding to features resisting grouping. However, no differences should be found between the enhanced responses of cells participating in different assemblies representing different objects. They should all be enhanced to a similar extent. If assemblies are distinguished in addition or alternatively by the temporal coherence of the responses of the constituting neurons a further set of predictions can be derived.

1. Spatially segregated neurons should synchronize their responses if activated by features that can be grouped together. This should be the case for features constituting a single object.

2. Synchronization should be frequent among neurons within a particular cortical area but it should also occur across cortical areas.

3. The probability that neurons synchronize their responses both within a particular area and across areas should reflect some of the Gestalt criteria used for perceptual grouping (Kofka 1935; Köhler 1969).

4. Individual cells must be able to rapidly change the partners with which they synchronize their responses if stimulus configurations change and require new associations.

5. If more than one object is present in a scene several assemblies should form. Cells belonging to the same assembly should synchronize their responses while no consistent temporal relations should exist between the discharges of neurons belonging to different assemblies.

6. Synchronization probability should at least in part depend on the functional architecture of reciprocal corticocortical connections and should change if this architecture is modified.

EXPERIMENTAL EVIDENCE FOR TEMPORAL RELATIONS

To test these predictions it is necessary to record simultaneously from spatially distributed neurons in the brain and to search for systematic temporal correlations among their responses (Gerstein et al. 1985). It is not sufficient to analyze synchronization probability of spontaneous activity as this would reveal only the architecture and coupling strength of connections and not the dynamic properties of the network that emerge only on stimulation. Thus, the correlation studies that have been performed with the goal of revealing anatomical connection patterns are relevant in the present context in as much as they provide data on the organization of coupling connections but they usually do not address the more dynamic, stimulus-dependent interactions that are predicted from the assembly hypothesis.

In the visual cortex correlations between the activities of simultaneously recorded cortical cells were found to be frequent, especially when they were closely spaced and located within single functional columns. The observed correlation patterns were indicative of constellations where cells receive either common excitatory or inhibitory input or where one cell excites or inhibits the other. (Toyama et al. 1981a,b; Michalski et al. 1983; Abeles and Gerstein 1988; Hata et al. 1988; Gochin et al. 1991). For cells located in different functional columns and hence being separated by several hundred micrometers along trajectories parallel to the pial surface correlations were more difficult to detect. This agrees with anatomical data that indicate that connections are densest between cells staggered within narrow cylinders orthogonal to the lamination and rapidly decrease along trajectories tangential to the lamination (for review see Douglas and Martin 1993). When interactions were found over larger tangential distances the cross-correlograms usually had a peak centered around zero delay that was interpreted as indicative of common excitatory input (Tso et al. 1986; Krüger and Aiple 1988; Gochin et al. 1991) or of common modulation of excitability (Aiple and Krüger 1988; Krüger and Aiple 1988). However, as detailed below, there may be interpretations other than common input for this type of interaction.

Data from the visual cortex of cats and monkeys suggested in addition that long-range interactions are confined to neurons that share similar preferences for the orientation and/or the spectral composition of stimuli (Tso et al. 1986; Tso and Gilbert 1988; Schwarz and Bolz 1991). This agrees with anatomical data that show that tangential intracortical connections are selective and link preferentially evenly spaced patches of cortical tissue that are closely related to functional columns (Rockland and Lund 1982; Gilbert and Wiesel 1989; but see Matsubara et al. 1985).

Systematic search for more dynamic stimulus-dependent interactions between spatially distributed cortical neurons had been initiated by the observation that adjacent neurons in the cat visual cortex can transiently engage in highly synchronous discharges when presented with their preferred stimulus (Gray and Singer 1987). Groups of neurons recorded simultaneously with a single electrode were found to discharge in synchronous "bursts" that followed one another at intervals of 15–30 msec. Typically, individual neurons would contribute only one or two spikes to such synchronous events but in the multicell recordings these episodes of synchronous discharge appeared as bursts. These sequences of synchronous rhythmic firing occur preferentially when cells are activated with slowly moving light stimuli. They last no more than a few hundred milliseconds and may occur several times during a single passage of the moving stimulus (figure 10.1). Accordingly, autocorrelograms computed from such response epochs often exhibit a periodic modulation (Gray and Singer 1987, 1989; Eckhorn et al. 1988; Gray et al. 1990; Schwarz and Bolz 1991; Livingstone 1991). During such episodes of synchronous firing a large oscillatory field potential is recorded by the same electrode, the negative deflections being coincident with the cells' discharges. The occurrence of such a large field response indicates that many more cells in the vicinity of the electrode than those actually picked up by the electrode must have synchronized their discharges (Gray and Singer 1989).

However, neither the time of occurrence of these synchronized response episodes nor the phase of the oscillations is related to the position of the stimulus within the neuron's receptive field. When cross-correlation functions are computed between responses to subsequently presented identical stimuli, these "shift predictors" reveal no relation between the temporal patterning of successive responses (Gray and Singer 1989; Gray et al. 1990). The rhythmic firing is thus not related to some fine spatial structure in the receptive fields of cortical neurons.

This phenomenon of local response synchronization has been observed with multiunit and field potential recordings in several independent studies in different areas of the visual cortex of anesthetized cats (areas 17, 18, 19, and PMLS) (Eckhorn et al. 1988, 1992; Gray and Singer 1989; Gray et al. 1990; Engel et al. 1991a), in area 17 of awake cats (Raether et al. 1989; Gray and Viana di Prisco 1993), in the optic tectum of awake pigeons (Neuenschwander and Varela 1990), and in the visual cortex of anesthetized (Livingstone 1991) and awake behaving monkeys (Kreiter and Singer 1992).

Subsequently, it has been shown with multielectrode recordings in anesthetized and awake cats (Gray et al. 1989; Raether et al. 1989; Engel et al. 1990; Kreiter and Singer, 1992; Kreiter et al. 1992) and anesthetized and awake monkeys (Kreiter and Singer 1992; Kreiter et al. 1992) that similar response synchronization can occur also between spatially segregated cell groups within the same visual area. When cells engage in such long distance synchronization the firing patterns of the local groups often exhibit the synchronous repetitive firing described above. Interestingly, the synchronization of responses over larger distances also occurs with zero phase lag. Hence, if the cross-correlograms show any interaction at all, they typically have a peak centered around zero delay. The half-width at half-height of this peak is on the order of 2–3 msec, indicating that most of the action po-

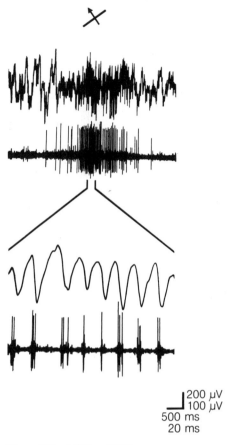

Figure 10.1 MUA and LFP responses recorded from area 17 in an anesthetized adult cat to the presentation of an optimally oriented light bar moving across the receptive field. Oscilloscope records of a single trial showing the response to the preferred direction of movement. In the upper two traces, at a slow time scale, the onset of the neuronal response is associated with an increase in high-frequency activity in the LFP. The lower two traces display the activity at the peak of the response at an expanded time scale. Note the presence of rhythmic oscillations in the LFP and MUA (35–45 Hz) that are correlated in phase with the peak negativity of the LFP. Upper and lower voltage scales are for the LFP and MUA, respectively. (Adapted from Gray and Singer 1989)

tentials that showed some consistent temporal relation had occurred nearly simultaneously. This peak is often flanked on either side by troughs that result from pauses between the synchronous bursts. When the duration of these pauses is sufficiently constant throughout the episode of synchronization, the cross-correlograms show in addition a periodic modulation with further side peaks and troughs. But such regularity is not a necessary requirement for synchronization to occur. There are numerous examples from anesthetized cats (see, e.g., Engel et al. 1991c; Nelson et al. 1992c) and especially from awake monkeys (Kreiter and Singer 1992) that responses of spatially distributed neurons can become synchronized and lead to cross-correlograms with significant center peaks without engaging in rhythmic activity that is sufficiently regular to produce a periodical modulation of averaged auto- and cross-correlograms. However, there are relations between oscillatory discharge patterns and response synchronization that will be discussed in detail in a later section.

The Dependence of Response Synchronization on Stimulus Configuration

As outlined above the hypothesis of temporally coded assemblies requires that the probabilities with which distributed cells synchronize their responses should reflect some of the Gestalt criteria applied in perceptual grouping. Another and related prediction is that individual cells must be able to change the partners with which they synchronize whereby the selection of partners should occur as a function of the patterns used to activate the cells. In this section experiments are reviewed that were designed to address these predictions. Detailed studies in anesthetized cats and recently also anesthetized and awake monkeys have revealed that synchronization probability for remote groups of cells is determined both by factors within the brain as well as by the configuration of the stimuli (Gray et al. 1989; Engel et al. 1990, 1991a,b,c; Kreiter et al. 1992; König et al. 1993). In general, synchronization probability within a particular cortical area decreases with increasing distance between the cells. If cells are so closely spaced that their receptive fields overlap, the probability is high that their responses will exhibit synchronous epochs if evoked with a single stimulus. This requires that the orientation and direction preferences of the cell pairs are sufficiently similar or that their tuning is sufficiently broad to allow for coactivation by a single stimulus. As recording distance increases synchronization probability becomes more and more dependent on the similarity between the orientation preferences of the neurons (Tso et al. 1986; Engel et al. 1990).

Concerning the dependence of synchronization probability on stimulus configuration single linearly moving contours have so far been found to be most efficient. Gray et al. (1989) recorded multiunit activity from two locations in cat area 17 separated by 7 mm. The receptive fields of the cells were nonoverlapping, had nearly identical orientation preferences, and were spatially displaced along the axis of preferred orientation. This enabled stimulation of the cells with bars of the same orientation under three different conditions: two bars moving in opposite directions, two bars moving in the same direction, and one long bar moving across both fields coherently. No significant correlation was found when the cells were stimulated by oppositely moving bars. A weak correlation was present for the coherently moving bars. But the long bar stimulus resulted in a robust synchronization of the activity at the two sites. This effect occurred in spite of the fact that the overall number of spikes produced by the two cells and the oscillatory patterning of the responses were similar in the three conditions.

In a related experiment Engel et al. (1991a) demonstrated that the synchronization of activity between cells in areas 17 and PMLS of the cat also depends on the properties of the visual stimulus. They recorded from cells in the two areas that had nonoverlapping receptive fields with similar orientation preference that were aligned colinearly. This made it possible to examine the effects of coherent motion on response synchronization between cells located in different areas. Little or no correlation was found when the cells were activated by oppositely moving contours but a robust synchronization occurred when the cells were coactivated by a single long bar moving over both fields (figure 10.2). (Engel et al. 1991a). These findings, combined with the earlier results, indicate that the global properties of visual stimuli can influence the magnitude of synchronization between widely separated cells located within and between different cortical areas. Single contours but also spatially separate contours that move coherently and therefore appear as parts of a single figure are more efficient in inducing synchrony among the responding cell groups than incoherently moving contours that appear as parts of independent figures.

These results indicate clearly that synchronization probability depends not only on the spatial segregation of cells and on their feature preferences, the latter being related to the cells' position within the columnar architecture of the cortex, but also and to a crucial extent on the configuration of the stimuli. So far, synchronization probability appears to reflect rather well some of the Gestalt criteria for perceptual grouping. The high synchronization probability of nearby cells corresponds to the binding criterion of "vicinity," the dependence on receptive field similarities agrees with the criterion of "similarity," the strong synchronization observed in response to continuous stimuli obeys the criterion of "continuity," and the lack of synchrony in responses to stimuli moving in opposite directions relates to the criterion of "common fate."

Experiments have also been performed to test the prediction that simultaneously presented but different contours should lead to the organization of two independently synchronized assemblies of cells (Engel et al. 1991c; Kreiter et al. 1992). If groups of cells with overlapping receptive fields but different orientation preferences are activated with a single moving light bar they synchronize their responses even if some of these groups are suboptimally activated (Engel et al. 1990, 1991c). However, if such a set of groups is stimulated with two independent spatially overlapping stimuli that move in different directions, they split into two independently synchronized assemblies, those groups joining the same synchronously active assembly that have a preference for the same stimulus (figure 10.3). Thus, the two stimuli evoke simultaneous responses in a large array of spatially interleaved neurons but these neurons become organized in two assemblies that can be distinguished because of the temporal coherence of responses within and the lack of coherence between assemblies. Cells representing the same stimulus exhibit synchronized response epochs while no consistent correlations occur between the responses of cells that are evoked

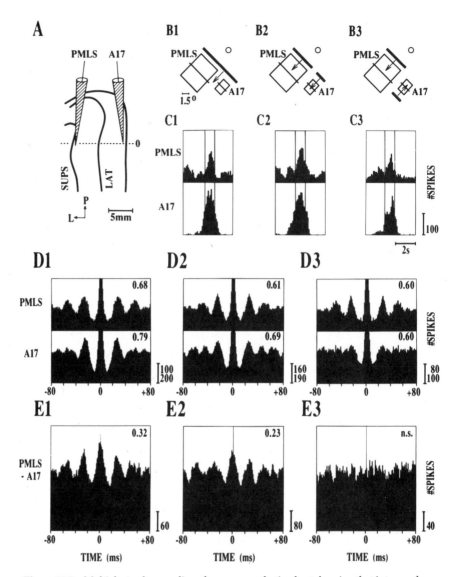

Figure 10.2 Multielectrode recordings from an anesthetized cat showing that interareal synchronization is sensitive to global stimulus features. (*A*) Position of the recording electrodes. A17, area 17; LAT, lateral sulcus; SUPS, suprasylvian sulcus; PMLS, posterior mediolateral suprasylvian sulcus; P, posterior; L, lateral. (*B1–B3*) Plots of the receptive fields of the PMLS and area 17 recording. The diagrams depict the three stimulus conditions tested. The circle indicates the visual field center. (*C1–C3*) Peristimulus time histograms for the three stimulus conditions. The vertical lines indicate 1-sec windows for which autocorrelograms and cross-correlograms were computed. Comparison of the autocorrelograms computed for the three stimulus paradigms. Note that the modulation amplitude of the correlograms is similar in all three cases (indicated by the number in the upper right corner). (*E1-E3*) Cross-correlograms computed for the three stimulus conditions. The number in the upper right corner represents the relative modulation amplitude of each correlogram. Note that the strongest correlogram modulation is obtained with the continuous stimulus. The cross-correlogram is less regular and has a lower modulation amplitude when two light bars are used as stimuli, and there is no significant modulation (n.s.) with two light bars moving in opposite direction. (From Engel et al. 1991a)

by different stimuli. The parameters of the individual responses such as their amplitude or oscillatory patterning were not affected by changes in the global configuration of the stimuli. Thus, it is not possible to tell from the responses of individual cells whether they were activated by a single contour or by two different stimuli. Even if one evaluated the extent of coactivation of the simultaneously recorded cells on a coarse time scale of several 100 msec as would be sufficient for the analysis of rate-coded populations, one would not be able to decide whether the cells had been activated by one composite figure whose features satisfy the preferences of the active cells or by two independent figures that excite the same set of cells. The only cue for this distinction was provided by the evaluation of synchronicity at a millisecond time scale.

The results of these experiments also prove that individual groups can change the partners with which they synchronize when stimulus configurations change. Cell groups that engaged in synchronous response episodes when activated with a single stimulus no longer did so when activated with two stimuli but then synchronized with other groups. One methodological caveat following from this is that cross-correlation analysis does not always reliably reflect anatomical connectivity (see also Aertsen and Gerstein 1985). In agreement with the predictions from the assembly hypothesis interactions between distributed cell groups were found to be highly dynamic, variable, and strongly influenced by the constellation of features in the visual stimulus.

Synchronization between Areas

Experiments have also been designed to test the prediction that cells distributed across different cortical areas should be able to synchronize their responses if they respond to the same contour. This prediction applies not only for interactions between cells distributed within different visual areas in the same hemisphere but also for cells in different hemispheres. The reason is that because of the partial decussation of the optic nerves, neurons responding to a figure extending across the vertical meridian are distributed across both hemispheres. As the responses of these cells have to be related to one another in the same way as those of cells located within the same hemisphere response synchronization should occur also across hemispheres and depend on stimulus configurations in the same way as intrahemispheric synchronization.

In the cat, interareal synchronization of unit responses has been observed between cells in areas 17 and 18 (Eckhorn et al. 1988, 1992; Nelson et al. 1992a), between cells in areas 17 and 19 and areas 18 and 19 (Eckhorn et al. 1992), between cells in area 17 and area PLMS, an area specialized for motion processing (figure 10.2) (Engel et al. 1991a; Munk et al. 1992), and even between neurons in area 17 of the two hemispheres (Engel et al. 1991b; Eckhorn et al. 1992; Nelson et al. 1992b). In the macaque monkey synchronous firing has been observed between neurons in areas V1 and V2 (Bullier et

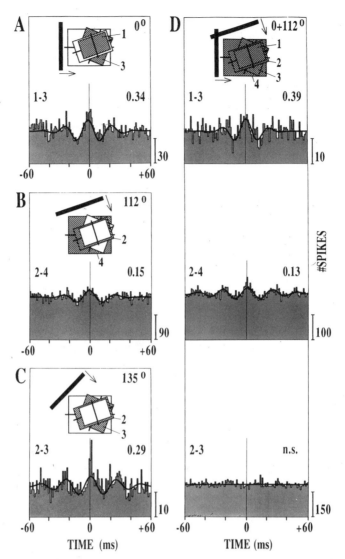

Figure 10.3 Stimulus dependence of short-range interactions. Multiunit activity was recorded from four different orientation columns of area 17 of cat visual cortex separated by 0.4 mm. The four cell groups had overlapping receptive fields and orientation preferences of 22° (group 1), 112° (group 2), 157° (group 3), and 90° (group 4), as indicated by the thick line drawn across each receptive field in (*A–D*). The figure shows a comparison of responses to stimulation with single moving light bars of varying orientation (*left*) with responses to the combined presentation of two superimposed light bars (*right*). For each stimulus condition, the shading of the receptive fields indicates the responding cell groups. Stimulation with a single light bar yielded a synchronization between all cells activated by the respective orientation. Thus, groups 1 and 3 responded synchronously to a vertically oriented (0°) light bar (*A*), groups 2 and 4 to a light bar at an orientation of 112° (*B*), and cell groups 2 and 3 to a light bar of intermediate orientation. (*C*) Simultaneous presentation of two stimuli with orientations of 0° and 112° respectively, activated all four groups (*D*). However, in this case the groups segregated into two distinct assemblies, depending on which stimulus was closer to the preferred orientation of each group. Thus, responses were synchronized between groups 1 and 3, which preferred the vertical stimulus, and between 2 and 4, which

Figure 10.3 *(continued)* preferred the stimulus oriented at 112°. The two assemblies were desynchronized with respect to each other, and so there was no significant synchronization between groups 2 and 3. The cross-correlograms between groups 1 and 2, 1 and 4, and 3 and 4 were also flat (not shown). Note that the segregation cannot be explained by preferential anatomical wiring of cells with similar orientation preference (T'so et al. 1986) because cell groups can readily be synchronized in all possible pair combinations in response to a single light bar. The correlograms are shown superimposed with a numerically fitted Gabor function. The number to the upper right of each correlogram indicates the relative modulation amplitude. n.s., not significant. Scale bars indicate the number of spikes (From Engel et al. 1991c)

al. 1992; Roe and Tso 1992). In all of these cases, whenever tested, synchronization depended on receptive field constellations and stimulus configurations, similar to the intraareal synchronization (Engel et al. 1991a,b). In the studies of Eckhorn et al. (1988, 1992) and Engel et al. (1991a) interareal and interhemispheric synchronous firing was found to occur primarily, if not exclusively, during coactivation of the cells by visual stimuli, and was particularly pronounced during periods of oscillatory firing (König 1994). In the studies of Nelson et al. (1992a) interareal synchronous firing was observed both during spontaneous activity and during the presentation of visual stimuli. The interactions span a wide "tripartite" range of temporal scales giving rise to correlograms having central peaks of narrow, medium, and broad width. The narrow (tight) coupling is most often seen between cells having overlapping receptive fields with similar properties. The broader coupling encompasses a much wider range of receptive field separations and differences in orientation preference (Nelson et al. 1992a). Synchronous firing between cells in V1 and V2 in the monkey shows similar features. Synchrony occurs between cells having both overlapping and nonoverlapping receptive fields (Bullier et al. 1992) and is most frequent between cells of similar color selectivity in the two areas (Roe and Tso 1992). Interestingly, whenever cells in V1 and V2 engage in synchronous discharges, on the average cells in V2 lead over cells in V1 by a few milliseconds (Bullier, personal communication). Synchronization of responses can thus occur over considerable distances and between cell groups located in different cortical areas and even hemispheres.

THE SYNCHRONIZING CONNECTIONS

It is commonly assumed in interpretations of cross-correlation data that synchronization of neuronal responses with zero-phase lag is indicative of common input (Gerstein and Perkel 1972). Because response synchronization occurred often in association with oscillatory activity in the range of 30–60 Hz, it has been proposed, that the observed synchronization phenomena in the visual cortex are due to common oscillatory input from subcortical centers (see chapter 6 by Llinás and Ribary, and Llinás and Ribary 1993). Oscillatory activity in the 30–60 Hz range has been described both for retinal ganglion cells and thalamic neurons (Doty and Kimura,

1963; Bishop et al. 1964; Fuster et al. 1965; Arnett, 1975; Ariel et al. 1983; Munemori et al. 1984; Ghose and Freeman 1990; Steriade et al. 1991; Ghose and Freeman 1992; Pinault and Deschênes, 1992a,b; Steriade et al. 1993). In both structures oscillatory activities have been observed in about 20% of the cells. They occurred during spontaneous activity and were often uninfluenced by visual stimulation or even suppressed (Ghose and Freeman 1992). These oscillatory patterns in afferent activity are likely to contribute to oscillatory responses in the visual cortex, but the possibility must also be considered that part of the thalamic oscillations are backpropagated from cortex by the corticothalamic projections.

Another question is whether these thalamic oscillations also play a role in stimulus-dependent synchronization of spatially distributed cortical neurons. Because the terminal arbors of thalamic axons span only 3–4 mm in the cortex (Ferster and LeVay 1978), the long distance correlations within areas but especially between areas and different hemispheres would require that thalamic activity becomes synchronized not only across different nuclei but even across the thalami of the two hemispheres to contribute effectively to long distance correlations at the cortical level. Large-scale synchronization of distributed thalamic neurons is common during sleep spindles (Steriade et al. 1990) but so far correlated 30–60 Hz oscillatory activity has been observed only between closely spaced cells (Arnett 1975).

If the synchronization phenomena observed at the cortical level were solely a reflection of common subcortical input, this would be incompatible with the postulated role of synchronization in perceptual grouping. The hypothesis requires that synchronization probability depends to a substantial extent on interactions between the neurons whose responses actually represent the features that need to be bound together. As thalamic cells possess only very limited feature selectivity one is led to postulate that corticocortical connections should also contribute to the synchronization process. This postulate is supported by the finding that synchronization between cells located in different hemispheres is abolished when the corpus callosum is cut (Engel et al. 1991b; Munk et al. 1992). This is direct proof (1) that corticocortical connections contribute to response synchronization and (2) that synchronization with zero-phase lag can be brought about by reciprocal interactions between spatially distributed neurons despite considerable conduction delays in the coupling connections. Thus, synchrony is not necessarily an indication of common input but may also be the result of a dynamic organization process that establishes coherent firing by reciprocal interactions.

Simulation studies are now available that confirm that synchrony can be established without phase lag by reciprocal connections even if they have slow and variable conduction velocities as long as the propagation delays do not exceed about one quarter of the oscillatory period (Schillen and König 1990; Schuster and Wagner 1990a,b; König and Schillen 1991). Use-dependent developmental selection of corticocortical connections could further contribute to the generation of architectures that favor synchrony.

During early postnatal development corticocortical connections are susceptible to use-dependent modifications and are selected according to a correlation rule (Löwel and Singer 1992; see below). This favors consolidation of connections whose activity is often in synchrony with the activity of their respective target cells. Hence, it is to be expected that connections are selected not only according to their feature-specific responses but also as a function of conduction velocities that allow for a maximum of synchrony.

However, the possibility to achieve synchrony through reciprocal cortical connections does not exclude a contribution of common input to the establishment of cortical synchronization. Especially if temporal patterns of responses need to be coordinated across distant cortical areas, bifurcating corticocortical projections or divergent corticopetal projections from subcortical structures such as the "nonspecific" thalamic nuclei, the basal ganglia, and the nuclei of the basal forebrain could play an important role. By modulating in synchrony the excitability of selected cortical areas they could influence very effectively the probability with which neurons distributed across these selected areas engage in synchronous firing. A contribution of diverging cortical backprojections to long-range synchronization is suggested by the observation that unilateral focal inactivation of a prestriate cortical area reduces intraareal and interhemispheric synchrony in area 17 (Nelson et al. 1992b). A contribution of thalamic mechanisms to the establishment of cortical synchrony has yet to be demonstrated.

EXPERIENCE-DEPENDENT MODIFICATIONS OF SYNCHRONIZING CONNECTIONS AND SYNCHRONIZATION PROBABILITIES

The theory of assembly coding implies that the criteria according to which particular features are grouped together reside in the functional architecture of the assembly forming coupling connections. It is of particular interest, therefore, to study the development of the synchronizing connections, to identify the rules according to which they are selected, to establish correlations between their architecture and synchronization probabilities, and, if possible, to relate these neuronal properties to perceptual functions.

In mammals corticocortical connections develop mainly postnatally (Innocenti 1981; Price and Blakemore 1985a; Luhmann et al. 1986; Callaway and Katz 1990) and attain their final specificity through an activity-dependent selection process (Innocenti and Frost 1979; Price and Blakemore 1985b; Luhmann et al. 1990; Callaway and Katz 1991). Recent results from strabismic kittens indicate that this selection is based on a correlation rule and leads to disruption of connections between cells which often exhibit decorrelated activity (Löwel and Singer 1992). Raising kittens with artificially induced strabismus leads to changes in the connections between the two eyes and cortical cells so that individual cortical neurons become connected to only one eye (Hubel and Wiesel 1965a). Cortical neurons split into two subpopulations of about equal size, each responding rather selectively to stimulation of one eye only. Because of the misalignment of the two eyes it is also to be expected that there are no consistent correlations between the activation patterns of neurons driven by the two eyes. Recently, it has been found that strabismus, when induced in 3-week-old kittens, leads to a profound rearrangement of corticocortical connections. Normally, these connections link cortical territories irrespective of whether these are dominated by the same or by different eyes. In the strabismics, by contrast, the tangential intracortical connections come to link with high selectivity only territories served by the same eye. These anatomical changes in the architecture of corticocortical connections are reflected by altered synchronization probabilities. In strabismic cats response synchronization no longer occurs between cell groups connected to different eyes while it is normal between cell groups connected to the same eye (König et al. 1990, 1993).

These results have several important implications. First, they are compatible with the notion that tangential intracortical connections contribute to response synchronization (see above). However, as strabismus also abolishes convergence of projections from the two eyes onto common cortical target cells, this result is also compatible with the view that synchrony is caused by common input. Second these results agree with the postulates of the assembly hypothesis that the assembly forming connections should be susceptible to use-dependent modifications and be selected according to a correlation rule. Third, the modifications of intracortical connections and synchronization probabilites add to the list of substrate changes that may be related to the specific perceptual deficits associated with early onset squint. Strabismic subjects usually develop normal monocular vision in both eyes, but they become unable to fuse signals conveyed by different eyes into coherent percepts even if these signals are made retinotopically contiguous by optical compensation of the squint angle (von Noorden 1990). Thus, in strabismics, binding mechanisms appear to be abnormal or missing between cells driven from different eyes. The lack of corticocortical connections and the lack of response synchronization could be among the reasons for this deficit in addition to the loss of binocular neurons.

These findings are, at least, compatible with the view that the architecture of corticocortical connections, by determining the probability of response synchronization, could set the criteria for perceptual grouping. Since this architecture is shaped by experience, this opens up the possibility that some of the binding and segmentation criteria are acquired or modified by experience.

CORRELATION BETWEEN PERCEPTUAL DEFICITS AND RESPONSE SYNCHRONIZATION IN STRABISMIC AMBLYOPIA

Further indications for a relation between experience-dependent modifications of synchronization probabilities and functional deficits come from a recent study of strabismic cats who had developed amblyopia. Strabismus, when induced early in life, does not only lead to a loss of binocular fusion and stereopsis but may also lead to amblyopia of one eye (von

Noorden 1990). This condition develops when the subjects solve the problem of double vision not by alternating use of the two eyes but by constantly suppressing the signals coming from the deviated eye. The amblyopic deficit usually consists of reduced spatial resolution and distorted and blurred perception of patterns. A particularly characteristic phenomenon in amblyopia is crowding, the drastic impairment of the ability to discriminate and recognize figures if these are surrounded with other contours. The identification of neuronal correlates of these deficits in animal models of amblyopia has remained inconclusive because the contrast sensitivity and the spatial resolution capacity of neurons in the retina and the lateral geniculate nucleus were found normal. In the visual cortex identification of neurons with reduced spatial resolution or otherwise abnormal receptive field properties remained controversial (for a discussion see Crewther and Crewther 1990; Blakemore and Vital Durand 1992). However, multielectrode recordings from striate cortex of cats exhibiting behaviorally verified amblyopia have revealed highly significant differences in the synchronization behavior of cells driven by the normal and the amblyopic eye, respectively. The responses to single moving bars that were recorded simultaneously from spatially segregated neurons connected to the amblyopic eye were much less well synchronized with one another than the responses recorded from neuron pairs driven through the normal eye (Roelfsema et al. 1993). This difference was even more pronounced for responses elicited by gratings of different spatial frequency. For responses of cell pairs activated through the normal eye the strength of synchronization tended to increase with increasing spatial frequency while it tended to decrease further for cell pairs activated through the amblyopic eye. Apart from these highly significant differences between the synchronization behavior of cells driven through the normal and the amblyopic eye no other differences were found in the commonly determined response properties of these cells. Thus, cells connected to the amblyopic eye continued to respond vigorously to gratings whose spatial frequency had been too high to be discriminated with the amblyopic eye in the preceding behavioral tests (see figure 10.4). These results suggest that disturbed temporal coordination of responses such as reduced synchrony may be one of the neuronal correlates of the amblyopic deficit. Indeed, if synchronization of responses at a millisecond time scale is used by the system for feature binding and perceptual grouping, disturbance of this temporal patterning could be the cause for the crowding phenomenon, as this can be regarded as a consequence of impaired perceptual grouping.

THE RELATIONSHIP BETWEEN SYNCHRONY AND OSCILLATIONS

Before reviewing the evidence on context-dependent synchronization across spatially segregated groups of neurons in structures other than the visual cortex it is necessary to examine the relationship between response synchronization on the one hand and oscillatory responses on the other,

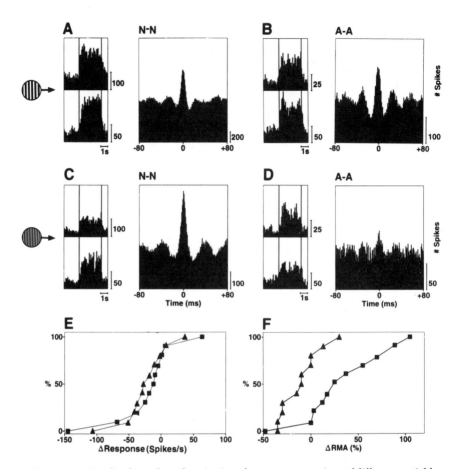

Figure 10.4 Amplitudes and synchronization of responses to gratings of different spatial frequencies recorded from cats with strabismic amblyopia. (*A–D*) Responses to low (*A* and *B*) and high (*C* and *D*) spatial frequency gratings, recorded simultaneously from two cell groups driven by the normal eye (N-sites) (*A* and *C*) and two cell groups driven by the amblyopic eye (A-sites) (*B* and *D*), respectively. The left and right panels show the response histograms and the corresponding cross-correlograms, respectively. Note that response amplitudes decrease at the higher spatial frequency in both cases, while the relative modulation amplitude increases for the *N–N* pair but decreases for the *A–A* pair. (*E*) Cumulative distribution functions of the differences between the amplitude of responses to low and high spatial frequency gratings of optimal orientation. N-sites, squares (*n*=53); A-sites, triangles (*n*=35). Abscissa, responses to high spatial frequency minus responses to low spatial frequency gratings. Note the similarity of the two distributions (*P* > 0.1). (*F*) Cumulative distribution functions of the differences between relative modulation amplitudes (DRMA of cross-correlograms obtained for responses to high and low spatial frequency gratings of *N–N* pairs (squares, *n*=24) and *A-A* pairs (triangles, *n*=11). DRMA values (abscissa) were calculated by subtracting the relative modulation amplitude obtained with the low spatial frequency from that obtained with the high spatial frequency. The difference between the DRMA distributions of *N–N* pairs and *A–A* pairs is highly significant. (From Roelfsema et al. 1993)

as the two phenomena are often but not necessarily always related. The occurrence of oscillatory responses does not logically imply that cells discharge in synchrony. Likewise, the nonoccurrence of oscillations does not exclude synchrony. Furthermore, it is useful to analyze to what extent different recording methods are appropriate for the assessment of synchrony or oscillatory behavior because there are a number of difficulties with the detectability of oscillatory firing patterns in single cell recordings and with the definition of oscillations.

No inferences can of course be drawn from single cell recordings as to whether the responses of the recorded cell are synchronized with others irrespective of whether the recorded cell is found to discharge in an oscillatory manner. The situation is different when multiunit recordings are obtained with a single electrode. In this case periodically modulated autocorrelograms are always indicative not only of oscillatory firing patterns but also of response synchronization, at least among the local group of simultaneously recorded neurons. The reason is that such periodic modulations can build up only if a sufficient number of the simultaneously recorded cells are oscillating synchronously and at a sufficiently regular rhythm. However, not observing periodically modulated autocorrelograms of multiunit recordings neither excludes that the recorded units oscillate, because nonsynchronized oscillations would not be observable, nor excludes that the recorded cells actually fire in synchrony, because they could do so in a nonperiodic way. The same arguments are applicable to field potential and even more so to EEG recordings. If they exhibit an oscillatory pattern this always implies that a large number of neurons must have engaged in synchronized rhythmic activity because otherwise the weak fields generated by activation of individual synapses and neurons would not sum to potentials recordable with macroelectrodes. But again, the reverse is not true: neither oscillatory discharge patterns nor response synchronization can be excluded if macroelectrode recordings fail to reveal oscillatory fluctuations.

Furthermore, it needs to be considered that single cell recordings may not be particularly well suited for the diagnosis of oscillatory activity. This is suggested by results from the visual cortex (Gray et al. 1990) and in particular from the olfactory bulb (Freeman and Skarda 1985). Individual discharges of single units may be precisely time-locked with the oscillating field potential, which proves that these discharges participated in an oscillatory process and occurred in synchrony with those of many other cells, without, however, showing any sign of oscillatory activity in their autocorrelation function. The reasons for this apparent paradox are sampling problems and nonstationarity of the time series. If the single cell does not discharge at every cycle and if the oscillation frequency is not perfectly constant over a period of time sufficiently long to sample enough discharges for an interpretable autocorrelation function, the oscillatory rhythm to which the cell is actually locked will not be disclosable. Thus, the less active a cell and the higher and more variable the oscillation frequency, the less legitimate it is to infer from nonperiodically modulated autocorrelograms that a cell is not oscillating. This sampling problem becomes more and more accentuated as the frequency of the oscillations increases. This explains why γ-band oscillations have been observed first with macroelectrodes and remain difficult to observe with microelectrodes unless one can record from several, synchronously active cells simultaneously. Finally, some ambiguities are associated with the term "oscillations." Most commonly, oscillations are associated with periodic time series such as are produced by a pendulum or an harmonic oscillator. But there are also more irregular or even aperiodic time series which are still called oscillatory. Such irregular oscillations typically occur in noisy linear or in nonlinear systems and cover a large spectrum of phenotypes from slightly distorted, periodic oscillations to chaotic oscillations to nearly stochastic time series. Oscillatory phenomena in the brain are rarely of the harmonic type and if so only over very short time intervals. Most often, oscillatory activity in the brain is so irregular that autocorrelation functions computed over prolonged periods of time frequently fail to reveal the oscillatory nature of the underlying time series.

These considerations need to be taken into account for the interpretation of the data reviewed below as these have been obtained with very different methods and often also different goals.

The evidence that in most structures investigated phases of response synchronization tend to be associated with episodes of oscillatory activity raises the question as to whether oscillations and synchrony are causally related.

One possibility is that oscillatory activity favors the establishment of synchrony and hence is instrumental for response synchronization. In oscillatory responses, the occurrence of a discharge predicts with some probability the occurrence of the next. It has been argued that this predictability is a necessary prerequisite to synchronize remote cell groups with zero phase lag, despite considerable conduction delays in the coupling connections (for review see Engel et al. 1992). This view is supported by simulation studies that have shown that zero-phase lag synchronization can be achieved despite considerable conduction delays and variation of conduction times in the synchronizing connections if the coupled cell groups have a tendency to oscillate (Schillen and König 1990, 1992; Schuster and Wagner 1990a,b; König and Schillen 1991). Another feature of networks with oscillatory properties is that network elements that are not linked directly can be synchronized via intermediate oscillators (König and Schillen 1991). This may be important, for instance, to establish relationships between remote cell groups within the same cortical area, or for cells distributed across cortical areas that process different sensory modalities. In both cases, linkages either via intermediate cortical relays or even via subcortical centers must be considered. The latter possibility is supported by the occurrence of γ-oscillations in a variety of thalamic nuclei (see above). These considerations suggest that oscillations, while not conveying any stimulus-specific information per se, may be instrumental for the establishment of synchrony over large distances. This conjecture is supported by the evidence that synchronization over larger distances within an area (> 2 mm) or between areas is in most cases associated with an oscillatory patterning of the local discharges (König et al. 1992a). However, it is also conceivable that oscillations occur as a consequence of synchrony. Simulation studies indicate that networks with excitatory and inhibitory feedback have the tendency to converge toward states where discharges of local cell clusters become synchronous (Sporns et al. 1991; Koch and Schuster 1992; Deppisch et al. 1992). Once such a synchronous voley has been generated, the network is likely to engage in oscillatory activity. Because of recurrent inhibition and because of Ca^{2+}-activated K^+ conductances (Llinás 1988a, 1990b), the cells that had emitted a synchronous discharge will also become simultaneously silent. On fading of these inhibitory events, firing probability will increase simultaneously for all cells and this, together with maintained excitatory input and nonlinear voltage-gated membrane conductances such as the low threshold Ca^{2+} channels (Llinás 1990b) will favor the occurrence of the next synchronous burst, and so on. Thus, oscillations are a likely consequence of synchrony and it actually becomes an important issue to understand how cortical networks can be prevented from entering states of global oscillations and, if they do, how these can be terminated (see, e.g., Freeman and Skarda 1985). These issues have recently been addressed in a number of simulation studies (Hansel and Sompolinsky 1992; König and Schillen 1991; König et al. 1992; Schillen and König 1991, 1993; Sporns et al. 1991).

EVIDENCE FOR RESPONSE SYNCHRONIZATION IN NONVISUAL STRUCTURES

Most experiments have concentrated on the analysis of oscillatory activity, but those that used multiunit or field potential methods provide information not only on the occurrence of oscillations but also allow one to make inferences on response synchronization. Only few studies are available that explicitly address the question of response synchronization with multielectrode recordings in structures other than the visual cortex. Such data are now available for the somatosensory and motor cortex (Murthy and Fetz 1992), the acoustic and the frontal cortex (Aertsen et al. 1991; Vaadia et al. 1991; Ahissar et al. 1992), and the pigeon optic tectum (Neuenschwander and Varela 1993). In every case evidence has been obtained for transient interactions between simultaneously recorded neurons. As in the visual cortex these episodes of manifest interactions were usually of short duration. In the acoustic and the somatomotor cortex as well as in the optic tectum the interactions resembled those in the visual cortex (i.e., the cells synchronized their responses with zero phase lag). In the sensorimotor cortex synchronization has been tested between different cortical areas and found between the arm representations in the somatosensory and the motor cortex and even between the motor cortices of the two hemispheres. In the other areas, only within-area correlations were sought. Some indications are available that the observed episodes of coupled discharges and synchrony are correlated with behavior. Synchronization between units in somatosensory and motor cortex is particularly pronounced while the monkey tries to solve a difficult reaching task but vanishes once the task is learned and reaching is executed without difficulty (Murthy and Fetz 1992). Synchronization between units in the frontal cortex has been reported to occur in contiguity with certain behavioral sequences in a complex delayed matching to sample task (Aertsen et al. 1991). And, most importantly, a recent study showed in the acoustic cortex of awake monkeys that learning a stimulus-stimulus association transiently increases the coupling between nearby cells responsive to the two stimuli with the effect that their responses become more synchronous (Ahissar et al. 1992).

EEG and field potential recordings from humans and higher mammals have provided abundant evidence for γ-band oscillations in a variety of nonvisual cortical and subcortical structures (for further review of the extensive literature see Basar 1980; Basar and Bullock 1992; Gray 1993; Singer 1993). In the mammalian olfactory system, for example, 40–80 Hz oscillatory activity is evoked during the inspiratory phase in both the olfactory bulb and piriform cortex (Adrian 1942; Freeman 1975). This activity is synchronous over a scale of several millimeters both within and betwen the two structures (Freeman 1975; Bressler 1984, 1987; Freeman and Skarda 1985). The patterns of activity that emerge during these coherent states correspond to specific odors, the animal's past experience with the odors and their behavioral significance (Freeman and Skarda 1985). The oscillatory activity in itself is not thought to convey any specific information, rather it is viewed as a basic neuronal mechanism for establishing synchrony among large populations of coactive cells. As discussed above, these data can be taken as evidence for the occurrence of synchrony, at least among the local cluster of neurons contributing to the electrical field that is picked up by the macroelectrode.

Similar field oscillations have been recorded from a variety of neocortical areas in cats and monkeys. This activity was particularly prominent when the subjects were aroused and in a state of focused attention (Dumenko 1961; Rougeul et al. 1979; Spydell et al. 1979; Bouyer et al. 1981; Montaron et al. 1982; Sheer 1984, 1989; Ribary et al. 1991; Gaal et al. 1992; Murthy and Fetz 1992; Tiitinen et al. 1993). These rhythmic activities are synchronous over relatively large areas of cortex (Bouyer et al. 1981), in cases of the sensorimotor cortex they occur in phase with similar activity in the ventrobasal thalamus (Bouyer et al. 1981) and are regulated by dopaminergic input from the ventral tegmentum (Montaron et al. 1982). Oscillatory field potential and unit activities in the range of 20–40 Hz have been observed in the motor cortex of alert monkeys (Gaal et al. 1992; Murthy and Fetz 1992). These signals are synchronous over widespread areas of the motor cortical map within and between the two cerebral hemispheres, between the motor and somatosensory cortices, and are enhanced in amplitude when the animals are preparing new and complicated motor acts (Murthy and

Fetz 1992). The rhythms are suppressed, however, during the execution of trained movements (Donoghue and Sanes 1991).

Oscillatory components in the β- and γ-frequency ranges have also been extensively documented in humans. In several early studies depth recordings of the local EEG from a number of cortical and subcortical sites revealed pronounced episodes of synchronous rhythmic activity in relation to particular behavioral states (Sem-Jacobsen et al. 1956; Chatrian et al. 1960; Perez-Borja et al. 1961). Surface EEG and MEG recordings have revealed γ-frequency components in the auditory evoked potential (Galambos et al. 1981; Galambos and Makeig 1988; Basar 1988; Pantev et al. 1991) and showed a broad distribution over the entire cerebral mantle (Ribary et al. 1991). During a number of different behavioral states the hippocampus is known to exhibit some of the most robust forms of synchronous rhythmic activity to be observed in the central nervous system. Foremost among these is the theta-rhythm, a sinusoidal-like oscillation of neuronal activity at 4–10 Hz that occurs during active movement and alert immobility. Theta-field potentials are often synchronous between the two hemispheres and over distances extending up to 8 mm along the longitudinal axis of the hippocampus (Bland et al. 1975). Local populations of cells also exhibit a high degree of synchronous firing during theta-activity (Kuperstein et al. 1986). In addition, two other hippocampal neuronal rhythms have been discovered, one having a frequency near 40 Hz that occurs during a variety of behavioral states (Buzsaki et al. 1983) and is both locally and bilaterally synchronous. Another more recently discovered signal having a frequency around 200 Hz is associated with alert immobility and the presence of sharp waves in the hippocampal EEG (Buzsaki et al. 1992). These events have been termed population oscillations because single cells do not exhibit high frequency periodic firing. Rather they fire at low rate in synchrony with the surrounding population of cells, which in the composite yields a periodic structure that is synchronous over distances up to 1.2 mm (Buzsaki et al. 1992).

There is thus ample evidence from brain structures other than the visual cortex that groups of cells engage in synchronous rhythmic activity in the γ-frequency range. The fact that this activity occurs in the awake brain and increases with attention and preparation of motor acts suggests that it is functionally relevant. An early Russian study described phase coherence for intracortically recorded field oscillations in the γ-band between visual, acoustic, and motor cortex of dogs. This coherence developed in a highly selective way in the course of Pavlovian conditioning while the dog acquired sensorimotor reactions to visual or acoustic stimuli (Dumenko 1961). When the dog was conditioned to withdraw the front paw after a visual warning stimulus, coherent oscillations became manifest over the motor representation of the front paw and the visual cortex. When the withdrawal reflex was shifted to the hind paw and the warning stimulus to a tone, coherent oscillations appeared over the acoustic cortex and the hind-paw representation.

EVIDENCE FOR β- AND γ-BAND OSCILLATIONS IN UNIT RECORDINGS FROM VISUAL AND NONVISUAL STRUCTURES

In contrast to the numerous field potential studies that have disclosed the presence of β- and γ-oscillations in many different cortical and subcortical areas, single unit analyses designed specifically for the search of such oscillatory discharge patterns have often failed to confirm their presence or have led to controversial results: So far, all investigators agree that oscillating unit activity in the γ-range occurs in the primary visual cortex of cats and monkeys, whether anesthetized or awake (Gray and Singer 1987, 1989; Eckhorn et al. 1988, 1992; Raether et al. 1989; Ghose and Freeman 1990, 1992; Livingston 1991; Schwarz and Bolz 1991; Jagadeesh et al. 1992; Gray and Viana di Prisco 1993), in cat area 18 (Eckhorn et al. 1988, 1992; Nelson et al. 1992a), and in area PMLS of cat visual cortex (Engel et al. 1991a). For area MT(V5) of the visual cortex of awake monkeys one positive report (Kreiter and Singer 1992) stands against two negative findings (Young et al. 1992; Bair et al. 1992). No evidence was found in temporal visual areas of the monkey (Tovee and Rolls 1992) but Nakamura et al. (1992) observed both low- and high-frequency oscillations associated with a recognition task in the temporal pole of *Macaca mulatta*. High-frequency oscillations in single cell activity have also been observed in somatosensory cortex where they were suppressed during sensory stimulation (Ahissar and Vaadia 1990), in the frontal cortex where they occurred in relation to preparatory phases of motion (Murthy and Fetz 1992; Donoghue and Sanes 1991; Gaal et al. 1992), and in prefrontal cortex where they were associated with particular behavioral sequences (Aertsen et al. 1991). Finally, synchronous multiunit responses very similar to those occurring in the cat visual cortex have been observed in the optic tectum of awake pigeons (Neuenschwander and Varela 1990). Thus, single unit data from a number of different labs agree with the evidence from field potential and EEG recordings that oscillatory activity in the γ-range is a common phenomenon in certain brain structures but there are also several negative findings that challenge this view. A possible reason for this discrepancy between single cell and field potential data is that single cell recordings are not well adapted to disclose the participation of a cell in an oscillatory process if oscillation frequency is high and irregular and the cell's discharge rate low (see above).

THE DURATION OF COHERENT STATES

It has been argued that synchronous oscillatory activity is unlikely to serve a function in visual processing because the time required to establish and to evaluate synchrony would be incompatible with the short recognition times common in visual perception (Tovee and Rolls 1992). The following considerations suggest that time constraints may not be that critical even if synchrony is used as a code. In a study by Gray et al. (1992a) recordings of field potential and unit activity were performed at two sites in cat visual cortex having a separation of at least 4 mm. Field potential responses were chosen for analysis in which the signals displayed particularly robust oscillations, a close correlation to the simultaneously recorded unit activity, and a statistically significant average cross-correlation. Under these conditions it became possible to determine (1) the onset latency of the synchronous activity; (2) the time-dependent changes in phase, frequency, and duration of the synchronous episodes within individual trials; and (3) the intertrial variation in each of these parameters.

The results, combined with previous observations (Engel et al. 1990), demonstrated that correlated responses in cat visual cortex exhibit a high degree of dynamic variability. The amplitude, frequency, and phase of the synchronous events vary over time. The onset of synchrony is variable and bears no fixed relation to the stimulus. Multiple epochs of synchrony can occur on individual trials and the duration of these events also fluctuates from one stimulus presentation to the next. Most importantly, the results demonstrated that response synchronization can be established within 50–100 msec, a time scale consistent with behavioral performance on visual discrimination tasks (Gray et al. 1992a).

Similar, rapid fluctuations between synchronous and asynchronous states have been observed in other systems, and recent methodological developments have made a quantitative assessment of these rapid changes possible. Using the joint-PSTH (Aertsen et al. 1989) and gravitational clustering algorithms (Gerstein et al. 1985; Gerstein and Aertsen 1985) it has been possible to examine the time course of correlated firing among pairs and larger groups of neurons, respectively (Aertsen et al. 1991). These findings clearly indicate that the formation of coherently active cell assemblies is a dynamic process. Patterns of synchronous firing can emerge from seemingly nonorganized activity within tens of milliseconds, and can change as a function of stimulus and task conditions within similarly short time intervals. These findings suggest that the temporal constraint imposed by perceptual performance can be met by the dynamic processes that underlie the organization of synchronously active cell assemblies.

Theoretical considerations point in the same direction. Assemblies defined by synchronous discharges need not oscillate at a constant frequency over prolonged periods of time. Rather, it is likely that neuronal networks that have been shaped extensively by prior learning processes can settle very rapidly into a coherent state when the patterns of afferent sensory activity match with the architecture of the weighted connections in the network. Such a good match can be expected to occur for familiar patterns that during previous learning processes had the opportunity to mould the architecture of connections and to optimize the fit. If what matters for the nervous system is the simultaneity of discharges in large arrays of neurons, a single episode of synchronous discharges in thousands of distributed neurons may actually be sufficient for recognition. Obviously, the nervous system can evaluate and attribute significance to coherent activity even if the synchronous event is only of short duration and not repeated because its parallel organization allows for simultaneous assessment of highly distributed activity.

Especially if no further ambiguities have to be resolved, or if no further modifications of synaptic connectivity are required, it would actually be advantageous if the system would not enter into prolonged cycles of reverberation after having converged toward an organized state of synchrony. Rather, established assemblies should be erased by active desynchronization as soon as possible to allow for the build-up of new representations. Thus, when processing highly familiar patterns or executing well-trained motor acts that raise no combinatorial problem the system would function nearly as fast as a simple feedforward network. Activity will be routed selectively through the network of tuned connections and a pattern of simultaneous discharges could emerge in the corresponding assembly of distributed cells with latencies that are only a little longer than the sum of the conduction and integration delays along the path of excitation. Convergence toward coherent states and hence "recognition" might actually occur even faster than one might predict from models that assume that retinal signals are relayed serially from one cortical area to the next and that recognition occurs only once cells in areas at the top of the processing hierarchy get driven.

In the present model it is assumed that the interconnected cortical areas are permanently active, collectively striving toward coherent states. In this case the role of retinal signals is not to provide the energy for the successive excitation of serially connected cells but to select paths of convergence toward coherent states by shifting the time of occurrence of discharges, most of which would have occurred anyway. The differential and flexible routing of activity that is required to organize the appropriate assemblies could thus be achieved in parallel within and among the different processing areas and within only a few reentrant cycles. Because the network is assumed to be engaged in an active exchange of signals even if "at rest" and because during the organization phase interactions only need to influence firing probability, excitatory postsynaptic potentials can effectively contribute to the organization without having to summate until they evoke strong discharges. Thus, not much time has to be reserved for temporal summation and hence the duration of reentrant cycles can be as short as the sum of the net conduction and synaptic transmission times. This possibility of rapid convergence toward coherent states and the option to maintain such states only for short durations is fully compatible with the hypothesis that representations consist of large assemblies of coherently active neurons. But it may become very difficult to experimentally identify the coherent states of assemblies if these last only for very short periods of time and if only a few neurons can be recorded simultaneously. Thus, as long as experimenters can assess the activity of only a few neurons at a time, coherence will be detectable only if it is maintained over sufficiently long periods. Such is likely to be the case when ambiguities have to be resolved and when novel patterns have to be learned.

But why then do episodes of prolonged coherence occur in anesthetized preparations and in response to rather simple stimulus configurations. In this case there are no ambiguities to be resolved and there is no learning. At present one can only speculate. One possibility is that anesthesia decreases the efficiency of feedback loops and associative connections (see, e.g., Cauller and Kulics 1991b) and that this reduces the complexity of the system. In the absence of feedback, neurons at peripheral processing levels can organize their responses only according to the criteria set by local connections and this is likely to result in rather stereotyped repetition of "attempts" to organize. Moreover, it is likely that anesthesia abolishes also the processes that would normally terminate states of synchrony once the system has successfully converged toward a coherent state and "recognition" has occurred (see below). This interpretation agrees with the consistent observation that episodes of response synchronization are shorter and less stereotyped in awake behaving animals (Kreiter and Singer 1992; Gray and Viana di Prisco 1993).

EYE MOVEMENTS AND SELECTIVE ATTENTION

So far, only grouping operations that do not require scanning eye movements or shifts of selective attention have been considered. However, under normal viewing conditions both processes certainly contribute. While familiar scenes, even if they are complex, are usually perceived readily within a few hundred milliseconds, recognition of unfamiliar objects or analysis of scenic details does require more time. Often it is even necessary to successively sample parts of the pattern by scanning eye movements. This requires that representations of components of the pattern are maintained (remembered) during successive eye movements to allow for the synthesis of successively perceived components. But it also requires that patterns at the more peripheral levels of analysis change from eye movement to eye movement. Thus, the possibility needs to be considered that the temporal scales at which activity patterns become organized differ at different levels of integration. At the levels directly involved in the segmentation of scenes and the appropriate grouping of features, organized states should last only briefly and they should definitely not outlast the duration of the retinal image. Moreover, they should be reset with each eye movement to reduce confusion between successive, often unrelated images. Both postulates agree with psychophysical evidence.

Patterns that are too complex to be remembered at a semantic level or that defy semantic description remain represented only little more than 100 msec after their offset. If interrupted for more than 150 msec, the approximate duration of visual persistence, modifications of the patterns go undetected when introduced during the interruption interval (Phillips and Singer 1974; Di Lollo and Wilson 1978). Likewise, saccadic eye movements also disrupt the ability to detect changes if these are introduced while the eyes move. Thus, disrupting the continuous flow of retinal activity by transiently obscuring the pattern or by making a saccadic eye movement appears to erase completely the organized state induced by the pattern that was present prior to the interruption. In case of eye movements it actually appears as if this resetting is an active process. The fact that the pontogeniculooccipital waves (PGO waves) that accompany saccadic eye movements are prominent in the lateral geniculate nucleus and in visual areas of the occipital lobe but not in more frontally located areas is compatible with such a view.

It has been suggested that the PGO waves or eye movement potentials reflect corollary activity that is generated in the brain stem in association with saccadic eye movements and serve to erase or reset activation patterns in peripheral visual centers each time an eye movement is executed (for review see Singer 1977, 1979). In the framework of the present model this resetting would have to consist of disrupting assemblies that have become organized in response to the pattern processed prior to the saccade. Hence this resetting should act by decorrelating previously synchronized activity. Evidence is indeed available that the activity that underlies PGO waves does have a desynchronizing effect both at the thalamic and the cortical level (for review see Singer 1977, 1979; Steriade and McCarley 1990). In addition it raises cortical excitability (for review see Singer 1979; Steriade 1991), which should in turn favor rapid self-organization of new assemblies in response to the pattern that is going to be processed after the saccade has occurred. In higher areas, by contrast, which are located more frontally and perform much more abstracted, semantic descriptions of patterns, such automatic eye movement-related resetting must not occur. Here, the time frames for the organization of assemblies should be set in a flexible way and adjusted according to the actual time it takes to organize coherent assemblies. If oscillatory activity is instrumental to establish coherence among distributed cells and to organize assemblies, one might then also expect that the frequency of these oscillations could be much slower in these higher areas and perhaps even varied as a function of the actually required integration times.

SYNCHRONIZATION AND ATTENTION

The hypothesis that information about feature constellations is contained in the temporal relation between the discharges of distributed neurons, and, in particular, in their synchrony, has also some bearing on the organization of attentional mechanisms. It is obvious that synchronous activity will be more effective in driving cells at higher levels than nonorganized asynchronous discharges.

Thus, those assemblies would appear as particularly salient and hence effective in attracting attention that succeed to make their discharges coherent with shorter latency and higher temporal precision than others. Conversely, responses of neurons reacting to features that cannot be grouped or bound successfully, and, hence, cannot be synchronized with the responses of other neurons, would have only a small chance of being relayed further and to influence shifts of selective attention. It is thus conceivable that of the many responses that occur at peripheral stages of visual processing only a few are actually passed on toward higher levels. These would either be responses to particularly salient stimuli causing strong and simultaneous discharges in a sufficient number of neurons or responses of cells that succeeded in being organized in sufficiently coherent assemblies. Thus, responses to changes in stimulus configuration or to moving targets have a good chance to be passed on even without getting organized internally because they would be synchronized by the external event. But responses to stationary patterns will require organization through internal synchronization mechanisms to be propagated.

This interpretation implies that neuronal responses that attract attention and gain control over behavior should differ from nonattended responses not so much because they are stronger but because they are better synchronized among one another. A neuronal network model using synchronization rather than rate modulation of discharges as a code for saliency in attentional processes has recently been realized by Niebur et al. (1993). Following the same reasoning, shifting attention by top-down processes would be equivalent with biasing synchronization probability of neurons at lower levels by feedback connections from higher levels. These top-down influences could favor the emergence of coherent states in selected subpopulations of neurons—the neurons that respond to contours of an "attended" object or pattern. Thus, the mechanism that allows for grouping and scene segmentation—the organization of synchrony—could also serve the management of attention. The advantage would be that nonattended signals do not have to be suppressed, which would hitherto eliminate them from competition for attention. Rather, cells could remain active and thus be rapidly recruitable into an assembly if changes of afferent activity or of feedback signals modify the balance among neurons competing for the formation of synchronous assemblies.

In a similar way shifts of attention across different modalities could be achieved by enhancing selectively synchronization probability in particular sensory areas and not in others. This could be achieved, for example, by modulatory input from the basal forebrain or nonspecific thalamic nuclei. If these projection systems were able to modulate in synchrony the excitability of cortical neurons distributed in different areas this would greatly enhance the probability that these neurons link selectively with each other and join into coherent activity. Such linking would be equivalent with the binding of the features represented in the respective cortical areas. Again, this view equates grouping or binding mechanisms with attentional mechanisms. The "attention" directing systems would simply have to provide a temporal frame within which distributed responses can then self-organize toward coherent states through the network of selective corticocortical connections. In doing so the attentional systems need not themselves produce responses in cortical neurons. It would be sufficient that they cause a synchronous modulation of their excitability. It is conceivable that the synchronous field potential oscillations that have been observed in animals and humans during states of focused attention are the reflection of such an attention mechanism (for review of the extensive literature see Singer 1993). The observations that these field potential oscillations are only loosely related to the discharge probability of individual neurons, are coherent across different cortical areas, are particularly pronounced when the subjects are busy with tasks requiring integration of activity across different cortical areas and stop immediately when the binding problem is solved—as witnessed by the execution of a well-programmed motor act—are in agreement with such an interpretation.

SUMMARY

In this section a scenario of cortical processes is developed in which response synchronization is used for scene segmentation, perceptual grouping, and the organization of sensory representations. The essential ingredients of this model are depicted schematically in figure 10.4. The different boxes stand for some of the numerous cortical areas devoted to the processing of retinal signals. The arrows between them symbolize the possibility of a reciprocal flow of signals between areas at similar and different levels of the processing hierarchy. For a detailed description of the connectivity pattern between different visual areas the reader is referred to Felleman and Van Essen (1991) and Young (1992). On presentation of a complex visual scene the following sequence of events is assumed to occur. Neurons in V1 that encounter a preferred feature in their receptive field start responding. At the very same time these responses become organized due to the action of the tangential connections within V1. Because of the specific architecture of these connections neurons coactivated by continuous contours or nearby contours with similar orientation, or neurons activated by colinear contour segments will tend to synchronize their activity. While this organization proceeds in V1 signals are passed on to other areas where similar organization processes are initiated. Of the many responses in V1 those that became synchronized best will be particularly effective in influencing neurons in higher areas. Therefore, response constellations that fit the grouping criteria set by the architecture of tangential connections in V1 will be passed on and processed further with higher probability than incoherent responses that also arrive from V1. Because the connections from V1 to the other areas convey already preprocessed activity and by divergence and convergence allow for remapping of neighborhood relations, the grouping criteria in these higher areas should differ from those in V1. Thus, it is assumed that in V5 those neurons have a tendency to synchronize their responses which code for the same direction of motion. Because the neurons in V5 have large receptive fields, and hence a great aperture, and are also sensitive to relative motion, this area can evaluate coherent motion both in relative and absolute terms over large distances.

While the responses in V5 become organized according to the grouping criteria set by the intrinsic interactions within V5 it is assumed that they influence via the backprojections the organization process in V1, adding the criterion of coherent motion to the grouping process in V1. This top-down influence is thought to bias synchronization probability between neurons in V1 either toward more or toward less synchrony, depending on stimulus configuration. Responses to contour elements that are far apart and have different orientations have a low probability of becoming synchronized by local interactions within V1. However, if these contour elements move coherently their coherence would be detected by neurons in V5 responses to these contours would synchronize in V4 and through the backprojections increase synchronization probability for the respective set of neurons in V1. Such top-down influences from motion-sensitive areas with large aperture could account for the observation that coherently moving line segments lead to synchronization of responses in area 17 even if the cortical representations of these line segments are much further apart than the maximal span of the tangential intracortical connections (see, for example, Gray et al. 1989). The finding that pharmacological inactivation of cells in motion-sensitive areas reduces considerably response synchronization to coherently moving contours in V1 supports this possibility (Nelson et al. 1992b). Conversely, responses to nearby contours of similar orientation that would have a tendency to become synchronized due to the local interactions in V1 may be prevented from synchronizing by top-down influences from motion-sensitive areas if the contours move in different directions and with different speed. Such differences in motion trajectories have been shown to prevent neurons in motion-sensitive areas from synchronizing (Kreiter and Singer 1992), and hence, activity in the backprojections would either not favor the occurrence of synchrony in V1 or even actively reduce its probability. Similar grouping operations are assumed to occur simultaneously in numerous other prestriate areas but according to different criteria.

Thus, while one area explores similarities in color space, another may search for related textures, and yet another for similarities in retinal disparity etc. The results of these evaluations, which can all occur in parallel, are sent back to V1 where they all contribute to the ongoing organization process. As a consequence, the synchronization probabilities among neurons in area 17 change and this in turn modifies the input configurations to prestriate areas. While the distributed search for the most probable grouping constellations proceeds, areas at the top of the processing hierarchy will also become involved. Because of the polysynaptic nature of the input to these areas they will probably become active only once activity in the preceding areas has become sufficiently coherent, but then responses should be organized in the higher areas according to the same general rules as at peripheral levels given the similarity of the intrinsic organization of the different cortical areas. The grouping criteria will be much more complex at these higher levels, however, because interactions now involve neurons that represent complicated constellations of features such as figural components and higher order geometric shapes (Tanaka et al. 1991; Gallant et al. 1993). Because these higher areas are connected to lower areas via massive backprojections, it must be assumed that once coherent patterns became organized at higher levels these influence in turn the organization of patterns at lower levels. These processes can all occur nearly simultaneously as the areas concerned are all interconnected either directly or via oligosynaptic pathways. Thus, the process of organizing the neuronal representation of a scene consists of parallel operations that occur nearly simultaneously at different levels of the processing hierarchy and according to similar rules. But because of differences in the way in which ascending activity from V1 is mapped into different areas, the evaluation criteria differ for each area and increase in complexity as one moves away from V1. In this model decisions required for successful perceptual grouping and scene segmentation are thus based on a highly distributed voting operation where each of the different areas contributes its "point of view" and where both bottom-up and top-down processes are intimately interleaved.

Each of the areas explores the feature space for which it is predisposed by its specific afferent and intrinsic connectivity, searches for coherence, and distributes the result of its computation simultaneously to all the areas to which it is connected. These messages are assumed to bias the probabilities with which neurons in the respective target areas are going to synchronize or desynchronize their discharges.

Successful segmentation could thus be viewed as the result of a self-organizing process that converges toward the state of maximal probability. If scenes contain little ambiguity with respect to the grouping criteria that are stored in the architecture of connections within and between areas, the organization process can be very rapid and in extreme cases it may not even require the contribution of backprojected activity. This could even be true for complex scenes if they contain mainly familiar objects. In this case the pattern of sensory activity would match directly with the functional architecture of coupling connections that has been shaped by previous learning and the system can converge nearly instantaneously into a coherent state. Under such conditions the system would function in a way that is not too different from a multilayered feedforward network. However, if the scene contains ambiguities allowing for several equally likely groupings or if it is highly unfamiliar, convergence may occur only after seconds. Such extreme processing times may actually be required for the segmentation of figures defined solely by similar disparity in random dot patterns or for the detection of figures hidden in background textures by camouflage as for example the well-known Dalmatian dog. In both cases it is helpful and reduces recognition time if one already knows what the figure is, a pragmatic proof of the notion that high level representations can directly influence figure-ground segmentation via top-down biasing of peripheral grouping criteria. In case of the Dalmatian dog, for example, recognition could be sped up either by top-down propagation if previous experience has already installed grouping criteria at the levels where figural attributes are bound together or if one provided additional cues that would facilitate grouping by bottom-up processes at peripheral levels. If the contour elements constituting the dog had any of the properties in common which V1 and prestriate areas can probably evaluate and relate to one another, such as disparity, color, motion, orientation, and texture, segmentation would occur much faster.

In this scenario a pattern is perceived as soon as segmentation is completed and neurons have become organized in distinct, coherently active assemblies. In that case, their output will be sufficiently coherent to allow for the propagation of signals to remote cortical areas and ultimately to effector levels. For this to occur it is necessary not only that enough cells coordinate their responses, but also that the spatial distribution of these coherently active cells matches the "receptive field" properties of cells at higher levels. Just as cells in V1 are selective for particular spatiotemporal patterns of retinal input, cells in higher cortical areas are likely to be activatable only by the appropriate spatiotemporal patterns that have organized in more peripheral cortical areas. But in contrast to the retinal and thalamic activation patterns, these cortical activation patterns are no longer a direct reflection of the retinal image but a result of a highly dynamic self-organizing process. The organization of the spatial and temporal structure of these patterns is initiated by the retinal input, but then it is extensively modified by dynamic interactions that are determined essentially by the functional architecture of connections linking cells within and between areas. The proposal is that this organization process converges toward coherent states in which responses that need to be related to one another are tagged by their synchrony.

Following the same line of reasoning it is also possible that access to the level of processing where representations reach consciousness is gated by coherence. As proposed by Crick and Koch (1990a) it is conceivable that only those activation patterns (assemblies) reach the threshold of conscious awareness that are sufficiently organized, that is coherent.

ACKNOWLEDGMENTS

I wish to thank Andreas Engel, Peter König, Andreas Kreiter, Siegrid Löwel, Peter Roelfsema, and Thomas Schillen for the many fruitful discussions that helped to clarify the issues addressed in this chapter, Christof Koch for constructive amendments to the manuscript, Renate Ruhl for preparing the figures, and Irmi Pipacs for editing the paper.

11 What Form Should a Cortical Theory Take?

Charles F. Stevens

Although neurobiology has accumulated an impressive body of information about neocortical structure and operation, the nature of the mathematical computations performed by cortex remains a mystery. A description of these computations is equivalent to developing a theory of cortical function, and the construction of such a theory is necessarily one of neurobiology's central problems. Although this chapter deals with cortical theory, its goal is much less ambitious than proposing what cortex computes. Rather it attempts one of the first steps in that direction: to outline what form a cortical theory should take.

Why try to define the form for such a theory when we surely are a long way from being able to develop an adequate theoretical structure? A strong form of the argument for this approach is as follows: A particular cortical region—primary visual cortex, for instance—performs some computations on its inputs to determine what information is sent to other areas. The types of computations one tends to think of are Fourier or Gabor transforms, calculation of cross-correlation functions, or deconvolutions, but the actual computations may not be ones that are currently familiar. As a prerequisite for understanding the role of a particular region in the overall cortical processing of information, then, we must identify the computations carried out by that region. And before these computations can be recognized, we must decide what sort of mathematical machinery is to be used for their characterization. For example, should we make a probabilistic description, or is a deterministic one adequate? Identifying the nature of the theory we seek is essential, because this determines, to a great extent, what sort of experiments are needed and what further theoretical approaches should be tried.

FOUR REQUIREMENTS FOR A THEORY OF CORTEX

The first step in defining the nature of a cortical theory is to identify some of the theory's requirements. Here four requirements for a theory of cortex are proposed and discussed.

Before a complete cortical theory could be developed, one must know: How many different types of cortex are there? That is, how many theories are required? Clearly, very many functionally distinct cortical regions can be recognized, over 30 in the visual system alone (Felleman and Van Essen, 1991). But functionally distinct areas may not be computationally unique. For example, the functional differences between areas (e.g., V1 vs. MT) might reside more in the nature and distribution of cortical inputs and on the disposition of cortical outputs than in the operations performed by the cortical circuits on the information they receive. Specifically, functionally distinct cortical regions, like V1 and MT, might perform identical mathematical operations on different sorts of inputs. This general notion has been proposed repeatedly (Lorente de Nó 1949; Creutzfeldt 1977; Powell 1981; Eccles 1984), and is supported by a variety of developmental, anatomical, and physiological observations.

If "type of cortex" refers to the mathematical computation performed, rather than to the some functional difference between areas revealed, say, by different receptive field structures, one might believe that only a single major computational type of cortex exists; this can be called the "unitary theory." Alternatively, activity-dependent rewiring of cortical circuits (see, for example, Shatz 1990) could modify the computations performed by even initially uniform cortices, so that the character of some mathematical operation might vary continuously across even an apparently uniform cortical region like primary visual cortex; this other limiting case could be termed the "continuous diversity theory." Most hopeful for neurobiology is the unitary theory: if this were true, understanding the basic computations carried out by any cortical area would then provide an answer for all of cortex. Unraveling cortical function in this limit would simply (!) amount to learning how inputs and outputs are mapped.

The *first requirement* for the framework of a cortical theory, then, is that it must be able to accommodate the spectrum of possibilities, from the unitary to the continuous diversity views.

How many inputs and outputs are present in cortex? The answer depends on how many classes of neurons are present. Currently available information (see, for example, Purves et al. 1992) indicates that the number of input and output types should be greater than one, but not a large number, perhaps 10 to 100. The cortical inputs and outputs must thus be described by vectors whose dimensions are to be determined experimentally. Note that inputs include information sent to one cortical region from another one and that outputs are defined in any convenient way. At this stage, a cortical theory must be able to accommodate an arbitrary number of inputs and outputs. This constitutes an additional part of the first requirement.

The *second requirement* is that the theory be an explicitly probabilistic one because synaptic transmission is a stochastic process (Katz 1969): Neurotransmitter is released at axon terminals in packets—called *quanta*—so that the total effect of a nerve impulse arrival is an integral multiple of the smallest effect, the one produced by a single quantum. The quanta, however, are released probabilistically according to a Poisson process (see Barrett and Stevens 1972) with a Poisson rate $\lambda(t)$ that depends on time t. Normally, $\lambda(t)$ is very small, but just after a nerve impulse arrives at the synapse, the release rate increases transiently. The net effect is that the size of the postsynaptic response due to the arrival of a nerve impulse at an axon terminal varies at random. The specific need for a probabilistic theory in brain arises from the following considerations.

The high signal-to-noise ratio of whole cell recording and the use of methods that cause localized release of neurotransmitter have permitted the characteristics of individual quanta to be determined for central neurons (Bekkers and Stevens 1989; Edwards et al. 1990; Bekkers et al. 1990; Manabe et al. 1992; Raastad et al. 1992; Silver et al. 1992). Further, this knowledge of quantal size, and its variation, has made possible a rigorous quantal analysis of central synapses (Bekkers and Stevens 1989, 1990). The conclusion of these investigations is that the probability of release at an individual synapse is generally very low, about 0.1 to 0.5.

Sometimes the number of synapses one neuron makes with another can be determined: in cortex, any particular neuron generally seems to receive only one or two synapses from any other neuron. In hippocampus, for example, Andersen (1990) estimated that a given axon usually makes only a single synapse (average estimated to be 1.3) on its target cell. The lateral geniculate axons that project to visual cortex also make only one or a few (up to about eight) synapses on their targets (Freund et al. 1985). Taken together, then, these observations indicate that when a pair of cells is connected, the communication link between them is quite unreliable for a single impulse arrival, although it is predictable in a statistical sense.

Experimental confirmation of this conclusion is available for hippocampus and primary visual cortex. When the intensity of a stimulus applied to axons that project onto CA1 neurons is reduced to low levels—intensities perhaps adequate for stimulating just a single axon — only a small fraction of stimuli (about 0.1 to 0.5) produces postsynaptic currents, that is, about five out of ten of the nerve impulses generated by a cell produce no postsynaptic response at a given target neuron. These "minimal" stimuli may actually stimulate more than one axon, and each axon may make more than one synapse with its target cell. In any event, two conclusions are secure for hippocampus: (1) the release probability of about 0.5 estimated in this way is an upper limit for the actual release probability and (2) the effect of one neuron on another is generally small and uncertain. This conclusion is supported in a general way by the correlational analysis by Tanaka (1983), which shows that even when geniculate and V1 receptive fields overlap, the extent to which a geniculate discharge predicts a cortical neuronal discharge is only about 0.1. Note that although the Tanaka result is difficult to connect directly to our argument because of the complexity of his experimental situation and the indirect nature of his measures of cell-to-cell connections, his observations do support the notion that a single input has only a relatively small effect on its target. Because transmission at a single synapse is so unreliable, and because each neuron makes only a few synapses with its target cell, neuron to neuron communication must also be quite unreliable: the random nature of synaptic transmission makes neuronal behavior uncertain and thus networks of such neurons must be described probabilistically. These statements apply, of course, to many, but probably not all, cortical neurons. Examples are known (Purkinje cells in the cerebellum, for example) in which one neuron makes thousands of synapses on its target cell, and statistical fluctuations in synaptic strength are very small.

Now we begin the background discussion for the third requirement, which is, we shall argue, that the theory should be continuous and should treat the cortex in a coarse-grained way. One might think that, ideally, a cortical theory should start from details of neuronal properties and the principles that determine connections of cortical circuitry, and then derive from these a description of the computations performed by each neuron and thus by the network as a whole. But such a cell-by-cell description is probably neither feasible nor desirable. A cubic millimeter of cortex—a good candidate for the size of a computational unit—contains 10^5 neurons, 10^9 synapses, and two miles of axons; each neuron receives about 10^4 synapses and communicates with about 10^4 other neurons (see White 1989; Stevens 1989; Braitenberg and Schüz 1991). Most of these connections are intracortical (Peters 1987). Furthermore, the average effect one cortical neuron has on another is quite small, ranging from less than 50 μV to several millivolts (Thompson et al. 1988; Mason et al. 1991). Because single neurons have small and uncertain effects on other neurons, the cortical description must be carried out in terms of neuronal populations rather than at the level of individual cells.

A consideration of cortical anatomy points to the nature of the neuronal populations that form the natural basis for a cortical theory. The argument will be made in terms of cat primary visual cortex layer 4, but the same conclusions are reached for any cortical region. Layer 4 neurons have a dendritic tree with a diameter of about 0.3 mm (Martin 1984). Pick a particular neuron as a reference and ask: how many layer 4 cells have dendritic trees that overlap that of the reference cell and thus potentially have access to the reference cell's synaptic input? Layer 4 is about 0.3 mm thick and cortex has a density of about 10^5 neurons per mm^3 (Powell 1981). All of the neurons in layer 4 that fall within a cylinder with a radius of about 0.3 mm will have overlapping dendritic trees. The number of neurons that overlap with the reference cell is thus $\pi(0.3)^2(0.3)(mm^3)$ times $(10^5)(cells/mm^3)$, or approximately 8000 neurons; within this population, a number of distinct neuronal types might be found. Each type could have different input patterns, but still a significant fraction of the population should represent essentially the same information.

In addition to the fact that overlapping dendritic trees tends to define equivalence classes of neurons (here, an equivalence class would be all of the neurons that would potentially have anatomical access to a particular set of axon terminals), a given axon generally arborizes over a considerable region of cortex with an arbor diameter of perhaps 0.5 mm (Martin 1984), and forms about 2000 boutons, each of which makes one or two synapses

(Freund et al. 1989). Thus, neurons of the same functional class and in the same cortical layer share nearly the same potential synaptic inputs whenever their cell bodies are separated by several hundred microns or less, and the degree of similarity in their inputs increases as the distance between cell bodies decreases. In cat primary visual cortex, about one third of the neurons whose receptive field overlaps with that of a particular geniculate neuron receive input from the geniculate cell (Tanaka 1983).

Altogether, these observations—together with the stochastic nature of neuronal behavior—suggest that the physiologically meaningful signal from cortex should be the average firing rates of a population of perhaps 100 to 1000 neurons near a particular cortical site. The behavior of cortex at a particular point would then be described by the firing in a population of neurons. The total firing that represents this population would be determined by a weighted average of the appropriate neurons in the cortical region that surrounded the point, perhaps with weights that are described by a spatial Gaussian. As one moved from one cortical location to an adjacent one, the neuronal population whose firing defined the state of the new cortical point would overlap with the previous population so that the variables describing cortical state would vary continuously with cortical position. Features of cortical structure such as ocular dominance columns are treated with a straight forward extensions of these notions in which the cortex is viewed as interleaved continuous regions.

The *third requirement* for a theory of cortex, then, is that it must be coarse-grained and treat cortical inputs and outputs as continuous variables that represent the summed behavior of appropriately sized and selected neuron populations.

Although individual neurons behave probabilistically, if the population of cells needed in this coarse-grained description were sufficiently large, a deterministic description would suffice. Indeed, deterministic theories probably will be adequate for many purposes, like the Hartline-Ratliff equation described below. But in certain situations—the treatment of activity-dependent modification of neuronal circuit connections discussed later, for example—the stochastic nature of brain operation will have to be treated explicitly. Furthermore, some of the essential calculations made by cortex, like the computation of cross-correlation functions, may well turn out to require a probabilistic description.

Finally, the prominent recurrent nature of lateral intracortical connections and relatively wide spatial distribution of cortical inputs mean that the cortical output at any one location must depend on both the input and output over relatively great expanses of cortex (for example, Gilbert and Wiesel 1979). That is, the output at any one point must be a functional of both inputs and outputs (for a brief description of functionals and the relevant literature, see Stevens 1987). This is the *fourth requirement*.

What are the chances that these are the right four requirements for a start on a theory for cortex? Not great, of course, but little explicit attention has been given to the types of theories we should attempt, and this issue seems to be an important one: the requirements selected should be examined and debated, alternatives explored, and questions raised should be addressed by experiments. If, for example, theories that use continuous mathematics are unsuitable, that would be important information.

SIMPLEST VERSION OF THE GENERAL APPROACH

According to the requirements outlined above, the goal of a cortical theory is to develop an equation, using continuous mathematics, for the probability functional of cortical outputs given the inputs. For simplicity in this initial description, the cortex will be considered to be one-dimensional, with cortical position specified by the variable x; the temporal coordinate will be suppressed. Further, only a single input $s(x)$ and a single output $f(x)$ will be used for this introductory treatment. The input and output at position x represent the firing rates of input and output neurons averaged over a small cortical volume. What is needed to describe this simple cortex, then, is the functional $P[f(x); s(x)]$ that specifies the probability of finding the output $f(x)$ (note that this function represents the entire output of the cortex) given that the input of the cortex is described by the function $s(x)$. This sort of formulation meets the four requirements: The macroscopic nature of the theory comes in the coarse-grained treatment used to define input and output, the inputs and outputs are treated as continuous functions, the requirement for a probabilistic formulation is explicit, and the effects of lateral connections reside in the fact that P is a functional.

A starting point for a useful description of cortex is the identity

$$P[f(x); s(x)] = e^{-S[f; s]}$$

where $S[f, s] = -\ln(P[f(x); s(x)])$, by definition. The motivation for this definition is as follows: Under some circumstances, for example, an input $s(x)$ that varies only very slightly across the cortex, a functional power series expansion in the arguments $f(x)$ and $s(x)$ should approach an accurate representation of cortical function. By transforming P is this way, we define a slowly varying functional that is more amenable to a power series approach; this is the case we want in the limit of inputs that are sufficiently close to a constant.

The functional S contains a complete description of the cortical operation. Cortical processing, however, occurs at specific locations, so S needs to be recast in a form that makes the local nature of the description explicit. Expand S in a functional power series (Volterra 1959)

$$S[f; s] = A_0[s] + \int dx A_1[s; x] f(x) \, dx + \int \int dx dx' A_2[s; x, x'] f(x) f(x') + \cdots$$

where the $A_k[s, x, x'...]$ are functionals of $s(x)$ and functions of x. Now rearrange the terms so that

$$S[f; s] = \int L[f(x); s(x), x] \, dx$$

where L is defined as

$$L[f; s, x] = f(x)\left[A_1(x) + \int dx' A_2[s; x, x']f(x') + \cdots\right]$$

Because L provides a local specification of cortical function, it will be called the *cortical characterization functional*. Note that the A_0 functional has been excluded because this term would appear in the normalization of the probability functional. Now the probability functional that describes the cortex is just

$$P[f(x); s(x)] \sim e^{-\int L[f; s, x] dx}$$

and the job of a cortical theory is to identify the cortical characterization functional L. So far, of course, the only physical content has been the four initial requirements in addition to the notion of a formulation in terms of local cortical properties. The extent to which this sort of formulation is useful depends on developing ways to determine L.

THE SIMPLEST CORTICAL CHARACTERIZATION FUNCTIONAL

An Approximate Cortical Characterization Functional

The preceding formulation was designed for a power series expansion approach. The simplest way to get closer to L, then, is to expand it in a functional power series and discard the higher order terms. Because the functionals to be expanded depend on both $f(x)$ and $s(x)$, it is easier to keep things straight if $S[f; s]$ itself, rather than L, is expanded; L can then be recognized in the resulting expressions. Expand S to second order in both of its arguments:

$$S[f; s] = A_0$$
$$+ \int dx A_1(x)f(x) + \int dx A_2(x)s(x)$$
$$+\frac{1}{2} \int \int dx dx' K(x, x')f(x)f(x')$$
$$+\frac{1}{2} \int \int dx dx' M(x, x')f(x)s(x')$$
$$+\frac{1}{2} \int \int dx dx' A_3(x, x')s(x)s(x')$$

here the A_k, K, and M are functions that arise in the Volterra expansion. Each of the A_k terms vanishes: they contribute to the probability of an output $f(x)$ without depending on that function, so they are included in the normalization of the probability. The final expression, to second order, for the probability functional P is thus

$$P[f(x); s(x)] \sim e^{-\frac{1}{2} \int \int dx dx' f(x)[f(x')K(x-x')+s(x')M(x-x')]}$$

Note the additional assumption that the cortical circuitry is spatially uniform across the particular cortical region being treated so that the integrals are convolutions. The cortical characterization functional L can be identified in this approximation as

$$L[f; s, x] = \frac{f(x)}{2} \int dx' \left[f(x')K(x - x') + s(x')M(x - x') \right]$$

The equation for P defines the probability of any output for a given input; K and M are functions that arise in the Volterra expansion. In this approximation describing cortical operation involves discovering the form of two functions, K and M.

Experiments generally measure the average output to a particular stimulus, so an expression for the average response must be extracted from the equation to provide a link between theory and experiment.

Because probabilities are specified, the equation for P also describes the output noise for no (or constant) input. In this preliminary treatment the time dependence of inputs and outputs has been suppressed, so the resting output noise would be spatial variations in neuronal firing $f(x)$ predicted for an unstimulated cortex. The statistics of these resting output fluctuations refer to an hypothetical ensemble of identical corticies. The average response and resting fluctuations are considered in turn.

Average Response

The average response $\bar{f}(x)$ for a given $s(x)$ is found, by definition, from the functional integral

$$\bar{f}(x) = \int Df e^{-\int dx L} f(x)$$

Here Df is a volume element in function space and is related to the Wiener measure (see the appendix of Stevens 1987 and the references cited there). Although this functional integration can be carried out for the simple L produced by the power series approach, an easier way to the desired result—and one that works in a wider variety of situations—is to find not the average, but rather the most probable response. In this particular situation, the average and the most likely responses happen to be identical (because, as will be seen, the fluctuations are Gaussian). The most likely response is found by discovering the output $\hat{f}(x)$ that maximizes the $e^{-\int dx L}$ term, which is equivalent to finding the $\hat{f}(x)$ that minimizes $\int dx L$. The extremum is found from the Euler–Lagrange equation

$$\frac{\delta}{\delta f(\xi)} \int L[\hat{f}; s, x] dx = 0$$

The functional differentiation gives, for the simple L obtained above,

$$\frac{\delta}{\delta f(\xi)} \int dx L = \int dx \left[2/\hat{f}(x)K(\xi - x) + s(x)M(\xi - x) \right] = 0$$

so that

$$2/ \int dx \hat{f}(x)K(\xi - x) = - \int dx s(x)M(\xi - x)$$

This equation relates the most likely response $\hat{f}(x)$ to the cortical input $s(x)$; note that both the functions K and M, which arise in the Volterra expansion, are needed to determine the response, and also that expansion of L to second order gives a linear relation between input and average output.

Since the most likely response $\hat{f}$ is described by linear equations, the appropriate characterization of a cortex in this limit is the Green function, defined to be $H(x)$; this function would, of course, define the receptive field structure of sensory cortical neurons. The Green function satisfies the equation [because $s(x)$ would be taken to be a delta function]

$$\int dx H(x)K(\xi - x) = -M(\xi)$$

so that (take the Fourier transform)

$$H(x) = \mathcal{F}^{-1}\left\{\frac{\tilde{M}}{\tilde{K}}\right\}$$

where $\mathcal{F}^{-1}\{\cdot\}$ denotes the inverse Fourier transform and the *tilde* indicates the Fourier transformed function.

Fluctuations

In addition to the average response, our probabilistic formalism also describes the cortex's random output fluctuations in the absence of an input. If $s(x) = 0$, the basic equation reduces to

$$P[f(x); 0] = e^{-\frac{1}{2}\int\int dx dx' \, f(x)f(x')K(x-x')}$$

This equation describes a Gaussian random process (Feynman and Hibbs 1965) with a covariance function $C(x)$ that is the functional inverse of K; that is,

$$\int dx' K(x')C(x - x') = \delta(x)$$

specifically, the covariance function is given by (take Fourier transforms of the preceding equation)

$$C(x) = \mathcal{F}^{-1}\left\{\tilde{K}^{-1}\right\}$$

Thus, the spatial fluctuations in the output should be Gaussian, and the statistical structure of these fluctuations is specified by the inverse of the function K.

Because the Green function and the covariance both involve K, the structure of the spontaneous fluctuations and the driven response are related by a sort of fluctuation-dissipation theorem. Specifically, if K is eliminated between the equations for the Green function and the covariance, the result is

$$H(x) = \int d\xi \, M(\xi)C(x - \xi)$$

This equation relates the average evoked response (through the Green function H) to the spontaneous output fluctuations about the mean (through the covariance function for the fluctuations C).

A Specific "Cortex"

How might the functions K and M be determined? These functions are, of course, dependent on the precise nature of the neural circuits that are being described and information about them must ultimately come from observations on cortical structure and function. The formulation developed here should apply to any essentially cortex-like network with a well-defined input and output. In particular, it should apply to the (one-dimensional) *Limulus* eye, a neuronal system whose descriptive equation is already known (Ratliff 1965). For the *Limulus* eye, the function M that describes the distribution of the input (light) should be a delta function, so that the term $-\int dx\, s(x)M(\xi - x) = ms(\xi)$, for some constant m. An ommatidium at position ξ in the eye is excited according to the input at that location $[ms(\xi)]$ and is subject to lateral inhibition by the surrounding cells. The magnitude of the inhibition at location ξ depends on the response of the inhibiting cell at x and on its distance $(x - \xi)$ from the neuron being inhibited according to the function $G(x - \xi)$. This means that the function K is given by

$$K(x - \xi) = \delta(x - \xi) - G(x - \xi)$$

with G specifying the lateral inhibitory interactions, so that the equation describing the eye's behavior would be the Hartline-Ratliff equation

$$\hat{f}(\xi) = ms(\xi) - \int dx\, \hat{f}(x)G(x - \xi)$$

Finally, anatomical constraints would make G a Gaussian: the distribution of inhibition in *Limulus* eye is rotationally symmetric and the formation of inhibitory connections in the x and y directions should be independent; the functional equation that results from these constraints has a Gauss function as its solution. In summary, the structure of the *Limulus* eye, together with the existence of rotationally symmetric lateral inhibition, serves to define the functions K and M, and the formalism then yields the Hartline-Ratliff equation.

For the special case of $M(x) = m\delta(x)$ (this is the situation for the Hartline-Ratliff equation), the Green function is just proportional to the covariance; this is the usual fluctuation-dissipation theorem.

Fluctuations in Development

The relation between the Green function and the covariance—this relation is a natural consequence of the inherently probabilistic nature of synaptic transmission—could be of importance for brain development (Mastronarde 1983). Activity-dependent modification of neural circuits has been proposed to be critical for the shaping of the final pattern of connections that determines what computation a circuit performs (Stent 1973; Changeux 1976; see Shatz 1990). The existence of these fluctuations at one level, with a correlation function that is related to the receptive field structure, would thus provide the appropriate activity during development in utero (when patterned input to cortex related to external stimulation should be minimal) for the selforganization of circuits at the next level.

EXTENSIONS OF THE SIMPLEST CASE

In the preceding sections, the cortex treated was one-dimensional, the time dependence of its operations was suppressed, cortical computations were supposed to be linear (functional expansion of L to second order), and only a single input and output cell type was permitted. The goal now is to remove these restrictions.

Including Additional Coordinates (Space and Time)

The extension to include two spatial dimensions and time is immediate; the cortical characterization functional now depends on two spatial dimensions (specified by the vector x) and on time (t):

$$P[f(\mathbf{x}, t); s(\mathbf{x}, t)] = e^{-\int d^2x \int dt\, L[f;s,x,t]}$$

so that L becomes a functional of $f(\mathbf{x})$ and $s(\mathbf{x})$ and a function of x and t. When L is approximated by carrying out a Volterra expansion and neglecting terms higher than second order, the probability functional becomes

$$P[f(\mathbf{x}, t); s(\mathbf{x}, t)] = e^{-\frac{1}{2}\int d^2x\, d^2x' \int dt\, dt' f(x,t)[f(x',t')K(x-x',t-t')+s(x',t')M(x-x',t-t')]}$$

The specialization of this equation to the *Limulus* eye gives

$$\hat{f}(\mathbf{x}, t) = ms(\mathbf{x}, t) - \int d^2x' \int dt\, G(\mathbf{x} - \mathbf{x}', t - t')\hat{f}(\mathbf{x}', t')$$

with the inhibitory influence function G known from experiment to be

$$G(\mathbf{x}, t) = ke^{-ax^2}e^{-bt}$$

k, a, and b are constants.

A Nonlinear Cortex

Up to this point the Volterra expansion has been carried out to second order in both f and s. Expansion to this order in the output $f(x)$ means the fluctuations are described by a Gaussian process, whereas the $s(x)$ expansion relates mean output to input by a linear operator. One might expect fluctuations still to be Gaussian even for a nonlinear cortex. To treat this case, the Volterra expansion must be carried out to higher order in s—here third order is used as an example, although any order is possible—but second order in f to maintain Gaussian fluctuations around the mean. If fluctuations happened to be non-Gaussian, the expansion could be carried

What Form Should a Cortical Theory Take?

out to higher order terms in f, but this would entail severe mathematical difficulties.

To simplify the equations, the cortex will again be one-dimensional and the following convention will be adopted: what was represented earlier as integrals, for example,

$$\int \int dxdx'\, K(x, x')f(x)f(x')$$

will now be expressed in operator notation: Kf^2. A third-order term involving an operator R would be

$$Rfs^2 = \int \int \int dxdydz\, R(x, y, z)f(x)s(y)s(z);$$

often the integrals will be convolutions, but they need not necessarily be. With this notation, a Volterra expansion of $S[f; s]$ to second order in f and third order in s is

$$S[f; s] = \tfrac{1}{2}Kf^2 - Mfs + Afs^2 + \tfrac{1}{2}Bf^2s$$

where A and B are operators that arise in the functional expansion, and terms that vanish into the normalization have not been included. Carry out the integration over one of the variables in the last two terms and define two new operators:

$$B' = Bs$$

$$A' = As$$

note that A' and B' are functionals of s. With this notation, S can be written

$$S[f; s] = \tfrac{1}{2}(K - B')f^2 - (M - A')s$$

Again,

$$P[f, s] = e^{-\tfrac{1}{2}(K - B')f^2 + (M - A')s}$$

represents a Gaussian process with a covariance function

$$C = (K - B')^{-1}$$

The inverse here is in the functional sense, and the covariance now is a functional of the input s, even if the input is uniform across the cortex. Thus, for a nonlinear cortex, the spatial correlations would change with the input's magnitude and pattern. As before, the most likely response $\hat{f}$ to a given input s can be found through the Euler–Lagrange equations and is

$$\hat{f} = [(M - A')C]\, s$$

The impulse response, therefore, is just

$$H = (M - A')C$$

this is a generalized fluctuation–dissipation relation, but now the impulse response (called the Green function in the earlier discussion of the linear cortex) is more complicated because A' and C are functionals of the input s. This would mean that the receptive field structure of a nonlinear sensory cortex could vary according to how it was measured (delta function input vs. noise, for example).

Multiple Inputs and Outputs

For simplicity, the treatment so far has considered only a single input and output. Here the theoretical framework for a cortical theory is extended to include multiple inputs and outputs. For example, primary visual cortex would require at least on- and off-center inputs for color coded cells, x and y cells, and left and right eye cells; the total number of inputs would thus be between one and two dozen. Because color and parvocellular and magnocellular pathways (each with on- and off-center varieties) project separately from primary visual cortex, at least about the same numbers of outputs would be necessary.

Use of the functional Fourier transform simplifies a treatment of multiple inputs and outputs. An N-dimensional functional Fourier transform of a probability functional $P[\mathbf{f}]$ is defined by

$$\Phi[\mathbf{w}] = \int \mathcal{D}^N f \, P[\mathbf{f}] e^{-i \int \mathbf{f} \cdot \mathbf{w} \, dx},$$

where $\mathbf{f}$ is an N-vector of input functions and $\mathbf{w}$ is a vector of transform functions. The input and output functions are treated here as if they depended only on a single spatial variable, but generalization to two spatial variables and time is immediate. Note that Φ, known as the *characteristic functional*, depends on the vector of transform functions $\mathbf{w}$. The characteristic functional is especially useful because the mean output, covariance of the output, etc. can be found from it immediately. For example,

$$i \left[\frac{\delta \Phi}{\delta w_j(\xi)} \right]_{\mathbf{w}=0} = \int \mathcal{D}^N f \, f_j(\xi) P[\mathbf{f}]$$

which is, by definition, the mean $\bar{f}_j(\xi)$ of the jth output. Similarly,

$$- \left[\frac{\delta^2 \Phi}{\delta w_j(\xi) \delta w_k(\xi')} \right]_{\mathbf{w}=0} = \int \mathcal{D}^N f \, f_j(\xi) f_k(\xi') P[\mathbf{f}]$$

the covariance function (assuming, for simplicity, a zero mean). Thus, moments of the outputs are readily found if Φ is known; fortunately, the characteristic functional for a Gaussian process, the type of process that results when the cortical characterization functional is expanded to second order in the outputs, is not difficult to calculate for multiple output and inputs functions.

What is required is a generalization of the single input-output cortical characterization functional

$$L[f; s, x] = \frac{f(x)}{2} \int dx' \, \left[f(x') K(x - x') + s(x') M(x - x') \right]$$

The functional L can be written in the shorthand operator notation employed above as

$$L[f's, x] = \frac{f(x)}{2} \left[Kf + Ms \right]$$

What Form Should a Cortical Theory Take?

and the S functional, in this same notation, is

$$S[f; x] = \frac{1}{2} \left[Kf^2 + Mfs \right]$$

To generalize to multiple inputs–outputs, consider vectors of output and input functions $\mathbf{f}$ and $\mathbf{s}$, and matrices $\mathbf{K}$ and $\mathbf{M}$ that contain operators. For example f_j would be the output from the jth class of cortical neuron and the j, kth operator (matrix element) would be interpreted, for a one-dimensional cortex with multiple inputs and outputs, as

$$f_j K_{jk} f_k = \int \int dx dx' \, K_{jk}(x - x') f_j(x) f_k(x')$$

The S functional for such a cortex is just

$$S[\mathbf{f}; \mathbf{s}] = \frac{1}{2} \left[\mathbf{fKf} + \mathbf{sMf} \right]$$

and the probability functional is

$$P[\mathbf{f}; \mathbf{s}] = e^{-\frac{1}{2}[\mathbf{fKf} + \mathbf{sMf}]}$$

The problem is to identify the covariance functions and mean associated with the probability functional.

The starting place is the corresponding characteristic functional

$$\Phi[\mathbf{w}] = \int \mathcal{D}^N f \, e^{-\frac{1}{2}[\mathbf{fKf} + \mathbf{sMf}] - i \int dx \, \mathbf{f} \cdot \mathbf{w}} \,.$$

The idea is to simplify this expression so that covariance and mean are apparent. This is done by completing squares. That is, the vector of output functions $\mathbf{f}$ is transformed to a new vector in such a way that the result contains only linear and squared terms in the transform variable $\mathbf{w}$ that can be immediately identified.

For a vector of constants $\mathbf{a}$, transform $\mathbf{f}$ according to $\mathbf{f} \to \mathbf{f}' = \mathbf{f} - \mathbf{a}$. When this change of variables is made, some terms in the expression for $\Phi[\mathbf{w}]$ contain only $\mathbf{f}'$ and the operator matrix $\mathbf{K}$, some do not contain the vector of functions $\mathbf{f}'$, and some — the cross-terms — contain combinations of $\mathbf{f}'$ with $\mathbf{s}$, $\mathbf{a}$, and $\mathbf{w}$. The vector $\mathbf{a}$ can be chosen to make these cross-terms vanish, a condition that makes $\mathbf{a} = -(i\mathbf{w} - \mathbf{sM})\mathbf{C}$, where $\mathbf{C} = \mathbf{K}^{-1}$. The combinations of elements in $\mathbf{C}$ that arise from $\mathbf{K}$ are most easily computed, when cortex is uniform (so convolutions can be used), with Fourier transforms. For example, C_{00} might be given (as it is for a cortex with just two outputs) by $K_{11}/(K_{00}K_{11} - K_{01}^2)$; this would mean that C_{00} is found from the inverse Fourier transform of the Fourier transformed K entries in the algebraic expression. The result of eliminating the cross terms by the appropriate selection of $\mathbf{a}$ is

$$\Phi[\mathbf{w}] = e^{-\frac{1}{2}[\mathbf{wCw} - 2is\mathbf{MCw} + s\mathbf{MCMs}]} \int \mathcal{D}^N f' e^{-\frac{1}{2}\mathbf{f}'\mathbf{Kf}'}$$

(Note that $\mathcal{D}^N f' = \mathcal{D}^N f$ because $\mathbf{a}$ is a constant with respect to differential.) The integral and the $\exp(-\frac{1}{2}\mathbf{sMCMs})$ term vanish in the normalization so the final expression for the characteristic functional is

$$\Phi[\mathbf{w}] = e^{-\frac{1}{2}[\mathbf{wCw} - 2is\mathbf{MCw}]}$$

The mean can be recognized as

$$\bar{f} = sMC$$

and the covariance functional is given by **C**. Thus, the same formalism can be easily generalized to corticies with multiple inputs and outputs, and a generalized fluctuation-dissipation relations still holds.

HOW CAN THE CORTICAL CHARACTERIZATION FUNCTIONAL BE FOUND?

If the formalism described here is an appropriate one, then the job of a cortical theorist is to specify the cortical characterization functional. A power series approach gives the form of the equations, but the number of inputs and outputs, and the nature of the functions that appear as a result of the Volterra expansion must be determined by biological and other constraints. In the simple *Limulus* eye example, anatomical and electrophysiological investigations revealed the number of inputs and outputs (one) and the existence of lateral inhibitory connections; symmetry conditions identify the inhibitory influence function as Gaussian.

Doubtless a cortical theory will involve the same sort of anatomical and physiological information combined with general constraints. Analysis of receptive field structure will partly specify the unknown functions (for example, Reid et al. 1991), but additional constraints—derived from principles like minimal redundancy (Atick and Redlich 1992), scale invariance, and other symmetries—will probably be required as well to fill in gaps left by incomplete information about cortical structure and function. The initial attempts may have to restrict the problem in some ways, for example, by considering an appropriately chosen subset of the inputs and outputs and by dealing with only a single cortical layer or sublayer (like primary visual cortex layer 4).

An alternative approach is to postulate the general nature of the cortical computation. For example, one might suppose that the job of cortex is to solve an underconstrained inverse problem (Poggio et al. 1985). Consider, again for simplicity, a one-dimensional cortex with time suppressed, and suppose that the desired output $f(x)$ is the one that minimizes $(Af - Bs)^2$ over the entire cortex for linear operators A and B. This problem might be ill-posed so that it must be regularized by adding $(Rf)^2$ for another linear operator R. The computation made by cortex then is to find the output f that satisfies the equation

$$\frac{\delta}{\delta f(\xi)} \int dx \left[(Af - Bs)^2 + (Rf)^2 \right] = 0$$

If the S functional is taken to be

$$S[f; s] = \int dx \left[(Af - Bs)^2 + (Rf)^2 \right]$$

then the most likely response $\hat{f}(x)$ will be the one that the cortex is supposed to compute. This means that the cortical characterization functional L can be immediately identified for this case as

$$L[f; s, x] = (Af - Bs)^2 + (Rf)^2$$

Examination of this last relation reveals that it has just of the same form as the expression for L developed earlier with the power series approach when the expansion was carried out to second order in both f and s. Specifically, if a uniform cortex is assumed so the integral operators are represented by convolutions, the S functional will be

$$S[f; s] = \int \int \int dx d\xi d\eta\, A(x - \xi)A(x - \eta)f(\xi)f(\eta)$$
$$- 2 \int \int \int dx d\xi d\eta\, A(x - \xi)B(x - \eta)f(\xi)s(\eta)$$
$$+ \int \int \int dx d\xi d\eta\, R(x - \xi)R(x - \eta)f(\xi)f(\eta)$$
$$+ (Bs)^2$$

Now, define

$$K(\xi - \eta) = \int dx\, A(x - \xi)A(x - \eta) + \int dx R(x - \xi)R(x - \eta)$$

and

$$M(\xi - \eta) = \int dx\, A(x - \xi)B(x - \eta)$$

With these definitions, the S functional is written, up to a functional that is independent of f and thus vanishes in the normalization of the probability functional, as

$$S[f; s] = \int \int d\xi d\eta K(\xi - \eta)f(\xi)f(\eta) - 2 \int \int d\xi d\eta\, M(\xi - \eta)f(\xi)s(\eta)$$

This is, of course, of just the same form as obtained earlier by expansion of S to second order in f and s. If the inverse problem to be solved involves just linear operators, it is thus equivalent to approximate (second-order) theory developed above. If the operators are nonlinear, however, then the situation would be more complex and the relationship to the power series approach would have to be established for each specific case. Whether the operators A, B, and R are linear or nonlinear, this approach permits the cortical characterization functional to be identified immediately.

To reiterate, the goal here has not been to formulate a theory of cortex but rather to identify the form that any such theory should take. Insofar as the arguments are correct, this initial step in considering cortical theories has defined the problem to be solved (specify the cortical characterization functional) and has indicated some of the paths that might be followed to do this.

The real challenge, of course, is to decide what requirements for a cortical theory are the correct ones and then to use the resulting theoretical framework to increase our understanding of how the brain works. Whether this will be possible is by no means obvious. Nevertheless, this effort is already producing beneficial results: our laboratory is carrying out experiments designed specifically to answer questions posed in the discussion of the requirements for the theory.

ACKNOWLEDGMENTS

The research reported here was supported by the Howard Hughes Medical Institute and NIH grant NS 12961. Most of this work was carried out at the Aspen Center for Physics and the Santa Fe Institute, and I am indebted to these institutions for making their facilities available to me.

12 Sequence Seeking and Counterstreams: A Model for Bidirectional Information Flow in the Cortex

Shimon Ullman

Considering the wide range of functions it performs, the mammalian neocortex is notably uniform in structure. Although cytoarchitectonic differences exist between neocortical areas (e.g., the striate cortex in certain primates, or the giant Betz cells in motor cortex), in terms of laminar organization, number of cells, cell types, and general connectivity patterns there are close similarities among different cortical areas in the same animal, and across species (Rockel et al. 1980; Van Essen 1985, Martin 1988; White 1989). In the words of Martin (1988), "it would take an expert to distinguish rat frontal cortex from sheep parietal cortex, or cat auditory cortex from monkey somatosensory cortex."

This structural uniformity has suggested the possibility of common computational principles that may be used, with suitable local variations, throughout the neocortex (Creutzfeldt 1978; Barlow 1985; Crick and Asanuma 1986; Sejnowski 1986). Several proposals have been made regarding the possible general operation of the neocortex (Marr 1970; Creutzfeldt 1978; Edelman 1978; Barlow 1985; Grossberg 1988; Mumford 1991, 1992; Poggio 1990; Ullman 1991).

In this chapter, a model for some general aspect of information flow in the neocortex is proposed. The proposed computation is quite general in nature, but the focus of the discussion will be on vision and the visual cortex. The first part of the chapter outlines the general computation proposed by the model, and the second its biological implementation. The model is used to account for known features of cortical circuitry, and to derive a number of new predictions.

SEQUENCE SEEKING AND COUNTERSTREAMS

A general task frequently faced by the brain is one of establishing a link between two different representations. For example, in visual recognition, the task involves establishing a connection between an incoming pattern and stored object representations in visual memory. The two will often fail to match exactly, due to changes in size, position, viewing direction, etc. A common view is therefore that prior to the matching the input is processed through a sequence of stages that includes, for example, edge detection, extracting features of varying complexity, normalization for size, position, and orientation. The model below modifies this general view in two directions. First, it proposes a bidirectional search, where the matching can occur at intermediate levels rather than some "topmost" level. Second, rather than following a single path, multiple processing alternatives are explored in parallel.

Bidirectional Search

In applying a sequence of transformations to match an incoming pattern P with stored patterns M_i, the transformations could be applied to P, or to M_i, or to both. A simple transformation, such as overall shift or scaling, is best applied to the input pattern, because then it will be applied to a single pattern. Other transformations are specific to a stored model (e.g., how a face may transform by facial expressions) and cannot be applied to the image in a "bottom-up" manner. An attractive solution is to apply a bidirectional search, which is also economical in terms of the number of patterns explored. More generally, the suggestion is to use two streams of processing, an ascending one starting at the input, and a descending one starting at the stored models. From a biological standpoint, these will correspond to the "forward" and "backward" connections between cortical areas.

Exploring Multiple Alternatives

A large number of alternative routes may have to be explored before a link is successfully established between a "source" and a "target" representation. To achieve fast computation, it will be necessary to explore simultaneously a large number of alternative routes. In many models of visual process-ing, the input pattern undergoes a single sequence of processing stages. In contrast, in the sequence-seeking scheme an input pattern gives rise to multiple sequences of transformations and mappings that are explored in parallel. The terms "transformations" and "mappings" should be taken here in a broad sense; they may include geometric transformations such as changes in size, position, and orientation, the recovery of different prop-erties such as color, motion, texture, and 3D shape, as well as exploring alternative ways of representing the pattern (e.g., in terms of its parts and its abstract shape properties).

Linking the Ascending and Descending Streams, the Counterstreams Structure

The bidirectional processing is diagrammed schematically in figure 12.1a. The basic operation in this scheme is to seek a sequence of processing steps linking a source pattern (S in figure 12.1a) in one area with stored repre-sentations (such as M_1, M_2) in another. The nodes in this schematic figure represent patterns of activity (e.g., subpopulations of neurons acting together, possibly with some degree of synchrony) (Abeles 1991; Engel et al. 1992), and the arrows indicate how patterns activate subsequent patterns (e.g., S can activate A_2, A_3, and A_5). Since different patterns may share neurons, implementation constraints will place some limitations on the coactivation of patterns; for example, patterns (B_2, B_3, B_4) may be prohibited from being coactive. In expanding the sequence down from M_1, only a subset of these patterns will be activated initially, and will later decay and be replaced by others.

The search is bidirectional, and a linking sequence is successfully established when the two searches meet somewhere in this large network of interconnected patterns. How can a successful link of patterns be found by the system? The proposed scheme (figure 12.1b) has two main components. First, the ascending and descending streams proceed along separate, complementary pathways. Second, when a track is being traversed in one stream, it is assumed to leave behind a primed trace in the complementary stream, making it more readily excitable, as explained further below. The scheme shown schematically in figure 12.1b is similar to that shown in figure 12.1a except that each node is now split into two complementary nodes (e.g., B_2 in figure 12.1a is now split into B_2 on the ascending pathway and its complementary pattern $\bar{B}_2$ on the descending one).

The full bidirectional search now proceeds as follows. A number of sequences originating at S begin to be activated along the ascending pathway. At the same time, sequences originating at M_1 and M_2 begin to expand downward along the descending pathway. Not all of the possible sequences are expanded simultaneously, but whenever a track (subsequence) is being traversed, the complementary track remains in a primed state, ready to be activated. Suppose that by the time S has activated A_2 along the ascending stream, the track $\bar{M}_1 \rightarrow \bar{B}_3 \rightarrow \bar{A}_2$ had already been traversed in the descending stream. Due to the primed traces, this will result in the immediate activation of the complete sequences $S \rightarrow M_1$ and $\bar{M}_1 \rightarrow \bar{S}$, establishing a complete link between the source and target patterns. This will also select M_1 as the stored pattern corresponding to the input image S. (A selection among models may be required if more than a single model is matched with the input.)

Two properties of this linking process are worth noting. First, a link between the ascending and descending streams can take place at any level. Second, to establish a link, the ascending and descending patterns need not arrive at a given node simultaneously; a meeting is also possible between an active pattern and a pattern that had been active some time before and decayed, but left a primed trace in the complementary stream.

In terms of connectivity, the excitatory connections between patterns are reciprocal, obeying the following general rule: whenever A is connected to B, there is a back-connection from $\bar{B}$ to $\bar{A}$, with cross-connections (of the priming type) between A and $\bar{A}$ and B and $\bar{B}$ (figure 12.1c). The reciprocity of the connections is an inherent aspect of the model, and it is also

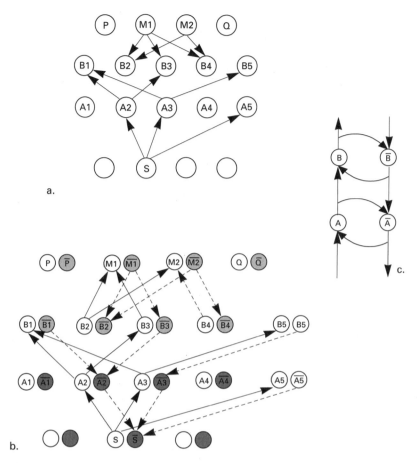

Figure 12.1 (*a*) The sequence-seeking computation seeks a sequence of mappings linking a source pattern (*S*) in one area with stored representations (M_1, M_2) in another. Nodes represent patterns of activity and arrows indicate how patterns activate subsequent patterns. In expanding sequences only a subset of patterns will be activated initially, and will later decay and be replaced by others. The processing is bidirectional, and a linking sequence is successfully established when the two searches meet somewhere in this large network of interconnected patterns. (*b*) Similar to (*a*), except that each node is split into two complementary ones. The ascending and descending streams proceed along complementary pathways. When a track is being traversed in one stream, it leaves behind a primed trace in the complementary stream. (*c*) The basic unit of the counterstreams structure. Patterns *A*, *B* on the ascending, $\bar{A}$, $\bar{B}$ on the descending path. Horizontal arrows denote connections of the priming type. This repeating unit is embedded in a network of richly interconnected patterns.

a distinguishing feature of cortical connectivity. Note that although the counterstreams structure uses "forward" and "backward" connections, it does not necessarily imply a hierarchical structure; it can incorporate a more general structure as long as the above connectivity rule is obeyed. (Inhibitory connections also play a role, but will not be discussed.)

The basic design of the sequence-seeking model is relatively straightforward, comprising two complementary networks going in opposite directions, with interaction between them primarily (but not exclusively) in the form of enhancing patterns across the two streams. Compared with other models, the scheme places more emphasis on the parallel exploration and selection of multiple alternatives, rather than relaxation and iterative computations. Timing considerations (Maunsell and Gibson 1992; Thorpe et al. 1991; Rolls et al. 1991) appear to place rather stringent restrictions on the use of multiiteration relaxation processes in tasks such as visual recognition. A visual cortical area may introduce an average delay of about 10–15 msec, and there are about six stations spanning the hierarchy from V1 to anterior IT. This suggests that visual processing should usually require a limited number of sweeps through the system. It is desirable, therefore, especially for a highly parallel system, to explore multiple alternatives simultaneously, rather than explore and refine them in sequence.

This is the skeleton of the computation, a number of elaborations and properties of the basic process are discussed below.

Express Lines

In expanding the descending sequences, how can the initial selection of models be performed? To cut down the number of competing sequences in the descending stream, it would be useful to expand with higher priority alternatives that appear more promising. For example, in attempting to recognize an object, some models (a face, say) may become more likely than others on the basis of partial analysis, although it may not yet be possible to identify the individual face. It would be advantageous under these circumstances to expand many face-related sequences, possibly at the expense of others. Such an effect can be obtained by using "express lines," directly connecting low-level to higher-level nodes along the ascending stream. The express lines will activate (directly or indirectly) patterns on the descending path. This will initiate an expansion of sequences from the selected patterns. This selection of higher-level patterns can be viewed as invoking a hypothesis suggested by the data, but which has yet to be confirmed. A link to the ascending stream will still be required to confirm the hypothesis. Note that, unlike the priming interaction between streams, in the case of express lines the ascending stream can directly activate the descending one. Express lines could also use inhibition rather than facilitation; if the partially expanded sequences in the ascending stream render some higher-level nodes unlikely, inhibitory "express lines" could be used to suppress their expansion in the descending stream.

The express lines provide one mechanism for "indexing" into the large number of models stored in memory. "Indexing" is a term used in computational vision for the initial selection of a general class, or classes of models, that might correspond to the input image. The express lines play a role in this process by the selection of likely models on the descending stream. This initial selection is not limited to the activation of models at a single "topmost" level; models at different levels along the descending stream can also be indexed and serve as the starting point for descending subsequences. For example, in addition to the selection of a complete face model, intermediate models of face parts can also be activated. Anatomically, such express lines may correspond to direct connections from low to high visual areas (such as the connections from area V4 to AIT, or from V3 and VP to area TF; Felleman and Van Essen 1991).

Another mechanism for model selection is provided by the effects of expectation and context. Knowledge about the current situation can lead to the activation or priming of a subset of models that will then become preferential sources for descending sequences. The set of active models will then be modified and refined throughout the sequence-seeking process, as described below.

The Effect of Context

Context can have a powerful influence on the processing of visual information (as well as in other perceptual and cognitive domains). A pair of similar, elongated blobs in the image may be ambiguous, but in the appropriate context (e.g., under the bed) they may be immediately recognized as a pair of slippers.

Context effects can operate in the framework of the sequence-seeking scheme by a prior priming of some of the nodes. The effect will be similar to the mutual priming of the ascending and descending streams but with longer time scales. (Priming between the streams may last for tens to hundreds of milliseconds, context effect should last for considerably longer, up to minutes or hours.) Sequences passing through the primed nodes will then become facilitated. In the above example, the location of the blobs under the bed will prime patterns representing objects that are commonly found in that location, making slippers a likely interpretation.

The general notion of priming internal representations is a common one, but its effects in the framework of the sequence-seeking scheme are particularly broad. When certain nodes are activated (e.g., by noticing and identifying the bed in the image) they will initiate sequences of their own, and an entire set of patterns may end up in a primed state. (Context may possibly involve also inhibitory effects, making some of the paths less favorable.) Later on, a large number of possible sequences passing through a primed node will be facilitated, compared with the nonprimed sequences. Context effects are thus not limited to directly increasing or decreasing the likelihood of a single match, but they can have indirect, widespread effects, by facilitating otherwise less favorable sequences. A context pattern *A* may help to bring about the activation of *B*, not as a result of direct association, but because *A* may have a sequence leading to some intermediate pattern *C*, and, later on, an activated pattern may have another sequence leading to *B* via the primed pattern *C*.

These general characteristics capture some of the fundamental aspects of context effects in humans. Humans' perception and cognition appear to have an almost uncanny capacity (which is extremely difficult to reproduce in artificial systems) for bringing in relevant context information in a broad and flexible manner. It seems that broad, indirect, context effects can be reproduced by the sequence-seeking computation.

Learning Sequences

A simple and local learning rule is sufficient in the counterstreams structure to reinforce selectively complete successful sequences. Every pattern node in a successful sequence will receive both a direct activation and a priming signal from the complementary track, while patterns on dead-end tracks will receive one or the other but not both. The approximate temporal coincidence of the two signals can be used to preferentially strengthen the successful sequence. This rule is local, since it depends on the activation of a single pattern. Yet, it is sufficient to reinforce preferentially successful sequences forming an uninterrupted link between source and target patterns. Following practice, out of the huge number of possible sequences, those that proved useful in the past will be explored with higher priority is future use of the network.

In the process of reinforcing successful sequences, changes due to learning are distributed throughout the system, and are not confined to high level centers specializing in learning (Sejnowski 1986). Recent studies of learning certain perceptual skills suggest that low level visual areas, including primary visual cortex, are indeed involved in the modifications that take place during the learning process (Karni and Sagi 1991).

In addition to the learning of complete sequences, as above, the system may also be engaged in the learning of the individual mappings, that is, the basic steps that make up the sequences (Poggio 1990). This aspect of the learning is, however, outside the scope of the current discussion, since the focus is not on the specifics of individual processes, but on their overall common structure.

Refining the Expansion

The matching between streams is not an all-or-nothing event, but a graded one: some sequences will lead to better matches than others, and then serve as starting points for exploring additional sequences, that will lead in turn to an improved match. This process has some features in common with a family of optimization and search procedures known as "genetic algorithms" (Holland 1975). Recent evaluations have shown such methods to behave quite efficiently (Brady 1985; Peterson 1990). Our own simulations in the context of pattern matching have also shown that computations based on sequence-seeking compare favorably with alternative methods, such as gradient descent and simulated annealing.

General Aspects of Sequence Seeking

The discussion above of the sequence-seeking process used as an example the domain on visual recognition. However, the process of establishing a sequence of transformations, mappings, or states, linking source and target representations, could provide a useful general mechanism for various aspects of perception as well as for nonperceptual functions. For example, the planning of a motor action can be cast at some level in terms of seeking a sequence of possible moves linking an initial configuration with a desired final state. Movement trajectories will be based in a sequence-seeking scheme on a stored repertoire of elementary movements. These basic movements could then be transformed (scaled, stretched, rotated, etc.) and concatenated together to generate more complex movements. In analogy with sequence seeking in vision, motion planning will involve the application of transformation and the generation of compound sequences. Similarly, more general planning and problem solving can also often be described in terms of establishing a sequence of transformations, mappings, or intermediate states, linking some source and target representations. The general aspects of the sequence-seeking process therefore provide a useful computation that could be applied, with appropriate modification, to a large variety of different tasks.

BIOLOGICAL EMBODIMENT

The sequence-seeking model requires two streams going in opposite directions with the appropriate cross-connections. A schematic diagram proposing how the counterstreams structure may be embedded in cortical connections is shown in figure 12.2a. The proposed implementation is presented in schematic outline only, focusing on a number of central aspect, and without discussing details or possible variations of the model.

The ascending stream goes through layer 4 to a subpopulation of the superficial layers, denoted in the figure as AS (for ascending superficial), and then projects to layer 4 of the next cortical area (II in the figure). The descending stream goes through a different subpopulation of the superficial layers (DS, for descending superficial) to DI (for descending infra), a subpopulation of the infragranular layers (often in layer 6), and from there to DS of a preceding area. The connections can also leap over one step (or occasionally more) in the stream (e.g., AS directly to AS on the ascending stream, and DS → DS or 6 → 6 on the descending stream) (thin lines in figure 12.2b).

Layer 5 is left out of the diagram because, according to the model, it (or a part of it) is involved primarily not in the main streams, but with their control, via subcortical structures. There are two reasons for assuming that layer 5 (or parts of it, e.g., 5b of the macaque's V1) may be involved in control functions. First, its orderly connections to subcortical structures (e.g., from visual cortex to the pulvinar and the superior colliculus, structures implicated in controlling attention and eye movements) that are reciprocally connected in turn in a topographic manner to multiple visual areas. Second, the firing pattern of a population of pyramidal cells in this layer that "can initiate synchronized rhythms and project them on neurons in all layers" (Silva et al. 1991).

Note that the counterstreams structure suggests a natural organization in about 5–6 main layers: one or two performing control functions, two (an input and an output layer) for the ascending and two for the descending streams. The division between the roles of the different layers may in reality be less clear cut, however, the main goal of the diagram is to emphasize the common underlying structure according to the model, rather that to account for possible variations.

Connections of V1: Data and Predictions

Figure 12.2b,c shows an expanded version of the diagram, applied to cortical area V1 (which is somewhat special, but for which the data are more comprehensive than for other visual areas), and its connections to the LGN and cortical area V2 (V1 is also connected to other visual areas, not shown in the diagram). Figure 12.2b shows the connections in the macaque of the magnocellular stream and figure 12.2c of the parvocellular stream (Rockland and Lund 1983; Lund 1988a,b; Martin 1988). The diagram shows the main connections; additional secondary ones will not be considered. The connections are drawn in a manner suggested by the model, and they includes both known connections (thick arrows) and connections predicted by the proposed scheme but for which empirical evidence is partial or lacking (thin arrows).

The pattern of connections in the two streams appears to be in general agreement with the counterstreams structure and figure 12.2a. This structure can be used as a repeating circuit to utilize the cortex inherent parallelism and combine ascending and descending information flows. If the general hypothesis regarding the counterstreams structure is broadly correct, then a number of predictions can be made regarding the main connectivity patterns within and between areas. One general prediction is the possible distinction between the AS and DS subpopulations. This separation reflects the most straightforward implementation of the scheme, however, some alternatives can exist without violating the constraints of the model.

A separation between the ascending and descending populations is evident in the connections involving layer 4: the ascending streams termi-

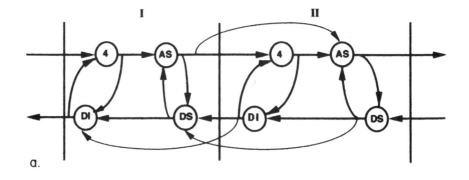

a.

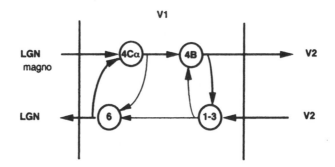

b.

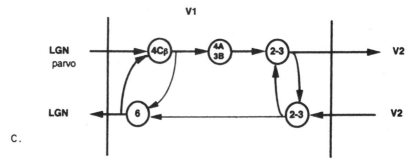

c.

Figure 12.2 (*a*) How the basic counterstream structure may be embodied in cortical connectivity. The structure contains two interconnected streams, and ascending and a descending
one. The ascending path goes through layer 4 and the ascending superficial population (AS)
to the next area. The descending path goes from the descending superficial (DS) population
to DI (descending infra) back to the first area. Thin arrows show pathways that "leap over" a
step in the stream. Inhibitory and long-range intraareal connections are not shown. See text
for more details. (*b*) The main connections according to the model along the magno stream
from the LGN via V1 to V2. V1 is also connected to other visual areas, not shown in the
diagram. The connections are drawn in a manner suggested by the model and (*a*). Thick
arrows, established connections; thin arrows, connections predicted by the model. (*c*) The
main connections according to the model along the parvo stream from the LGN via V1 to V2.
Thick arrows, established connections; thin arrows, connections predicted by the model.

nate in layer 4, the descending streams always avoid it. In the superficial layers the situation is more difficult to assess, and the available evidence is at present restricted. In the magnocellular projection from V1 to V2 the forward projection originated mainly in 4B, while the back projection in mainly to other layers (2b). It is further expected that even when the superficial layers provide both the source and the target of connections to another area, there will in fact often be a separation to the AS/DS subpopulations (presumably a vertical, rather than horizontal separation; Wong-Riley 1978; Zeki and Shipp 1988). If these populations exist, they should be connected in a reciprocal manner. A related expectation derived from the model is the existence of priming-type synaptic interactions, that is, excitatory synaptic input that by itself may not be very effective in driving the target cells, but that facilitates the effects of subsequent inputs to these cells.

An example of a more specific prediction is that in the magnocellular stream the model suggests reciprocal interconnections between layer 4B (playing the part of AS in the model), and layers 1–3, the recipients of descending projections from V2 (DS in the model). The projection from 4B to the superficial layers is well established. It is also known (Lund 1988a) that 4B pyramidal cells send apical dendrites to the superficial layers where the connection may take place. It has been noted in this regard (Lund 1988a) that the significant 4B projection has a surprisingly limited effect on properties of superficial layers units (such as directional selectivity). In figure 12.2b, this connection has mainly a priming role, and therefore the lack of direct effect is not unexpected. A related prediction is the expectation that the same superficial cells connected to 4B will also be the recipients of descending projections from V2.

The model also includes a reciprocal connection between layer 4 and the LGN-projecting cells in layer 6. The projection from 6 to 4 is well-established in both cat (McGuire et al. 1984) and monkey (Lund 1988a), and there is support for the opposite connection as well (Lund and Boothe 1975). It is also interesting to note in this regard that the population of layer 6 cells projecting back to the LGN was found (in the cat) to be the same cells that are also connected to layer 4C, by axonal collaterals and dendritic arbors (Katz et al. 1984), in accordance with the connectivity in figure 12.2b,c.

The connections between layers 4 and 6 are expected in the model to have a priming effect (not necessarily the only effect, see Martin 1988), and this notion has some physiological support. It was found (Ferster and Lindström 1985) that using electrical activation of layer 6 cells by antidromic activation increased the probability of layer 4 firing, and most cells fire multiple spikes in response to each stimulus. Under the opposite conditions, when layer 6 is inactivated, the main observed effect was the reduction in excitability of layer 4 cells (Grieve et al. 1991).

From an anatomical standpoint, EM reconstructions (McGuire et al. 1984) have shown terminations of layer 6 axons on smooth and sparsely spiny cells. The model suggests also a projection onto layer 4 spiny cells, and this remains to be clarified in future anatomical studies.

Lateral Connections between Areas

Connections between cortical areas (not only visual, also somatosensory and motor) can be classified into "forward," "backward," and "lateral" connections, on the basis of the laminar distribution of their source and destination (Rockland and Pandya 1979; Maunsell and Van Essen 1983; Friedman 1983; Van Essen 1985; Zeki and Shipp 1988; Andersen et al. 1990; Boussaud et al. 1990; Felleman and Van Essen 1991). Lateral connections terminate in all layers, and their origin is bilaminar, from the supra and infra layers. The lateral pattern is relatively complex and sometimes perplexing (Felleman and Van Essen 1991). It is therefore interesting that a number of its main features can be derived almost directly from the model. The counterstreams structure does not have a distinct, third type of connections, but it allows forward and backward connection simultaneously in both directions, and it can include lateral connections by assuming that they are the "union" of ascending and descending connections. If this view is correct, then the main connections participating in the lateral connection can be inferred from the basic scheme (figure 12.2a). According to the model, they include the direct connections: AS → 4, and 6 → DS, as well as the connections that leap over one stage in the diagram, namely, AS → AS, DS → DS,DI, and DI → DI.

The origin of the projections according to the model would be bilaminar, and the terminations would span all layers, in agreement with the observed pattern. This can also explain several difficulties such as the problem of irregular terminations (Felleman and Van Essen 1991), that occurs, e.g., when some of the terminations are restricted to layer 4 of the target area while others show columnar terminations. This was termed F/C (i.e., a mixture of "four" and "columnar") paradoxical termination, since termination in layer 4 is usually a signature for ascending connections, while a columnar termination signifies lateral connections. Usually these connection types are distinct, but some interconnections exhibit a mixed type. In the counterstreams structure, the main point to note is that the lateral connections from the superficial layers of area A to target area B are composed of two subprojections: AS → 4 (ascending) and DS → DS, DI (descending). (In addition, there is a descending connection DI → DI.) Anterograde labeling of the upper layers of area A can therefore show a mixed pattern of terminations: 4 alone (from AS of A), or a columnar termination (from AS and DS). This is in agreement with the F/C paradoxical termination (Felleman and Van Essen 1991). It can also (by labeling the DS alone) show a bilaminar pattern of connections; this can account for the other types of irregular terminations. If this account is correct, it also provides support for the existence of the separate AS and DS subpopulations.

Possible Priming Mechanisms

Synaptic interactions in the model include priming-type effects between the complementary streams. Although this has not been studied directly, some known or physiologically plausible mechanisms may play a role in such priming interactions.

One such mechanism has been investigated by Miller et al. (1989). In this study, responses of cells in the cat's visual cortex to visual stimulation were profoundly suppressed by the blocking of NMDA receptors (by using APV). A possible mechanism proposed by Miller et al. (1989) by which NMDA receptors could control the responsiveness of cells is that such receptors, when activated in neocortex pyramidal cells, cause a slow, long-lasting EPSP that rises to a peak in 10–75 msec. They suggest that this slow EPSP could provide a base on which subsequent subthreshold input would become suprathreshold.

Another mechanism has been proposed by Koch (1987) and by empirical studies in the LGN (Esguerra et al. 1989; Sherman et al. 1990). The proposed mechanism makes use of the capacity of NMDA receptors to increase the cell's response in a nonlinear fashion, as a function of the depolarization in the postsynaptic cell. The proposal, in the context of the LGN (Koch 1987), is that the descending stimulation from the cortex can cause long-lasting subthreshold depolarization, and that the ascending stimulation involves receptors of the NMDA type. If ascending stimulation arrives while the units are still in a depolarized state, the response will be enhanced significantly. A similar mechanism based on the nonlinear properties of the NMDA-type receptor could be used for priming between the streams.

The long-lasting depolarization could be contributed by postsynaptic responses with slow time course, similar to the persistent Na^+ channel, or the I_T calcium channel (McCormick 1990). Synaptic mechanisms of this type have been implicated in cortical cells (Hirsch and Gilbert 1991; Amitai et al. 1993). A similar effect can also be contributed by the activation of distal parts of the dendritic tree. Simulations of pyramidal cells (Stratford et al. 1989) have shown that such stimulation can have a significant temporal extent.

Priming can thus be obtained by long-lasting depolarization, caused by the properties of ionic channels, the NMDA receptors, or the stimulation of distal dendritic branches, combined with subsequent input, added either linearly or nonlinearly. Other mechanisms, not considered here, might participate as well. Although the details are not known, it appears that synaptic mechanisms for priming connections are physiologically plausible, and it will be of interest to try to test them empirically.

Effects of the Feedback Projection

According to the sequence-seeking scheme, the physiological effects of the descending projections can assume two different forms: either the priming and modulation of the ascending stream or the direct activation of a lower area.

Both effects have been observed in physiological studies, modulatory (Nault et al. 1990; Sandell and Schiller 1982), as well as direct excitatory effects (Mignard and Malpeli 1991; Cauller and Kullics 1991b). Further predictions of the model regarding the modulatory effects include (1) the facilitation by the back projections will not require strict temporal coincidence, (2) similar modulatory effects are also likely to be induced by ascending signals on descending ones, (3) the two effects may be segregated into two distinct subpopulations: in figure 12.1c $\bar{B}$ can be directly driven along the descending stream, but patterns such as B on the ascending stream are expected to show modulatory effects.

In summary, the computation proposed by the sequence-seeking model is a bidirectional search performed by top-down and bottom-up streams seeking to meet. Key properties of the scheme include the simultaneous exploration of multiple alternatives, the relatively simple, uniform, and extensible structure, the flexible use of "bottom-up" and "top-down" sequences that can meet at any level, and the learning of complete sequences by a simple local reinforcement rule. The implementation in the counterstream structure proposes a computational account for several basic features of cortical circuitry, such as the predominantly reciprocal connectivity between cortical areas, the forward, backward, and lateral connection types, the regularities in the distribution patterns of interarea connections, the organization in 5–6 main layers, and the effects of back projections.

ACKNOWLEDGMENTS

I would like to thank C. Gilbert, C. Koch, J. Maunsell, D. Mumford, T. Poggio, and W. Richards for their helpful comments. This work was supported in part by NSF grant IRI-8900207.

13 Dynamic Routing Strategies in Sensory, Motor, and Cognitive Processing

David C. Van Essen, Charles H. Anderson, and
Bruno A. Olshausen

INTRODUCTION

Our understanding of how the cerebral neocortex carries out its essential functions has progressed through several important stages over the past three decades. In the 1960s, Hubel and Wiesel provided the first detailed characterization of orientation selectivity, direction selectivity, and binocular interactions in visual cortex. Equally important, they proposed simple and attractive models that could account for these properties by invoking "bottom-up" convergence of ascending inputs in a strictly serial processing hierarchy (Hubel and Wiesel 1965b).

In the 1970s these ideas were modified by the discovery of parallel pathways within the visual system and by evidence for the importance of intracortical inhibition and other aspects of local intrinsic circuitry of the cortex (Stone et al. 1979; Orban 1984). This picture subsequently became enriched by the realization that intrinsic connections extend rather widely within each cortical area (Gilbert 1983) and also that there are massive feedback pathways projecting from higher to lower cortical areas (Rockland and Pandya 1979; Maunsell and Van Essen 1983). The function of these long-distance and feedback pathways is not fully understood, but it is widely presumed that they subserve the strong modulatory effects known to arise from outside the classical receptive field (Allman et al. 1985a; Knierim and Van Essen 1992). This represents a form of "top-down" processing that makes cortical responses to sensory stimuli highly dependent on the context in which they are presented.

Recently, we and others (Poggio 1984; Baron 1987; Anderson and Van Essen 1987, 1993; Van Essen and Anderson 1990; Olshausen et al. 1992, 1993) have argued that a critical component is missing from this view of cortical processing, namely, an explicit mechanism for dynamically regulating the flow of information within and between cortical areas. The need for such a mechanism arises from computational considerations relating to (1) the vast amounts of data continuously impinging on the nervous system, (2) the finite computational resources that can be dedicated to any given task, and (3) the need for highly flexible linkages between a large number of physically separate modules. In our previous articulations of this hypothesis we have focused on the visual system, but we believe that the concepts and issues are equally applicable to other systems as well. To provide an intuitive illustration of this point, we begin by considering two examples, one relating to sensory processing and the other to motor processing. While superficially rather different, we will argue that these two types of processing may share important similarities in their "deep structure" at the computational level and at the level of neurobiological implementation.

Visual Attention

First, consider the phenomenon of directed visual attention, which has a number of characteristics illustrated by the following hypothetical task. Suppose that an observer stares at a fixation spot on an otherwise blank screen. A simple cuing stimulus is flashed on the screen, and it is followed after a brief interval by an array of letters arranged concentrically about the fixation point. If the cue is a small spot occurring in the location of the leftmost target letter, then attention will be directed to that location (figure 13.1a); the observer will be able to identify that letter as a C but will be unable to reliably identify the remaining letters in the array (cf. Nakayama and Mackeben 1989; Kröse and Julesz 1989). If the cue is switched to the rightmost letter, the observer's attention will likewise shift so as to allow identification of the letter T (figure 13.1b). A very different result will occur, though, if the cuing spot is expanded to cover the entire target array (figure 13.1c). In attending to the overall pattern, the observer will immediately recognize the ring-shaped geometric configuration of letters, but the spatial resolution within this broader "window of attention" will be too coarse to allow reliable identification of any of the individual letters, let alone the entire word that they form (cf. Van Essen et al. 1991).

Our analysis of this phenomenon starts with the argument that visual attention is a process that has evolved to subserve general purpose object recognition. The ability to recognize a wide range of highly complex patterns (e.g., faces) is too computationally demanding to have the requisite neural machinery replicated separately for each location in the visual field. Instead, this machinery is relegated to a small number of modules situated at high levels of the visual hierarchy in inferotemporal cortex. In our view, visual attention is a mechanism for dynamically regulating information flow so as to bring information from the appropriate region of the visual field and in an appropriate format to the appropriate high-level object recognition center. Five general characteristics of spatially directed visual attention warrant explicit mention in connection with this hypothesis.

- Attention can readily be directed to different locations and to different spatial scales (Sperling and Dosher 1986; Eriksen and Murphy 1987; Julesz 1991).

- Attentional shifts can be initiated by bottom-up cues and/or top-down influences. When initiated by bottom-up cues (as in the above example),

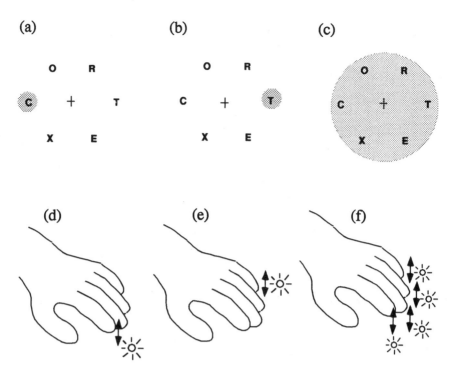

Figure 13.1 Directed visual attention (*a–c*) and directed motor control (*d–f*). The process of shifting attention to different locations and to different scales (shaded regions) is analogous in important ways to the process of directing a basic motor routine (e.g., up-down movements signified by arrows) to different digits or combinations of digits in response to instructions embodied, say, by the blinking light icons.

attentional shifts occur with a finite temporal delay (50–100 msec) and tend to persist at any given location for a relatively brief period (Nakayama and Mackeben 1989; Kröse and Julesz 1989). In this respect, they are analogous to saccadic eye movements, except that they occur on a faster time scale.

• Attention is directed not simply to the initial cue, but to whatever image data lie within the window of attention once it has been shifted. For example, a highly salient cue might be followed by a very subtle spatial pattern whose characteristics can nonetheless be scrutinized via attentive processing (Nakayama and Mackeben 1989; Kröse and Julesz 1989; Julesz 1991). In this respect, the cue evidently serves simply as a gating mechanism to regulate the flow of image data.

• Visual attention acts as an informational bottleneck that reduces to manageable levels the amount of image data reaching high-level cortical centers involved in pattern recognition. By our estimate, less than 0.1% of the information carried in the optic nerve at any given moment passes through the attentional bottleneck (Van Essen et al. 1991).

• For complex patterns to be recognized, it is important that information about spatial relationships be preserved within the window of attention. However, because of the narrowness of the information bottleneck, the spatial resolution within the window is limited to the equivalent of about 30 × 30 resolution elements.

Directed Motor Control

An analogous set of issues arise in certain aspects of motor function, which we shall refer to as "directed motor control." The essential notion is that our motor system is capable of selecting among a large repertoire of stereotyped motor routines, each of which can be distributed to different body parts according to specific instructions and cues. An instructive example of this is that of an orchestral conductor. In general, a given section of music has a steady beat that must be communicated to the orchestra by rhythmic movements of the baton. The conductor has a wide repertoire of specific movement routines (simple up-down strokes, triangular strokes, and a variety of more complex patterns), from which one must be selected to convey the desired beat. In addition, any particular set of strokes can be transmitted via movements of the hand, the arm, a single finger, or whatever combination the conductor chooses from moment to moment.

To sharpen the analogy with directed visual attention, note that the cues for switching from one type of movement to another can be dictated by bottom-up sensory cues as well as the top-down cognitive control that would occur in the case of the orchestral conductor. For example, suppose that an observer is instructed to move an appendage up and down at a steady rate, in response to a set of lights situated adjacent to the fingertips (figure 13.1, bottom row). Thus, according to which particular lights were blinking, the observer would wiggle the index finger (figure 13.1d), the little finger (figure 13.1e), or all fingers simultaneously (figure 13.1f).

We contend that motor coordination is too computationally demanding to have the circuitry for *de novo* generation of each possible motor routine separately represented for each of the many dozens of appendages in the body. Instead, we support the notion that there exist only one or a few central representations of each motor routine (wiggles, circles, figure-8 movements, etc.) that can be generated via top-down (cognitive) processing (Keele 1981, 1986). Directed motor control is a process for dynamically regulating information flow so as to distribute information about a desired motor routine to the appropriate target appendage(s). More specifically, there are five characteristics of directed motor control that share important similarities with those outlined above for visual attention.

- A given motor routine can be distributed to target appendages at different locations and to different spatial scales (e.g., one digit or many).

- Directed motor control can be initiated by a variety of bottom-up cues and/or top-down influences. A given motor routine can be sustained more or less indefinitely if desired, though; it does not show the strong transient component characteristic of some aspects of visual attention.

- The cue used for initiation need not be a direct replica of the trajectory that is to be carried out in any given motor routine. Rather, the cue can be an essentially arbitrary abstract stimulus that serves simply as a gating mechanism to regulate the flow of motor signals.

- Directed motor control represents a strategy that reduces to manageable levels the amount of learned information that needs to be stored in the circuits involved in generating specific motor routines.

- For a given motor routine to be executed, it is important that the spatiotemporal sequence of information associated with a given pattern be preserved as it is distributed to the appropriate target appendages.

The common theme in both of these examples is that a sophisticated system for dynamically controlling information flow may be essential for the brain to carry out its duties. The notion that gating mechanisms are important in CNS function is by no means new. For example, gating has long been suspected to play a critical role in the modulation of pain sensitivity (Melzack and Wall 1965; Fields and Basbaum 1978). However, the evidence for it in cortical function has been fragmentary and controversial. To guide further experimentation, there is a clear need for detailed, neurobiologically plausible models. In this chapter, we will review the key features of our model of directed visual attention, emphasizing recent progress in making it an autonomous, self-contained process. Then we will discuss ways in which this model might be adapted to account for motor functions. Finally, we will briefly consider the relevance of these strategies for cognitive processing.

A MODEL OF DIRECTED VISUAL ATTENTION

The goal of our model is to provide a neurobiologically plausible mechanism for shifting and rescaling the representation of an object from the retinal reference frame into an object-centered reference frame. Information in the retinal reference frame is represented on a neural map (for instance, the topographic representation in V1), and we hypothesize that information in the object-centered reference frame is also represented on a neural map that preserves some degree of information about local spatial relationships. This is illustrated in the simplest possible scheme in figure 13.2a, which shows a purely pixel-based (but object-centered) representation providing the input to a high-level associative memory. However, we do not presume that only "pixels" are being routed into the high level areas. Rather, each sample node in the high level map may be thought of as a feature vector representing various local image properties, such as orientation, texture, and depth (figure 13.2b). For simplicity, our current computer model simulates the routing of only pixel-based data. However, it should be relatively straightforward to incorporate into future models the additional processing needed for dynamic routing of more complex feature vectors.

To topographically map an arbitrary section of the input (retina) onto the output (object-centered reference frame), each neuron in the output

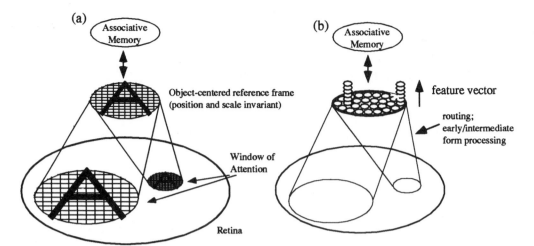

Figure 13.2 Shifting and rescaling the window of attention. The image within the window of attention in the retina is remapped onto an array of sample nodes in an object-centered reference frame. (*a*) In the simplest scheme, each "pixel" in the object-centered reference frame represents image luminance. (*b*) More realistically, each pixel should presumably correspond to a feature vector that integrates over a somewhat larger spatial region and represents orientation, depth, texture, etc.

stage needs to have dynamic access to a large number of neurons in the input stage. In the brain, this access must necessarily be obtained via the physical hardware of axons and dendrites. Since these pathways are physically fixed for the time scale of interest to us (< 1 sec), there needs to be a way of dynamically modifying their strengths. We propose that the efficacy of transmission of these pathways is modulated by the activity of *control neurons* whose primary responsibility is to dynamically route information through successive stages of the cortical hierarchy.

A Simple Dynamic Routing Circuit

Figure 13.3a illustrates a simple, 1D dynamic routing circuit. Each node in the circuit forms a linear weighted sum of its inputs, and the weights are dynamically modified by a set of control neurons to set the position and scale of the window of attention. The hierarchical connection scheme shown has the attractive property of keeping the fan-in (number of inputs) on any node fixed to a relatively low number while allowing the nodes in the output layer access to any part of the input layer. This property will be important in scaling-up the model.

An example of how the weights might be set for different positions and sizes of the window of attention is shown in figure 13.3b,c. When the window is at its smallest size (i.e., at the same resolution as the input stage, figure 13.3b) , the weights are set so as to establish a one-to-one correspondence between nodes in the output and the attended nodes in the input. When the window is larger, the weights must be set so that

Van Essen, Anderson, and Olshausen

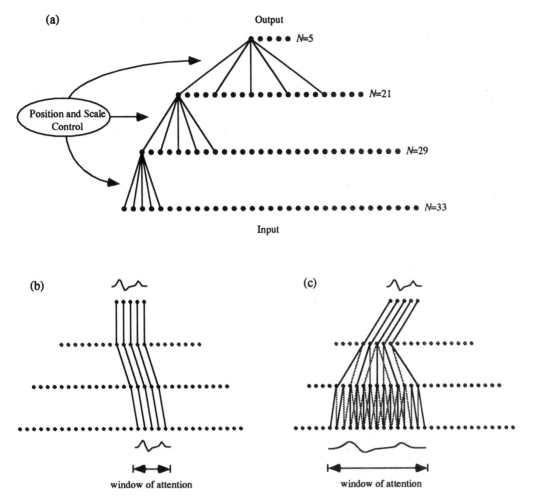

Figure 13.3 A simple, one-dimensional dynamic routing circuit. (*a*) Connections are shown for the leftmost node in each layer. The connections for the other nodes are the same, but merely shifted. *N* denotes the number of nodes within each layer. A set of control units (not explicitly shown) provides the necessary signals for modulating connection strengths so that the image within the window of attention in the input is mapped onto the output nodes. (*b*, *c*) Some examples of how the weights would be set for different positions and sizes of the window of attention. The gray-level of each connection denotes its strength. Essentially, the sets of weights feeding into each node are samples of a Gaussian interpolation function that is shifted and rescaled according to how the window of attention is shifted and rescaled.

Dynamic Routing Strategies in Sensory, Motor and Cognitive Processing multiple inputs converge onto a single output node, resulting in a lower-resolution representation of the contents of the window of attention on the output nodes. One can see from this illustration, then, that the challenge in controlling the routing circuit lies in properly setting the weights to yield the desired position and size of the window of attention. Note that in general there are many possible solutions in terms of the combinations of weights that could achieve any particular input-output transformation.

Control

Since the purpose of attention is to focus the neural resources for recognition on a specific region within a scene, it would make sense for the attentional window to be automatically guided to salient, or potentially informative areas of the visual input. Salient areas can often be defined on the basis of relatively low-level cues — such as pop-out due to motion, depth, texture, or color (e.g., Koch and Ullman 1985). Here, we utilize a very simple measure of salience based on luminance pop-out in which attention is attracted to "blobs" in a low-pass filtered version of a scene. (A blob may be defined simply as a contiguous cluster of activity within an image.) Attention can also be directed via voluntary or cognitive (top-down) influences, but these are not incorporated into our current model.

We propose the following simple but useful strategy for an autonomous visual system (see figure 13.4):

1. Form a low-pass filtered version of the scene so that objects are blurred into blobs.

2. Select one of the blobs from the low-pass image — whichever is brightest or largest — and set the position and size of the window of attention to match the position and size of the blob.

3. Feed the high-resolution contents of the window of attention to an associative memory for recognition.

4. If a match with one of the memories is close enough (by some as yet unspecified criterion), then consider the object to have been recognized; note its identity, location, and size in the scene. If there is not a good match, then consider the object to be unknown; either learn it or disregard it.

5. Now inhibit this part of the scene and go to step 2 (find the next most salient blob).

The next three sections describe details for carrying out steps 2, 3, and 5. Step 1 is trivial, whereas step 4 is a high-level problem beyond the scope of this paper (see chapter 7 by Mumford).

Focusing the Window of Attention on a Blob   To formulate a solution for controlling the routing circuit, we will simplify matters even further and consider controlling a simple two-layer, one-dimensional network consisting of an input layer and an output layer only (figure 13.5). The output

Van Essen, Anderson, and Olshausen

1. Blur objects into blobs.

2. Focus the window of attention on a blob.

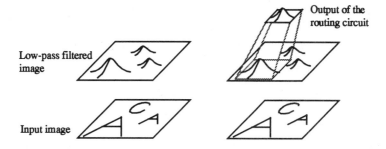

Low-pass filtered image

Input image

Output of the routing circuit

3. Feed the high resolution contents of the window of attention to an associative memory.

4. Note the location, size, and identity of the object.

5. Move on to the next blob and repeat.

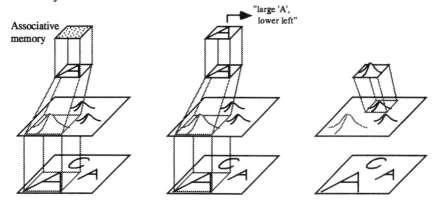

Associative memory

"large 'A', lower left"

Figure 13.4 A simple attentional strategy for recognizing objects in a scene. Objects are preattentively segmented via lowpass filtering. Once an object has been localized, the contents of the window of attention are fed to an associative memory for recognition. This process is then repeated ad infinitum.

units, I^{out}, compute their activation from the input units, I^{in}, via a simple linear summation

$$I_i^{out} = \sum_j w_{ij} I_j^{in} \tag{13.1}$$

and the weights are dynamically modified by a set of control neurons, c, via

$$w_{ij} = \sum_k c_k \Gamma_{ijk} \tag{13.2}$$

where Γ_{ijk} denotes the weight with which control neuron c_k modulates the strength of synapse (i, j) between neuron j in the input layer and neuron i in the output layer. For ease of notation, this equation describes the general

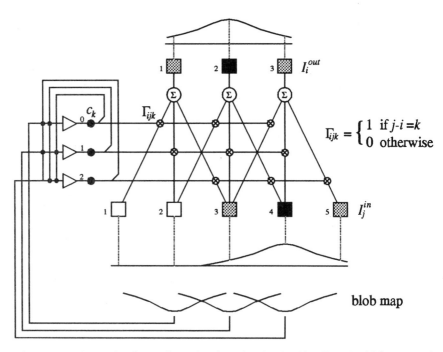

$$\Gamma_{ijk} = \begin{cases} 1 & \text{if } j\text{-}i = k \\ 0 & \text{otherwise} \end{cases}$$

Figure 13.5 A very simple one-dimensional routing circuit with a Gaussian blob presented to the input units. Each control unit corresponds to a different position of the window of attention: left (c_0), center (c_1), or right (c_2). For example, to accomplish the remapping shown, the values on the control units should be $c_2 = 1$ and $c_0 = c_1 = 0$. The circuitry for autonomous control is shown on the left, with each control unit receiving input from a Gaussian receptive field in the input layer; the control units then compete among each other via negatively weighted interconnections, so that only the control unit corresponding to the strongest blob in the input prevails.

case in which each control neuron may modulate any or all synapses (i, j). In fact, each control neuron actually modulates only a local group of synapses, and so Γ_{ijk} will be nonzero for only a few combinations of i, j, and k. In the particular circuit of figure 13.5, the Γ_{ijk} are set simply so that each control unit c_k corresponds to a global position of the window of attention, but in general this need not be the case.

To focus the window of attention on a blob in the input, the network's "goal" will be to fill the output units with a blob while maintaining a topographic correspondence between the input and output nodes (figure 13.4, step 2). Since the dynamic variables in this network are the c_k, we need to formulate a dynamic equation for the c_k that accomplishes this objective. We can achieve this by letting c_k follow the gradient of an objective function, E_{blob}, that provides a measure of how well a blob is focused on the output units.

$$E_{\text{blob}} = -\sum_i I_i^{\text{out}} G_i \tag{13.3}$$

where G denotes the desired blob shape (e.g., in the circuit of figure 13.5, $G_i = e^{-(i-1)^2}$). The second part of the objective (maintaining topography)

Van Essen, Anderson, and Olshausen

may be accomplished by letting c_k follow the gradient of a *constraint* function that favors control states corresponding to translations or scalings of the input-output transformation.

$$E_{\text{constraint}} = -\frac{1}{2} \sum_{k,l} c_k \, T^C_{kl} \, c_l \qquad (13.4)$$

where the constraint matrix $\mathbf{T}^c$ is chosen so as to appropriately couple the control neurons. For example, in the particular circuit of figure 13.5, $\mathbf{T}^c$ is set to accomplish a winner-take-all function ($T^c_{kl} = -1$, $k \neq l$) since each c_k corresponds to a global position of the window of attention.

A dynamic equation for c_k that simultaneously minimizes both of these objective functions (E_{blob} and $E_{\text{constraint}}$) is given by

$$c_k = \sigma(u_k) \qquad (13.5)$$

$$\frac{du_k}{dt} + \alpha u_k = \eta \sum_i \sum_j G_i \, \Gamma_{ijk} \, I^{\text{in}}_j + \beta \sum_l T^c_{kl} \, c_l \qquad (13.6)$$

where α, η, and β are constants that determine the rate of convergence of the system, and σ is a sigmoidal squashing function. The neural circuitry required for computing equations (13.5) and (13.6) is shown in figure 13.5. The first term on the right of equation (13.6) is computed by correlating the Gaussian, G, with a shifted version of the input (the amount of shift depends on the index k). The second term is computed by forming a weighted sum of the activities on the other control units. These two results are then summed together and passed through a leaky integrator and squashing function to form the output of the control unit, c_k. Thus, the c_k essentially derive their inputs directly from a "blob map," and then compete among each other so that the c_k corresponding to the strongest blob prevails.

This circuit could easily be modified to allow for different sizes of the window of attention by adding another set of control units for each desired size of the window of attention. For example, a control unit corresponding to a large window of attention would be connected to the weights, w_{ij}, so as to converge multiple inputs into a single output node, as in figure 13.3c. Such a control unit would then have a larger Gaussian receptive field in the input (or, correspondingly, it would receive its input from a coarser-grained blob map). The control units for each different size and position would then compete among each other to constrain the attentional window to be of a single size and position.

Note that since equations (13.5) and (13.6) are nonlinear in c_k, there exists the potential for getting stuck in local minima. This is not a serious problem for the circuit of figure 13.5, however, because the control neurons are so tightly constrained due to their global connectivity (each control neuron corresponds to a global position of the window of attention, so the winner-take-all circuit ensures a single, affine transformation). On the other hand, in larger circuits where the control neurons must be connected to local groups of synapses instead of globally, the existence of local minima will present a significant problem. This problem can be overcome by utilizing a coarse-to-fine control architecture, in which routing is at first performed by a small number of control neurons on a low-pass filtered version of the image. This smaller set of control neurons can then be used to constrain the activities of the fine-grained, locally connected control neurons that are routing the high-resolution information.

Recognition Once the window of attention has focused on a blob, the underlying high-resolution information can also be fed through the routing circuit and into the input of an associative memory for recognition. However, it is likely that the initial estimation of position and size made by routing the blob was only approximately correct, and this may cause problems for matching the high-resolution information. Thus, it would be desirable to have the associative memory help adjust the position and scale of the attentional window while it converges. How, then, shall the associative memory be incorporated into the control of the routing circuit?

If a Hopfield associative memory (Hopfield 1984) is used for recognition, we can replace the blob search objective function, E_{blob}, with the associative memory's objective function, E_{mem}. Normally, the only dynamic variables in a Hopfield associative memory are the output voltages, V_i, which evolve by simply following a monotonically increasing function of the gradient of the energy.

$$V_i = g_i(u_i^m) \tag{13.7}$$

$$C_i \frac{du_i^m}{dt} = -\frac{\partial E_{\text{mem}}}{\partial V_i}$$

$$= \sum_j T_{ij} V_j - \frac{u_i^m}{R_i} + I_i^{\text{mem}} \tag{13.8}$$

where T_{ij} denotes the connection strength between neurons i and j, C_i and R_i are constants that determine the integration time constant of each neuron, and g is a squashing function (usually tanh). Since the inputs of the associative memory, I_i^{mem}, are to be obtained directly from the outputs of the routing circuit, the c_k now become additional dynamic variables incorporated into the associative memory's energy function. By letting the c_k follow the gradient of the energy, along with the V_i, the combined associative memory/routing circuit should relax to the closest stored pattern and to the correct position and size of the window of attention simultaneously. A dynamic equation for c_k that simultaneously minimizes both the associative memory's energy and the constraint function for preserving topography is given by

$$c_k = \sigma(u_k) \tag{13.9}$$

$$\frac{du_k}{dt} + \alpha u_k = \eta \sum_i \sum_j V_i \Gamma_{ijk} I_j^{\text{in}} + \beta \sum_l T_{kl}^c c_l \tag{13.10}$$

A neural circuit for computing equations (13.9) and (13.10) is shown in figure 13.6. The first term on the right-hand side of equation (13.10) is

Van Essen, Anderson, and Olshausen

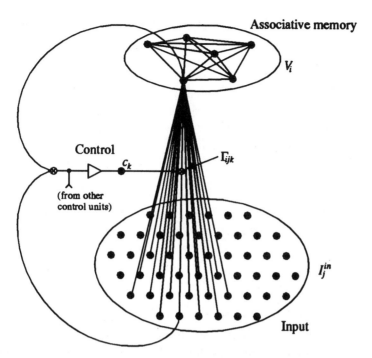

Figure 13.6 An autonomous routing circuit for recognition. Each node of the associative memory receives its external input from an output node of the routing circuit. Hence, each node of the associative memory has dynamic connections to many input nodes. The outputs of the associative memory are then fed back and correlated with the inputs to drive the control units.

computed by correlating the inputs, I_j^{in}, and memory outputs, V_i, whose connection pathways are influenced by control unit c_k (specified by Γ_{ijk}). The other terms are computed as before. Thus, the main qualitative difference between this circuit and the "blob finder" (figure 13.5) is that the control is guided by the interaction between top-down and bottom-up signals rather than purely bottom-up sources.

Again, since the equations (13.9) and (13.10) are nonlinear in c_k, the potential exists for the routing circuit to get stuck in local minima. This problem can be overcome in a similar manner as outlined previously by having the associative memory and routing circuit work at varying levels of resolution. Matching would first be performed at low resolution, and this information would then be used to constrain the matching at progressively higher resolutions.

Shifting Attention to the Next Object   Once an object has been recognized, the window of attention should move on to another interesting part of the scene. One way this could be accomplished would be for the control units to be self-inhibited through a delay. Thus, when a group of control units is active for some time (long enough for recognition to take place) it should begin to shut off. This would allow other blobs or interesting items to compete successfully for control of the window of attention (see also Koch and Ullman 1985).

Computer Simulation   Figure 13.7 shows the results of a computer simulation of a simple attentional system for recognizing objects, based on the principles elucidated above. The network begins in blob search mode, attempting to fill the output of the routing circuit with something interesting. In figure 13.7a the network has settled on the *A*, since it has the greatest overall brightness in the input. (Since the shapes used in this example are so compact and simple we have bypassed the step of prefiltering them into blobs. Thus, during blob search, an object is low-pass filtered by the routing circuit itself.) After settling on a potentially interesting object, the network is switched into recognition mode and the output of the routing circuit is fed to an associative memory. Two patterns—*A* and *C*—have been previously stored in the associative memory. The blurred version of the object initially drives the inputs of the associative memory to begin the pattern search. If the position of the window of attention is slightly off, the blurred version of the object is not affected much and still sends the memory searching in the correct direction. As the associative memory converges, control units compute the correlation between outputs and inputs and set their activation correspondingly. This tends to maximize the similarity between the outputs of the memory and the outputs of the routing circuit, which will also refine the position of the attentional window so that the high-resolution components can be properly matched (figure 13.7b). After allowing a fixed amount of time for the associative memory to converge (another time constant or two), the simulation states the position and presumed identity of the object. The current control state is then self-inhibited and the network switches back into blob search mode. This then puts the next interesting object at a competitive advantage in attracting the window of attention so that it may also be recognized (figure 13.7c,d).

This simulation demonstrates the operation of simple, neural-like circuits for routing and control in both "preattentive" (blob search) and "attentive" (recognition) modes. Although we have greatly oversimplified matters for the purpose of explanation, the same basic principles can be extended to larger, hierarchical circuits. We now turn to the issue of how such circuits may possibly be implemented in the brain.

Neurobiological Substrates and Mechanisms

Figure 13.8a illustrates the major visual processing pathways of the primate brain. Information from the retinogeniculostriate pathway enters the visual cortex through area V1 in the occipital lobe and proceeds through a hierarchy of visual areas that can be subdivided into two major functional streams (Ungerleider and Mishkin 1982). The so-called "form" pathway leads ventrally through V4 and inferotemporal cortex (IT) and is mainly concerned with object identification, regardless of position or size. The so-

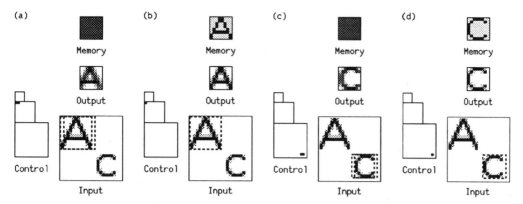

Figure 13.7 Computer simulation of a simple attentional system for recognizing objects. The input to the routing circuit consists of a 22 × 22 pixel array of sample nodes and the output of the routing circuit is an 8 × 8 array of sample nodes. There are three sets of control units, each one corresponding to a different size of the window of attention (small [8 × 8], medium [11 × 11], and large [16 × 16]). Each control neuron within a set corresponds to particular position of the window of attention. The Hopfield associative memory network ("Mem output," see figure 13.6) is composed of 64 units, fully interconnected and arranged into an 8 × 8 grid (i.e., one node for each output of the routing circuit). The dashed outline denotes the position and size of the window of attention. (*a*) In blob search mode, the network settles on the *A*, since it has the greatest overall brightness. (*b*) The network is then switched into recognition mode and settles on the identification of the object. The position and size of the object are encoded in the activities of the control neurons. After a fixed amount of time, the current control state is self-inhibited and the network is switched back into blob search mode so that the next most interesting object may be recognized (*c*, *d*).

called "where" pathway leads dorsally into the posterior parietal complex (PP), and seems to be concerned with the locations and spatial relationships among objects, regardless of their identity. The pulvinar, a subcortical nucleus of the thalamus, makes reciprocal connections with all of these cortical areas (cf. Robinson and Petersen 1992). The following subsections describe how we envision the dynamic routing circuit mapping onto this collection of neural hardware.

Cortical Areas   Figure 13.8b shows the scaled-up routing circuit that we propose as a model of attentional processing in visual cortex. The different stages of the network correspond to the major cortical areas in the "form" pathway. There are two stages for V1: V1a corresponding to layer 4C, and V1b corresponding to superficial layers, because V1 has about twice the density of neurons per unit surface area as the rest of neocortex (O'Kusky and Colonnier, 1982). The remaining areas—V2, V4, and inferotemporal cortex (IT)—occupy one stage apiece. Each node represents, in the simplest sense, a sample of image luminance. More realistically, each node would correspond to a feature vector that is represented by the activity profile of a large group (hundreds or thousands) of neurons in each visual area. For example, in V1, each group would include cells selective for various orientations, spatial frequencies, etc. in a small region of visual space. It

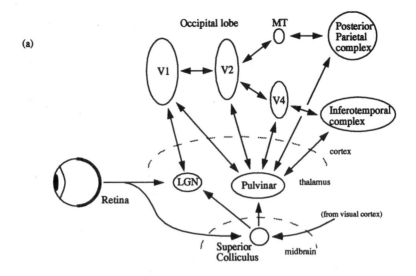

(a)

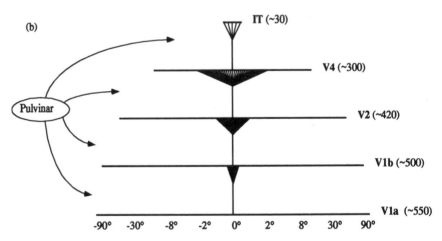

(b)

Figure 13.8 Neurobiological substrates of the model. (*a*) The major visual processing pathways of the primate brain. Some connection pathways (e.g., V4–PP) are not shown to avoid clutter. (*b*) Proposed neuroanatomical substrates for dynamic routing. The label beside each layer indicates the corresponding cortical area and the number of sample nodes in one dimension. The number of sample nodes in two dimensions is approximately the square of this number (of course, there would be many neurons at each node in any model that represented complex features rather than pixels.) At the bottom is shown a scale of the approximate eccentricity of the input nodes to the circuit. Connections are shown for the middle node in each layer. (Individual nodes are indistinguishable here because of their density.) Control signals originate from the pulvinar to effectively gate the feedforward synapses.

Van Essen, Anderson, and Olshausen

is impractical at this stage to include these characteristics explicitly in our model, but we contend that these details can be safely neglected for now without losing the predictive value of the model.

The size of each stage scales roughly with the relative size of each corresponding cortical area. (A notable exception, of course, is the IT complex, due to the fact that much of the neural resources in IT are probably devoted to recognition rather than representing the contents of the window of attention itself.) In addition, the fan-in per node ($\sim$1000:1) and resulting receptive field sizes are consistent with neuroanatomical and neurophysiological data (Gattass et al. 1985; Cherniak 1990; Douglas and Martin 1990). The number of nodes in the output ($\sim$ 30 nodes in diameter) corresponds to the spatial resolution of the window of attention in our model, which seems to be consistent with several lines of psychophysical evidence (Campbell 1985; Van Essen et al. 1991).

A number of physiological experiments indicate that the posterior parietal complex (PP) may be representing the locations of potential attentional targets in the visual scene (Mountcastle et al. 1981; Robinson et al. 1991; Steinmetz et al. 1992). For this reason, we propose that PP may act as a "saliency map" (e.g., Koch and Ullman 1985), analogous to the blob map utilized in the simple attentional system described previously. The superior colliculus may also supplement PP in this role by acting as a crude saliency map, but with a quicker response time due to its direct retinal input. These neurons would then drive the control neurons that compete for control of the window of attention.

Subcortical Areas We hypothesize that the pulvinar plays an important role in providing the control signals required for the routing circuit (see also chapter 5 by Koch and Crick). The pulvinar is reciprocally connected to all areas in the form pathway, thus making it a plausible candidate for modulating information flow from V1 to IT. In addition, the pulvinar receives projections from both PP and superior colliculus, which are known to encode the direction of saccade targets and may also be involved in setting up attentional targets (Posner and Petersen 1990). Finally, neurophysiological studies (Petersen et al. 1985), lesion studies (Desimone et al. 1990), and PET studies (LaBerge and Buchsbaum 1990) of the pulvinar suggest that it plays an important role in visual attention.

A subcortical nucleus such as the pulvinar also has the important property of being spatially localized while at the same time being able to communicate with vast areas of the visual cortex. The relative proximity of pulvinar neurons to each other would facilitate the competitive and cooperative interactions among the control neurons that are necessary to enforce the constraint of having a single window of attention. Intrapulvinar communication could possibly be subserved by interneurons within the pulvinar (Ogren and Hendrickson 1979) or through the reticular nucleus of the thalamus (Conley and Diamond 1990).

Although it is difficult to estimate exactly how many control neurons would be required for a cortical routing circuit, we estimate that roughly 10^6 neurons would suffice, which is in the range of the total number of neurons we estimate for the pulvinar (see Olshausen et al. 1993).

Gating Mechanisms   Neural gating mechanisms are believed to play an important role in many aspects of nervous system function. For example, the extent to which a noxious stimulus is perceived as painful varies greatly as a function of one's emotional state and other external factors. This is subserved, at least in part, by gating mechanisms in the spinal cord, where descending fibers from the raphe nuclei form part of a control system that modulates pain transmission via presynaptic inhibition in the dorsal horn (Fields and Basbaum 1978). Gating mechanisms are also thought to play an important role in sensorimotor coordination; for example, central pattern generators in the spinal cord are known to gate sensory inputs according to the phase of the movement cycle in which the input occurs (Sillar 1991). A somewhat different form of gating seems to take place in the LGN, in which thalamic relay cells seem to exhibit two distinct response modes: a *relay* mode, in which cells tend to more or less faithfully replicate retinal input, and a nonrelay *burst* mode, in which cells burst in a rhythmic pattern that bears little resemblance to the retinal input (Sherman and Koch 1986). In this instance, the reticular nucleus of the thalamus is thought to be the source of the signal that switches the LGN into the nonrelay burst mode.

Although there is as yet no explicit evidence for gating mechanisms in the visual cortex, there are several possible biophysical mechanisms that would allow control neurons to gate synapses along the V1-IT pathway. Presynaptic inhibition, as in the spinal cord, would provide the most localized gating effect. However, to date there exists no morphological evidence for this type of mechanism (Berman et al. 1992). Postsynaptically, a control neuron could decrease or possibly nullify the efficacy of a corticocortical synapse via shunting inhibition. Evidence for this type of mechanism playing a role in orientation or direction tuning is mixed, with some for (Volgushev et al. 1992; Pei et al. 1992) and some against (Douglas et al. 1988). Another possible postsynaptic gating mechanism could be realized via the combined voltage- and ligand-gated NMDA receptor channel, which has been shown to play an important role in normal visual function (Nelson and Sur 1992; Miller et al. 1989). In this case, a neuron could effectively boost the gain of a corticocortical synapse by locally depolarizing the membrane in the vicinity of the synapse. Also, there exist voltage-gated Ca^{2+} channels in dendrites that could provide nonlinear coupling between inputs (Llinás, 1988a). All of these mechanisms, and possibly others, offer a multiplicative-type effect that is suitable for gating information flow through the cortex (see also Koch and Poggio 1992).

From a computational viewpoint, gating inputs within the dendrites provides a much higher degree of flexibility than would merely gating the outputs of pyramidal cells. Since the output of a pyramidal cell may branch to several cortical areas and make synaptic connections to a multitude of neurons, any modulation of the cell's output will simply be duplicated at all these input points. Gating inputs within the dendrites, on the other hand, allows many intermediate results, $\sum_k c_k \Gamma_{ijk} I_j^{in}$, to be computed within the postsynaptic membrane and then summed together within a single cell. This results in a computational structure far richer and more compact (Mel 1992), and provides a higher degree of flexibility in remapping visual information. We believe the demonstrable computational advantage of dendritic gating mechanisms for visual processing motivates the need to specifically look for such mechanisms experimentally.

Predictions

Neurophysiology   The most obvious prediction of the dynamic routing circuit model is that the receptive fields of cortical neurons should change their position or size as attention is shifted or rescaled. This effect should be especially pronounced in higher cortical areas. Moran and Desimone (1985) found that receptive fields of neurons in areas V4 and IT of primate visual cortex seem to be dynamically modulated so that unattended stimuli have a reduced effect on the cells response, even though they lie within the classical receptive field. This result is consistent with the prediction of the model, as explained graphically in figure 13.9. The stronger prediction of the model—that receptive fields should shift and rescale proportional to the position and size of the attentional window—needs to be tested by explicitly mapping out receptive fields under different attentional conditions.

Since we hypothesize that pulvinar neurons control the remapping process, we would predict that lesions to the pulvinar should dramatically degrade attentional and pattern recognition abilities. Neurophysiological data thus far indicate that pulvinar lesions do indeed degrade attentional capabilities, but tests of pattern recognition capabilities (e.g., Chalupa et al. 1976) have used such simple stimuli that it is difficult to discern to what extent detailed spatial recognition is affected. One would also expect to find some form of enhancement in the response of pulvinar neurons projecting to those areas of the cortex within the topographic vicinity of the attentional beam. Petersen et al. (1985) reported such an enhancement effect for neurons in the dorsomedial portion of the pulvinar (which is connected with PP), but not in the inferior or lateral portion (which is connected to V1-IT). The lack of enhancement here may be due to the fact that the task used in this experiment was very simple (detecting the dimming of a spot of light).

Neuroanatomy   Our particular routing circuit model predicts that the size of the cortical region from which a cell receives its input should increase by roughly a factor of two at each stage in the hierarchy of visual areas in the form pathway (see figure 13.8). While there exists some evidence in support of this prediction—for example, projections from V4 to IT are more diffuse than projections from V1 to V2 (Van Essen et al. 1986; DeYoe and Sisola 1991; see also Rockland 1992b)—more accurate and higher resolution

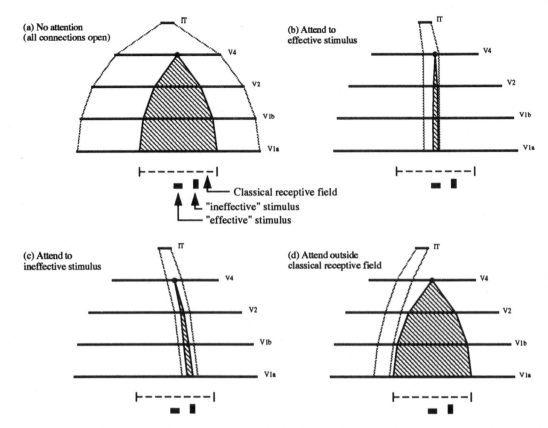

Figure 13.9 The dynamic routing circuit interpretation of the Moran and Desimone (1985) experiment. The node in layer V4 indicates the cell under scrutiny. The hatched region indicates those connections to the cell that are enabled; the others are disabled. The bounds of the window of attention in each area are shown by the stippled lines. (*a*) In the nonattentive state, all connections will be open and the effective stimulus can excite the cell anywhere within its classical receptive field. (*b*) When attending to the effective stimulus, the cell's response should be unaltered since the neural pathways to the stimulus are still open. (*c*) When attending to the ineffective stimulus, the cell's response should decrease substantially since the neural pathways to the effective stimulus are gated out. (*d*) When attending outside the cell's classical receptive field, there is no need to gate the cell's inputs since it is no longer taking part in the process of routing information within the window of attention.

data are needed to confirm or contradict the specific architecture of the proposed routing circuit. The model also predicts that pulvinar afferents should terminate in the cortex in such a way that they could effectively modulate intercortical synaptic strengths. Neuroanatomical studies thus far seem to be in agreement with this prediction (e.g., Trojanowski and Jacobson 1976), but it would be of interest to know if the pulvinar afferents make contact with inhibitory interneurons or directly onto the dendrites of pyramidal cells. If the latter is true, it would be of interest to know whether these synapses are made near corticocortical synapses. Finally, the model predicts that there should exist lateral interconnections among pulvinar neurons to constrain their activity to be consistent with a single position and

Van Essen, Anderson, and Olshausen

size of the window of attention. This is partially supported by the existence of interneurons within the pulvinar (Ogren and Hendrickson 1979), but it remains to be seen if the axons of projection neurons have collaterals that spread horizontally within the pulvinar, or to what extent the reticular nucleus of the thalamus may subserve intrapulvinar communication.

Psychophysics   The number of sample nodes in the top layer of the routing circuit implies that the spatial resolution of the window of attention is limited to a diameter of about 30 pixels. Although this estimate is consistent with several lines of psychophysical evidence, including studies of spatial acuity, contrast sensitivity to gratings, and recognition (Campbell 1985; Van Essen et al. 1991), none of these studies was actually directed at studying visual attention. Most of the experiments had long display times that could conceivably have allowed several attentional fixations (although we doubt that this would have been a major contaminating factor in most cases). One possible approach to testing this prediction more thoroughly would be to test pattern discrimination ability as a function of the position, size, and resolution of the object. Assuming a subject could be properly pre-cued to attend to a certain position and size of the visual field, and that display times were limited to the order of 50 msec, we would predict that performance would drop off sharply once the task-specific spatial frequency content of the stimulus exceeded approximately 15 cycles across the object.

The model also suggests that once a location has been attended to in the visual field, it should be difficult to stay there or immediately revisit it since the the control neurons corresponding to that part of the visual field are currently inhibited from firing. This is consistent with the psychophysical observation that involuntary attentional fixations tend to be transient (Nakayama and Mackeben 1989) and appear to be inhibited from return (Posner and Cohen 1984).

Recognition of Highly Complex Patterns   The current version of our dynamic routing model is capable of subserving translation and scale-invariant pattern discrimination when tested with small numbers of relatively simple stimuli (see figure 13.7). These rudimentary capabilities are a far cry from those of our own visual system, which can effortlessly discriminate, for example, among hundreds or thousands of human faces, independent of size, position, and viewing angle. Besides quantitatively scaling to a higher density representation, two types of qualitative refinement, both alluded to already, will be needed for our model to more closely approach human performance. First, there needs to be a much more sophisticated control structure that provides for warping of image representations, in addition to the translational and scaling capacities already achieved in our current model. Second, substantial processing of form and textural cues should be carried out at early and intermediate stages, rather than having only pixel-based information transmitted to the high-level associative memory. For example, it might be sensible to have cells at the high levels

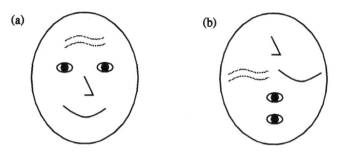

Figure 13.10 The importance of preserving spatial relationships. The two objects (*a*) and (*b*) contain the same spatial features but result in very different percepts.

of the routing circuit, just before the associative memory stage (cf. figure 13.2), that have attained selectivity for complex local features, such as the shapes of particular parts of the face (e.g., the eyes, nose, or mouth). Indeed, cells with characteristics of this type have been reported in posterior inferotemporal cortex of anesthetized monkeys (Fujita et al. 1992). A crucial (but untested) aspect of our model is that such cells should also be tuned for the location of features within the window of attention, thereby preserving an explicit representation of local spatial relationships.

To illustrate how this strategy might work, consider the example of two readily distinguishable cartoon faces that are made up of identical elements, one in a natural configuration (figure 13.10a) and the other in a spatially scrambled configuration where the positions of the eyes, nose, and mouth are interchanged (figure 13.10b). When one attends to the face as a whole, we propose that each pattern activates very different populations of the aforementioned feature-selective neurons in posterior inferotemporal cortex, even though they consist of identical components. For example, the normal face would activate cells that are selective for eyes situated in the upper part of the window of attention, whereas the scrambled face would activate a different set of cells selective for eyes situated in the middle and lower part of the window. (The former cells might be more numerous than the latter as a result of biased exposure in normal visual experience.) In each case, the inputs would feed into an associative memory, where they would elicit different patterns of activity and hence different visual percepts. We contend that the preservation of information about local spatial relationships up to (but not including) the associative memory stage provides an efficient basis for discriminating among a wide variety of complex natural objects.

One possible concern about this proposed strategy is the requirement for a large number of cells, because these would need to be tuned for many different positions of each featural type relative to the boundaries of the window of attention. This difficulty could be minimized by having reasonably broad tuning for position as well as featural characteristics. In any event, similar issues must be addressed by any model that is capable of complex pattern recognition. For example, one alternative would be to

Van Essen, Anderson, and Olshausen

have cells that are explicitly tuned for the distance between particular features as well as the orientation of the axis between them (F. Crick, personal communication). To discriminate between figure 13.10a and 13.10b using this strategy, it would be important to have cells that are selective both for particular feature conjunctions and for certain positional relationships (e.g., a nose-like feature that is a particular distance directly above a mouth-like feature). We contend that this strategy would lead to an even more severe combinatorial explosion in the numbers of cells needed to encode all of the requisite combinations of featural and positional cues.

Comparison with Other Network Models

Control vs. Synchronicity   Recently, there has been widespread interest in the possibility that synchrony of neural firing could serve as a code for linking features common to a given object. Synchronous activation could operate by transient increases in connection strengths (Crick 1984; von der Malsburg and Bienenstock 1986). In this way, temporal information might be used to solve the "binding problem" (see chapter 10 by Singer) and thereby mediate aspects of figure/ground segregation, attention, and perhaps even consciousness (see chapter 5 by Koch and Crick). A potential weakness common to these approaches is that information about what is being connected to what at any instant in time is not explicitly encoded anywhere in the system. In our model, this information is encoded explicitly in the activities of the control neurons, which then allows it to be utilized advantageously in a number of ways.

One way that information about connectivity can be utilized is in constraining the active connections between retinal and object-based reference frames to be in accordance with a global shift and scale transformation. This constraint is incorporated in our model via the competitive and cooperative interactions among the control neurons. During object recognition, this constraint drastically reduces the number of degrees of freedom in matching points between the retinal and object-centered reference frames, because once a few point-to-point correspondences have been established, the number of potential matches between other pairs of points is greatly reduced. Researchers in machine vision have termed this the *viewpoint consistency constraint*, and it has proved to be a powerful computational strategy for object recognition systems (Hinton 1981b; Lowe 1987).

Another advantage of having information about active connection states readily available is that the ensemble of control neurons together forms a neural code for the current position and size of the window of attention. Therefore, the position and size of an object can be inferred by simply reading-out the state of the control neurons. In addition, it would also be possible for the control neurons to warp the reference frame transformation to form object representations that are invariant to distortion (e.g., handwritten digits), in which case information about the particular shape of the object (e.g., its slant or style) could also be preserved. Note that such information is typically lost in networks that utilize feature hierarchies of complex cells (Fukushima 1980, 1987; LeCun et al. 1990) or Fourier transforms (e.g., Pollen et al. 1971; Cavanagh 1978, 1985) for forming position-, scale-, and/or distortion-invariant representations.

One final advantage of having control explicitly represented is that it allows attention to be easily directed "at will," or by other modalities, since those areas of the brain that have access to the control neurons (such as parietal cortex) can directly influence where attention is directed. This also provides a convenient format for mediating the access to control among various competing demands.

An interesting issue is whether the control strategy advocated in our model could be implemented using oscillations or temporal synchrony rather than the multiplicative synaptic interactions suggested in an earlier section. While we cannot rule this out, we find it difficult to envision precisely how such a model could route information flow within the cortex while simultaneously preserving local spatial relationships within the attentional window. Alternatively, oscillations might provide, by analogy to digital computers, a clocking signal to control the precise timing of switches in information flow.

Control-Based Network Models   A number of other network models of attention have also utilized the concept of control neurons for directing information flow. Niebur et al. (1993), Ahmad (1992), Tsotsos (1991), and Mozer and Behrmann (1990), among others, have proposed various schemes for selecting and routing information from a select portion of the visual scene. However, none of these models explicitly preserves spatial relationships within the window of attention, which we consider to be a critical component of the routing process. Hinton (1981a), Hinton and Lang (1985), and Sandon (1990) have proposed control-based models that do preserve spatial relationships within the window of attention and share the same basic flavor as the model presented here (i.e., remapping object representations from retinal into object-centered reference frames). Although these latter models attempt to model psychophysical data, we feel that they lack the necessary level of neurobiological detail to give them strongly predictive value in biology.

Recently, Postma et al. (1992) proposed a neural model based on the original shifter circuit proposal (Anderson and Van Essen 1987) to account for translational invariance in visual object priming (Biederman and Cooper 1992). This model shares many similarities to the model presented here, including top-down, or template-driven control, although it differs in the specifics of the control structure.

Ullman (see chapter 12 by Ullman) has proposed that pattern recognition is achieved by a "counterstreams" strategy, in which information about stored patterns flows top-down at the same time that information about currently viewed patterns flows in the bottom-up direction. Multiple coexisting representations flow in each direction, and recognition is manifested by a winner-take-all competition to find the best match between patterns propagating in the two directions. His model shares with ours the notion that explicit control processes are needed to regulate information flow. However, these processes are different in many respects, particularly with regard to the multiplicity of actively propagated stimulus representations.

Lastly, there are important lines of convergence and divergence in comparing our model to the ideas of Mumford (see chapter 7 by Mumford) that are based on Grenander's Pattern Theory (Grenander 1976–81). He invokes the need for "domain warping" that can be mediated by neural shifter circuits analogous to those we have proposed.

A MODEL OF DIRECTED MOTOR CONTROL

Computational Framework

In the introduction, we argued that directed motor control and directed visual attention share common computational underpinnings, in terms of requirements for a high degree of anatomical convergence and divergence and for mechanisms for dynamically controlling information flow. These commonalities are illustrated schematically in figure 13.11. As noted already, visual attention involves the transient selection of a tiny subset of the information being transmitted along the optic nerve. In the preceding section, our model included only a single high-level center mediating all aspects of pattern recognition. In reality, though, there may well be distinct neural populations responsible for qualitatively different types of pattern recognition. For example, the population of cells responsible for recognizing faces may be largely or entirely different from those responsible for recognizing alphanumeric characters or from those responsible for recognizing different flowers and trees. If so, there needs to be an output selection process for determining the target population to which attended information is sent in addition to the aforementioned input selection process for directing the position and scale of the window of attention (figure 13.11a; see also figure 5 in Anderson and Van Essen 1987). For simplicity, we assume here that the different target populations can be represented by separate entries at the top, even if they happen to overlap physically within the cortex.

For directed motor control, there needs to be a cognitively driven input selection process to generate the appropriate motor routine and an output selection process that determines to which appendage or combination of appendages this pattern is directed. Consider, once again, the example of the orchestral conductor discussed in the introduction. The basic rhythm (say, for a brisk march having two beats per measure) is presumably represented in some central pattern generator (an internal metronome, in effect) that can be activated and modulated by auditory, visual, and other cues (figure 13.11b). For any given rhythm, the conductor must transform this beat into a particular motor routine (e.g., up-down or left-right strokes, or circular strokes) and then direct this routine to the appropriate digit(s) or

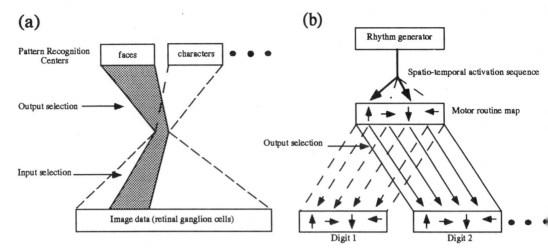

(a)

Pattern Recognition Centers

faces   characters • • •

Output selection →

Input selection →

Image data (retinal ganglion cells)

(b)

Rhythm generator

Spatio-temporal activation sequence

↑ → ↓ ←   Motor routine map

Output selection →

↑ → ↓ ←   ↑ → ↓ ← • • •

Digit 1   Digit 2

Figure 13.11 Control of information flow for visual attention (*a*) and directed motor control (*b*). The direction of information flow is bottom-up for sensory processing and top-down for motor function, but in both cases we propose the need for separate control mechanisms to regulate the flow dynamically. The attentional selection process tends to be relatively transient, whereas directed motor routines can be very sustained.

other appendage(s) (cf. Keele 1981). Note that information flow is directed downward in this scheme, because it involves top-down communication from high-level centers to ones that are closer to the periphery.

A Simplified Vector-Code Scheme

The motor system is extremely complex and is arguably more diverse than the visual system in the layout of its major components, which include cerebellar as well as cerebral cortical areas and numerous subcortical nuclei in the forebrain (basal ganglia), thalamus, midbrain, and brainstem. It has been intensively studied using anatomical, physiological, and computational approaches in attempting to decipher how information flows and is processed in this complex network (cf. Houk et al. 1993; Hoover and Strick 1993). For our purpose in this preliminary analysis, it is sufficient to focus on a highly simplified scheme in which the representation at both middle and lower levels in figure 13.11b involves a vector-code strategy similar to that proposed by Georgopolous and colleagues for primary motor cortex (Georgopolous et al. 1988, 1989). In this scheme, each neuron "votes" for a particular direction of movement for the portion of the limb that it influences, and the strength of its vote is proportional to its firing rate. We further assume that there is an orderly representation of movement direction across the cortical surface. In the example illustrated, neurons on the left represent upward movement, and neurons further to the right represent progressively more clockwise directions of movement. Thus, the up-down movements schematized in figure 13.1 would be represented as a motor routine in which activity would alternate between neurons repre-

Van Essen, Anderson, and Olshausen

senting upward and downward movement. To have a particular digit or other appendage execute this movement, this stereotyped spatiotemporal pattern would be selectively routed to the neural populations in motor cortex that control the relevant digits. By having direction of movement be encoded in the same way at both levels in this scheme, the routing process is simplified so that neurons at the lower level can project in a 1:1 mapping to neurons within the representation for each appendage. Note that for conceptual clarity, we have treated the representation of each digit or other appendage as a separate neural ensemble. In reality, it is likely that these are often overlapping, interleaved ensembles of neurons (Schieber 1990, 1992). However, we will ignore these complexities in our zeroth-order model, for the same reason that our attention model involves major simplifications that we regard as not crucial to the initial formulation even though they in due course will necessitate expansion and refinement of the model.

In considering how this scheme might be implemented in the primate motor system, it is important to identify the cortical and/or subcortical structures in which different motor routines are represented and the major anatomical pathways along which this information is transmitted. Unfortunately, these basic issues are not well understood. One interesting observation is that primary motor cortex is at a lower level than other motor areas in an anatomically based hierarchy that derives from the laminar patterns in which different pathways originate and terminate (Felleman and Van Essen 1991). A priori, there is no compelling reason why this need be the case, as opposed to a situation in which primary motor cortex was at the pinnacle of a sensory input to motor output hierarchy. The observed relationship fits nicely with our presumption that directed motor control involves top-down flow of information, as already indicated in figure 13.11. However, it raises the puzzling possibility that the spatiotemporal patterns associated with different motor routines might be communicated between cortical areas via anatomically descending (feedback) pathways, whose terminations preferentially avoid the middle layers of cortex. This would contrast with the basic pattern of information flow in visual, auditory, and somatosensory systems, which is characterized by ascending inputs that preferentially terminate in the middle layers of cortex (see Felleman and Van Essen 1991). On the other hand, it would correspond to the pattern of information flow in olfactory cortex and related areas, where ascending pathways tend to terminate in layer 1 and other superficial layers (Haberly and Price 1978; Carmichael 1993).

An intriguing alternative is that information about a given motor routine might be communicated indirectly between cortical areas, via the well-known route involving the basal ganglia and thalamus (Alexander et al. 1986), rather than via direct corticocortical connections. Since thalamic motor nuclei (VA and VL) terminate mainly in the middle layers of motor cortex (Sloper and Powell 1979), this would represent a better match to the laminar patterns associated with ascending information flow in sensory cortex. In short, our current level of understanding is insufficient to determine whether the information for encoding a given motor routine (1) flows through cortical areas and is gated by thalamic projections, (2) flows through corticosubcortical loops and is gated by descending cortico-cortical pathways, or (3) flows in a very different way that does not involve the explicit gating mechanisms proposed in our model.

COGNITIVE PROCESSING

In this chapter, we have concentrated on sensory and motor processing, where the conceptual issues can be most sharply formulated and their pos-sible neurobiological substrates most rigorously analyzed. However, we suspect that similar processing strategies apply also to language process-ing and to other cognitive functions as well. This supposition derives from thinking about examples such as the following.

Suppose that an observer is asked to remember the identity of an object (say, an apple) whose picture is briefly flashed on a screen. After a brief delay, the observer is asked to report the identity of the object by (1) stating its name verbally, (2) writing its name on a sheet of paper, or (3) signing its name in American Sign Language. In another variation of this task, a multilingual observer could be asked to state the name of the object in different languages (e.g., English, French, or Polish). In situations of this type, we presume that information about the object's identity is stored in a restricted number of central representations and that this information can be routed to a variety of different output modalities and in a variety of different formats. For the purposes of this argument, it does not matter whether the object's identity is stored as a visual memory, in a particular language-specific format, or as some completely abstract cognitive con-struct. Whatever the format of the representation, the brain must be able to access the information, rapidly translate it into an appropriate format, and transmit it to the appropriate motor structures. We have already ar-gued that the sensory and motor components of such tasks are likely to entail dynamic routing strategies. It requires only a modest conceptual leap to suppose that analogous routing strategies may be used to control the flow of information in whatever central structures are used to represent semantic information and other high-level abstractions that are the coinage of cognitive function.

CONCLUDING REMARKS

Our approach to formulating models of cortical function has been guided by a number of computational and systems engineering considerations. Some of these can be illustrated by the following analogy. We regard the brain as a system designed to treat information as an essential commodity, much as an efficient factory is designed for optimal handling of the phys-ical materials that traffic across its floors. In both cases, the raw materials that enter the system generally represent only a small fraction of the final

Van Essen, Anderson, and Olshausen

product. The production process involves careful selection of useful materials, discarding of excess or unnecessary materials, and transforming and repackaging of the desired materials in an appropriate configuration for the particular applications for which the product is intended. For efficient function, the flow of material must be carefully monitored and controlled. This requires specialized systems that are explicitly designed for this purpose, rather than for construction and fabrication processes per se.

In the nervous system, we believe that an important general strategy for attaining these objectives is the systematic use of multiplicative operations, whereby one set of inputs from a class of "control neurons" dynamically modulates the connections between two other groups of neurons. In this chapter our emphasis has been on the modulation of feedforward pathways, although in the future the coordinated modulation of the corresponding feedback pathways will be included in our discussions. Conventional neural network models typically rely on computations that are dominated by linear combinations of synaptic feedforward inputs followed by a nonlinear operation. This simple neural network structure has proven to be too rigid and unwieldy when applied to large problems. More complex nonlinearities have been introduced to achieve flexibility in neural computational systems using, for example, dynamic links (von der Malsburg and Bienenstock 1986) or oscillations (Koch and Crick, chapter 5; Singer, chapter 10). While these models have encompassed feedback projections, they have kept the basic input-output structure and have effectively introduced what we would call control functions in an implicit fashion, rather than the explicit fashion we favor. We suggest that models that do not distinguish control functions from information flow and processing will not scale well with increased problem complexity.

Explicit three-way interactions provide a contextual framework for modifying the interpretation of information, which is an essential ingredient for cognitive processing. Hierarchical interpretative systems can be designed in a cleaner more efficient fashion using a substrate of modules specialized in terms not only of memory and processing, but control as well. Implementing complex circuits of the type described in this chapter requires their physical structure to be largely laid down by genetic factors. Numerous anatomical and physiological facts support the picture of a rather well-defined structure for the neocortex and its connections to subcortical bodies that is replicated across individuals and not modified on a gross scale by experience. This would suggest there are preprogrammed strategies for learning how to interact with and adapt to the environment. Thus, while the details of the information about complex objects contained within the inferotemporal cortex differs between individuals, the strategies for acquiring that information, and the way it is stored and retrieved, is presumed to be very similar. Our brains did not evolve as general purpose computational systems, but rather to compete and thrive in the rich, but restricted context of human society and the world we live in.

References

Abeles, M. (1991) *Corticonics: Neural Circuits of the Cerebral Cortex.* Cambridge University Press, Cambridge.

Abeles, M. and Gerstein, G.L. (1988) Detecting spatiotemporal firing patterns among simultaneously recorded single neurons. *J. Neurophysiol.* 60: 909–924.

Abelson, H. and Sussman, G.J. (1985) *Structure and Interpretation of Computer Programs.* MIT Press, Cambridge, MA.

Adrian, E.D. (1935) *The Mechanism of Nervous Action.* University of Pennsylvania Press, Philadelphia.

Adrian, E.D. (1941) Afferent discharges to the cerebral cortex from peripheral sense organs. *J. Physiol. (Lond.)* 100: 159–191.

Adrian, E.D. (1942) Olfactory reactions in the brain of the hedgehog. *J. Physiol.* 100: 459–473.

Aertsen, A.M.H.J. and Gerstein, G.L. (1985) Evaluation of neuronal connectivity: Sensitivity of cross-correlation. *Brain Res.* 340: 41–354.

Aertsen, A.M.H.J., Gerstein, G.L., Habib, M.K., and Palm, G. (1989) Dynamics of neuronal firing correlation: Modulation of effective connectivity. *J. Neurophysiol.* 61: 900–919.

Aertsen, A., Vaadia, E., Abeles, M., Ahissar, E., Bergmann, I.I., Carmon, B., Lavner, Y., Margalit, E., Nelken, I., and Rotter, S. (1991) Neural interactions in the frontal cortex of a behaving monkey. Signs of dependence on stimulus context and behavioral states. *J. Hirnforschung* 32: 735–743.

Agmon, A. and Connors, B.W. (1992) Correlation between intrinsic firing patterns and thalamocortical synaptic responses of neurons in mouse barrel cortex. *J. Neurosci.* 12: 319–329.

Ahissar, E. and Vaadia, E. (1990) Oscillatory activity of single units in the somatosensory cortex of an awake monkey and their possible role in texture analysis. *Proc. Natl. Acad. Sci. USA* 87: 8935–8939.

Ahissar, E., Vaadia, E., Ahissar, M., Bergmann, H., Arieli, A., and Abeles, M. (1992) Dependence of cortical plasticity on correlated activity of single neurons and on behavioral context. *Science* 257: 1412–1415.

Ahmad, S. (1992) VISIT: A neural model of covert visual attention. In: *Advances in Neural Information Processing Systems* 4, Moody, J.E., Hanson, S.J., and Lippman, R.P. (eds.). Morgan Kaufmann, San Mateo, CA.

Aiple, F. and Krüger, J.(1988) Neuronal synchrony in monkey striate cortex: Interocular signal flow and dependency on spike rates. *Exp. Brain Res.* 72: 141–149.

Alexander, G.E., DeLong, M.R., and Strick, P.L. (1986) Parallel organization of functionally segregated circuits linking basal ganglia and cortex. *Annu. Rev. Neurosci.* 4: 357–381.

Allman, J. (1987) Evolution of the brain in primates. In: *The Oxford Companion to the Mind*, Gregory, R.L. (ed.). Oxford University Press, Oxford.

Allman, J., Miezin, F., and McGuinness, E. (1985a) Stimulus specific responses from beyond the classical receptive field: Neurophysiological mechanisms for local-global comparisons in visual neurons. *Annu. Rev. Neurosci.* 8: 407–430.

Allman, J.M., Miezin, F., and McGuinness, E. (1985b) Stimulus specific responses from beyond the receptive field: Neurophysiological mechanisms for local-global comparisons of visual motion. *Annu. Rev. Neurosci.* 3: 532–548.

Allport, D.A. (1989) Visual attention. In: *Foundations of Cognitive Science*, Posner, M.I. (ed.). Bradford Books/MIT Press, Cambridge, MA, pp. 636–682.

Aloimonos, Y. and Rosenfeld, A. (1991) Computer vision. *Science* 253: 1249–1254.

Amitai, Y., Friedman, A., Connors, B.W., and Gutnick, M.J. (1993) Regenerative activity in apical dendrites of pyramidal cells in neocortex. *Cerebral Cort.* 3: 26–38.

Andersen, P. (1990) Synaptic integration in hippocampal CA1 pyramids. *Prog. Brain Res.* 83: 215–222.

Andersen, R.A. (1987) The role of the inferior parietal lobule in spatial perception and visual-motor integration. In: *The Handbook of Physiology, Section 1: The Nervous System, Volume IV, Higher Functions of the Brain, Part 2*, American Physiological Society, Bethesda, MD, pp. 483–518.

Andersen, R.A., Asanuma, C., Essick, G., and Siegel, R.M. (1990) Cortico-cortical connections of anatomically and physiologically defined subdivisions within the inferior parietal lobule. *J. Comp. Neurol.* 296: 65–113.

Andersen, R.A., Siegel, R.M., and Essick, G.K. (1987) Neurons of area 7 activated by both visual stimuli and oculomotor behavior. *Exp. Brain Res.* 67: 316–322.

Anderson, C.H. and Van Essen, D.C. (1987) Shifter circuits: A computational strategy for dynamic aspects of visual processing. *Proc. Natl. Acad. Sci. USA* 84: 6297–6301.

Anderson, C.H. and Van Essen, D.C. (1993) Dynamic neural routing circuits. In: *Proceedings of the Second International Conference on Visual Search*, Brogan, D., Taylor, and Francis (eds.). London, pp. 311–319.

Arbib, M.A. (1989) *The Metaphorical Brain 2: Neural Networks and Beyond*. Wiley, New York.

Ariel, M., Daw, N.W., and Rader, R.K. (1983) Rhythmicity in rabbit retinal ganglion cell responses. *Vision Res.* 23: 1485–1493.

Arnett, D.W. (1975) Correlation analysis in the cat dorsal lateral geniculate nucleus. *Exp. Brain Res.* 24: 111–130.

Artola, A. and Singer, W. (1987) Long-term potentiation and NMDA receptors in rat visual cortex. *Nature* 330: 649–652.

Artola, A. and Singer, W. (1990) The involvement of N-methyl-D-aspartate in rat visual cortex. *Eur. J. Neurosci.* 2: 254–269.

Atick, J.J. and Redlich, A.N. (1992) What does the retina know about natural scenes? *Neural Comp.* 4: 196–210.

Baddeley, A. (1986) *Working Memory*. Oxford University Press, Oxford.

Baddeley, A. (1990) *Human Memory*. Allyn and Bacon, Boston.

Bair, W., Koch, C., Newsome, W., Britten, K., and Niebur, E. (1992) Power spectrum analysis of MT neurons from awake monkey. *Soc. Neurosci. Abstr.* 18: 11–12.

Bair, W., Koch, C., Newsome, W., and Britten K. (1993) Temporal structure of spike trains from MT neurons in the awake monkey. In: *Computation and Neural Systems '92*, Eeckman, F.H. and Bower, J.M. (eds.). Kluwer Academic Publishers, Boston, pp. 495–502.

Bair, W., Koch, C., Newsome, W., and Britten, K. (1994) Power spectrum analysis of bursting cells in area MT in the behaving monkey. J. Neurosci., in press.

Bajcsy, R. (1988). Active perception. *Proc. IEEE* 76: 996–1005.

Ballard, D.H. (1986) Cortical connections and parallel processing: Structure and function. *Behav. Brain Sci.* 9: 67–120.

Ballard, D.H. (1991) Animate vision. *Art. Intell.* 48: 57–86.

Ballard, D.H. and Whitehead, S.D. (1990) Active perception and reinforcement learning. *Neural Comp.* 2: 409–419.

Ballard, D.H., Hayhoe, M.M., Feng, M.L., and Whitehead, S.D. (1992) Hand-eye coordination during sequential tasks. *Phil. Trans. Roy. Soc. London B* 337: 331–339.

Barlow, H.B. (1959) Sensory mechanisms, the reduction of redundancy, and intelligence. In: *The Mechanisation of Thought Processes.* Her Majesty's Stationery Office, London, pp. 535–539.

Barlow, H.B. (1972) Single units and cognition: A neurone doctrine for perceptual psychology. *Perception* 1: 371–394.

Barlow, H.B. (1985) Cerebral cortex as a model builder. In: *Models of the Visual Cortex*, Rose, D., and Dobson, V.G. (eds.). John Wiley, New York, pp. 37–46.

Barlow, H.B. (1989) Unsupervised learning. *Neural Comp.* 1: 295–311.

Barlow, H.B. (1990) A theory about the functional role and synaptic mechanism of visual after-effects. In: *Vision: Coding and Efficiency*, Blakemore, C.B. (ed.). Cambridge University Press, Cambridge.

Barlow, H.B. (1991) Vision tells you more than "What is Where." In: *Representations of Vision*, Gorea, A. (ed.). Cambridge University Press, Cambridge, pp. 319–329.

Barlow, H.B., and Földiák, P. (1989) Adaptation and decorrelation in the cortex. In: *The Computing Neuron*, Durbin, R., Miall, C., and Mitchison, G. (eds.). Addison-Wesley, Wokingham, UK, pp. 54–72.

Barlow, H.B. and Reeves, H.B. (1979) The versatility and absolute efficiency of detecting mirror symmetry in random dot displays. *Vision Res.* 19: 783–793.

Barlow, H.B. and Tolhurst, D.J. (1992) Why do you have edge detectors? In: *1992 Technical Digest Series*, Vol. 23. Optical Society of America, Washington, D.C., p. 172.

Baron, R.J. (1981) Mechanisms of human facial recognition. *Int. J. Man Mach. Stud.* 15: 137–178.

Baron, R.J. (1987) *The Cerebral Computer.* Erlbaum, Hillsdale, NJ.

Barrett, E.F. and Stevens, C.F. (1972) The kinetics of transmitter release at the frog neuromuscular junction. *J. Physiol.* 227: 691–708.

Bartlett, R.C. (1958) *Thinking.* Basic Books, New York.

Basar, E. (1980) *EEG-Brain Dynamics.* Elsevier, Amsterdam.

Basar, E. (1988) EEG-Dynamics and evoked potentials in sensory and cognitive processing by the brain. In: *Dynamics of Sensory and Cognitive Processing by the Brain*, Springer Series in Brain Dynamics, Vol. 1, Basar E. (ed.). Springer, Berlin, pp. 30–55.

Basar, E. and Bullock, T. (ed.) (1992) *Induced Rhythms in the Brain*. Birkhauser, Boston.

Basri, R. and Ullman, S. (1989) Recognition by linear combinations of models. Technical report, The Weizmann Institute of Science.

Baylis, G.C. and Rolls, E.T. (1987) Responses of neurons in the inferior temporal cortex in short term and serial recognition memory tasks. *Exp. Brain Res.* 65: 614–622.

Baylis, G.C., Rolls, E.T., and Leonard, C.M. (1985) Selectivity between faces in the responses of a population of neurons in the cortex in the superior temporal sulcus of the monkey. *Brain Res.* 342: 91–102.

Beer, R.J. (1990) *Intelligence as Adaptive Behavior*, Academic Press, New York.

Bekkers, J.M. and Stevens, C.F. (1989) NMDA and non-NMDA receptors are co-localized at individual excitatory synapses in cultured rat hippocampus. *Nature* 341: 230–233.

Bekkers, J.M. and Stevens, C.F. (1990) Presynaptic mechanism for long-term potentiation in the hippocampus. *Nature* 346: 724–729.

Bekkers, J.M., Richardson, G.B., and Stevens, C.F. (1990) Origin of variability in quantal size in cultured hippocampal neurons and hippocampal slices. *Proc. Natl. Acad. Sci. USA* 87: 5359–5362.

Belew, R.K. (1993) Interposing on ontogenetic model between genetic algorithms and neural networks. In: *Advance in Neural Information Processing Systems 5*, Hanson, S. J., Cowan, J. D., and Giles, C. L. (eds.). Morgan Kaufmann, San Mateo, CA, pp. 99–106.

Bergen, J.R. and Julesz, B. (1983) Parallel versus serial processing in rapid pattern discrimination. *Nature* 303: 696–698.

Berman, N.J., Douglas, R.J., and Martin, K.A.C. (1992) GABA-mediated inhibition in the neural networks of visual cortex. In: *Progress in Brain Research, Vol. 90*, Mize, R.R., Marc, R.E., and Silito, A.M. (eds.). Elsevier Science Publishers, Amsterdam.

Berns, G.S., Dayan, P., and Sejnowski, T.J. (1993) A correlational model for the development of disparity selectivity in visual cortex that depends on prenatal and postnatal phases. *Proc. Natl. Acad. Sci. USA* 17: 8277–8280.

Bichsel, M. (1991) Strategies of robust object recognition for the identification of human faces. Ph.D. thesis, Eidgenossischen Technischen Hochschule.

Biederman, I. (1987) Recognition by components: A theory of human image understanding. *Psychol. Rev.* 94: 115–147.

Biederman, I. and Cooper, E.E. (1992) Evidence for complete translational and reflectional invariance in visual object priming. *Perception* 20: 585–593.

Bishop, P.O., Levick, W.R., and Williams, W.O. (1964) Statistical analyses of the dark discharge of lateral geniculate neurons. *J. Physiol.* 170: 598–612.

Blake, A. and Yuille, A. (eds.) (1992) *Active Vision*, MIT Press, Cambridge, MA.

Blakemore, C. and Vital-Durand, F. (1992) Different neural origins for 'blur' amblyopia and strabismic amblyopia. *Ophthal. Physiol. Opt.* 12.

Bland, B.H., Andersen, P., and Ganes, T. (1975) Two generators of hippocampal theta activity in rabbits. *Brain Res.* 94: 199–218.

Bonds, A.B. (1992) Dual inhibitory mechanisms for definition of receptive field characteristics in cat striate cortex. In: *Advances in Neural Information Processing Systems, Vol. 4*, Moody, J.E., Hanson, S.J., and Lippmann, R. P. (eds.). Morgan Kaufmann, San Mateo, CA, pp. 75–82.

Boussaud, D., Ungerleider, L.C., and Desimone, R. (1990) Pathways for motion analysis: Cortical connections of the medial superior temporal and fundus of the superior temporal visual areas in the macaque. *J. Comp. Neurol.* 296: 462–495.

Bouyer, J.J., Montaron, M.F., and Rougeul, A. (1981) Fast frontoparietal rhythms during combined focused attentive behaviour and immobility in cat: Cortical and thalamic localizations. *Electroencephalogr. Clin. Neurophysiol.* 51: 244–252.

Braddick, O.J. (1980) Low-level and high-level processes in apparent motion. *Phil. Trans. R. Soc. B* 290: 137–151.

Brady, R.M. (1985) Optimization processes gleaned from biological evolution. *Nature* 317: 804–806.

Braitenberg, V. (1978) Cell assemblies in the cerebral cortex. In: *Lecture Notes in Biomathematics 21, Theoretical Approaches in Complex Systems*, Heim, R., and Palm, G. (eds.). Springer, Berlin, pp. 171–188.

Braitenberg, V., and Schütz, A. (1991) *Anatomy of the Cortex*. Springer-Verlag, New York.

Braun, J. (1994) Visual search among objects of unequal salience: Removal of selective attention mimics a lesion in extrastriate area V4. *J. Neurosci.*, in press.

Braun, J. and Sagi, D. (1992) Vision outside the focus of attention. *Percep. Psychophys.* 48: 45–58.

Bressler, S.L. (1984) Spatial organization of EEG from olfactory bulb and cortex. *Electroencephalogr. Clin. Neurophysiol.* 57: 270–276.

Bressler, S.L. (1987) Relation of olfactory bulb and cortex: I. Spatial variation of bulbocortical interdependence. *Brain Res.* 409: 285–293.

Bressler, S.L. and Freeman, W.J. (1980) Frequency analysis of olfactory system EEG in cat, rabbit and rat. *Electroencephalogr. Clin. Neurophysiol.* 50: 19–24.

Breuel, T.M. (1992) Geometric aspects of visual object recognition. Ph.D. thesis, Massachusetts Institute of Technology, Cambridge, MA.

Bricolo, E. and Bülthoff, H.B. (1992) Translation-invariant features for object recognition. *Perception* 21–S2: 59.

Britten, K.H., Shadlen, M.N., Newsome, W.T., and Movshon, J.A. (1992) The analysis of visual motion: a comparison of neuronal and psychophysical performance. *J. Neurosci.* 12: 4745–4765.

Brooks, R.A. (1989) A robot that walks: Emergent behaviors from a carefully evolved network. *Neural Comp.* 1: 253–262.

Brunelli, R. and Poggio, T. (1992) HyperBF networks for gender classification. In: *Proceedings of the Image Understanding Workshop*. Morgan Kaufmann, San Mateo, CA.

Bullier, MJ., Munk, M.H.L., and Nowak, L.G. (1992) Synchronization of neuronal firing in areas V1 and V2 of the monkey. *Soc. Neurosci. Abstr.* 11.7.

Bullock, T.H., Orkand, R., and Grinnell, A. (1977). *Introduction to Nervous Systems*. W. H. Freeman, San Francisco.

Bülthoff, H.H. and Edelman, S. (1992) Psychophysical support for a 2-D view interpolation theory of object recognition. *Proc. Natl. Acad. Sci. USA* 89: 60–64.

Bülthoff, H.H., Edelman, S., and Sklar, E. (1991) Mapping the generalization space in object recognition. *Invest. Ophthalmol. Vis. Sci. Suppl.* 32: 996.

Bundesen C. (1992) A theory of visual attention. *Psych. Rev.* 97: 523–547.

Burkhalter, A. (1993) Development of forward and feedback connections between areas V1 and V2 of human visual cortex. *Cerebral Cortex* 3: 476–487.

Buzsaki, G., Leung, L.S., and Vanderwolf, C.H. (1983) Cellular bases of hippocampal EEG in the behaving rat. *Brain Res. Rev.* 6: 139–171.

Buzsaki, G., Horvath, Z., Urioste, R., Hetke, J., and Wise, K. (1992) High-frequency network oscillation in the hippocampus. *Science* 256: 1025–1027.

Cajal, S.R. (1929) *Etude sur la Neurogénese de quelques Vertébrés*. Thomas, Springfield.

Callaway, E.M. and Katz, L.C. (1990) Emergence and refinement of clustered horizontal connections in cat striate cortex. *J. Neurosci.* 10: 1134–1153.

Callaway, E.M. and Katz, L.C. (1991) Effects of binocular deprivation on the development of clustered horizontal connections in cat striate cortex. *Proc. Natl. Acad. Sci. USA* 88: 745–749.

Campbell, F.W. (1985) How much of the information falling on the retina reaches the visual cortex and how much is stored in the visual memory? In: *Pattern Recognition Mechanisms*, Chagas, C., Gattass, R., Gross, and C. (eds.). Springer-Verlag, Berlin.

Campbell, K.B., and Bartoli, E.A. (1986) Human auditory evoked potentials during natural sleep: The early components. *Electroencephalogr. Clin. Neurophysiol.* 65: 142–149.

Caprile, B., and Girosi, F. (1990) A nondeterministic minimization algorithm. A.I. Memo 1254, Artificial Intelligence Laboratory. Massachusetts Institute of Technology, Cambridge, MA.

Carmichael, T. (1993) The orbital and medial prefrontal cortex of the macaque: Anatomical parcellation and evidence for sensory, limbic and premotor integration. Ph.D. thesis, Washington University.

Carpenter, G. and Grossberg, S. (1987) A massively parallel architecture for a self-organizing neural pattern recognition machine. *Comp. Vis. Graph. Image Proc.* 37: 54–115.

Castaigne, P., Buge, A., Escourolle, R., and Masson, M. (1962) Ramollissement pédonculaire médian, tegmento-thalamique avec ophtalmoplégie et hupersomnie. *Rev. Neurol.* 106: 357–367.

Cauller, L.J. and Kulics, A.T. (1991a) A comparison of awake and sleeping cortical states by analysis of the somatosensory-evoked response of postcentral area 1 in rhesus monkey. *Exp. Brain Res.* 72: 584–592.

Cauller, L.J. and Kulics, A.T. (1991b) The neural basis of the behaviorally relevant NI component of the somato-sensory evoked potential in awake monkeys: Evidence that backward cortical projections signal conscious touch sensation. *Exp. Brain Res.* 84: 607–619.

Cavanagh, P. (1978) Size and position invariance in the visual system. *Perception* 7: 167–177.

Cavanagh, P. (1985) Local log polar frequency analysis in the striate cortex as a basis for size and orientation invariance. In: *Models of the Visual Cortex*, Rose, D., and Dobson, V.G. (eds.). John Wiley, New York.

Chagnac-Amitai, Y., Luhmann, H.J., and Prince, D.A. (1990) Burst generating and regular spiking layer 5 pyramidal neurons of rat neocortex have different morphological features. *J. Comp. Neurol.* 296: 598–613.

Chalupa, L.M., Coyle, D., and Lindsley, D.B. (1976) Effect of pulvinar lesions on visual pattern discrimination in monkeys. *J. Neurophysiol.* 39: 354–369.

Changeux, J.P. (1976) Selective stabilization of developing synapses as a mechanism for the specification of neural networks. *Nature* 264: 705–711.

Changeux, J. and Danchin, A. (1976) Selective stabilization of developing synapses as a mechanism for the specification of neuronal networks. *Nature* 264: 705–711.

Chatrian, G.E., Bickford, R.G., and Uilein, A. (1960) Depth electrographic study of a fast rhythm evoked from the human calcarine region by steady illumination. *Electroencephalogr. Clin. Neurophysiol.* 12: 167–176.

Chelazzi, L., Miller, E.K., Duncan, J., and Desimone, R. (1993) A neural basis for visual search in inferior temporal cortex. *Nature,* 363: 345–347.

Chen, B.M. and Buchwald, J.S. (1986) Midlatency auditory evoked responses: Differential effects of sleep in the cat. *Electroencephalogr. Clin. Neurophysiol.* 65: 373–382.

Cherniak, C. (1990) The bounded brain: Toward quantitative neuroanatomy. *J. Cog. Neurosci.* 2: 58–68.

Chomsky, N. (1965) *Aspects of the Theory of Syntax.* MIT Press, Cambridge, MA.

Chomsky, N. (1980) *Rules and Representations.* Columbia University Press, New York.

Churchland, P.S. (1986). *Neurophilosophy: Toward a Unified Science of the Mind-Brain.* MIT Press, Cambridge, MA.

Colby, C.L. (1992) The neuroanatomy and neurophysiology of attention. *J. Child Neurol.* 6: S90–S118.

Conley, M. and Raczkowski,D. (1990) Sublaminar organization within layer VI of the striate cortex in *Galago. J. Comp. Neurol.* 302: 425–436.

Connors, B.W. and Gutnick, M.J. (1990) Intrinsic firing patterns of diverse neocortical neurons. *Trends Neurosci.* 13: 99–104.

Compton, P.E., Grossenbacher, P., Posner, M.I., and Tucker, D. (1991) A cognitive-anatomical approach to attention in lexical access. *J. Cog. Neurosci.* 3: 303–312.

Conley, M. and Diamond, I.T. (1990) Organization of the visual sector of the thalamic reticular nucleus in galago. *Eur. J. Neurosci.* 2: 211–226.

Cooper, L.A. and Shepard, R.N. (1975) Chronometric studies of the rotation of mental images. In: *Visual Information Processing,* Chase, W.G. (ed.). Academic Press, New York, pp. 75–176.

Corbetta, M., Miezin, F., Dobmeyer, S., Shulman, G., and Petersen, S. E. (1993) Attentional modulation of neural processing of shape, color, and velocity in humans. *Science* 248: 1556–1559.

Corkin, S. (1984) Lasting consequences of bilateral medial temporal lobectomy: Clinical course and experimental findings in H.M. *Sem. Neurol.* 4: 249–259.

Craik, K.J.W. (1943) *The Nature of Explanation.* Cambridge University Press, Cambridge.

Creutzfeldt, O.D. (1977) Generality of the functional structure of the neocortex. *Naturwissenschaften* 64: 507–517.

Creutzfeldt, O.D. (1978) The neocortical link: Thoughts on the generality of structure and function of the neocortex. In: *Architectonics of the Cerebral Cortex,* Brazier, M.A.B. and Petsche, H. (eds.). Raven Press, New York.

Crewther, D.P. and Crewther, S.G. (1990) Neural sites of strabismic amblyopia in cats: spatial frequency deficit in primary cortical neurons. *Exp. Brain Res.* 79: 615–622.

Crick, F. (1984) Function of the thalamic reticular complex: The searchlight hypothesis. *Proc. Natl. Acad Sci. USA* 81: 4586–4590.

Crick, F. and Asanuma, C. (1986) Certain aspects of the anatomy and physiology of the cerebral cortex. In: *Parallel Distributed Processing, Vol. 2,* McClelland, J.L. and Rumelhart, D.E. (eds.). MIT Press, Cambridge, MA.

Crick, F. and Koch, C. (1990a) Towards a neurobiological theory of consciousness. *Sem. Neurosci.* 2: 263–275.

Crick, F. and Koch, C. (1990b) Some reflections on visual awareness. *Cold Spring Harbor Symp. Quant. Biol.* 55: 953–962.

Crick, F. and Koch, C. (1992) The problem of consciousness. *Sci. Am.* 267: 153–159.

Cutting, J.E. (1986) *Perception with an Eye for Motion*. MIT Press. Cambridge, MA.

Damasio, A.R. (1989a) Time-locked multiregional retroactivation: A systems-level proposal for the neural substrates of recall and recognition. *Cognition* 33: 25–62.

Damasio, A.R. (1989b) The brain binds entities and events by multiregional activation from convergence zones. *Neural Comp.* 1: 123–132.

Damasio, A.R. (1989c) Concepts in the brain. *Mind Language* 4: 24–28.

Damasio, A.R. (1989d) Multiregional retroactivation. *Cognition* 12: 263–288.

Damasio, A.R. (1990) Category-related recognition defects as a clue to the neural substrates of knowledge. *Trends Neurosci.* 13: 95–98.

Damasio, A.R. (1994) Descartes' error. In: *Emotion, Knowledge, Decision-Making, and the Human Brain*. Putnam, New York.

Damasio, H. and Frank, R. (1992) Three-dimensional in vivo mapping of brain lesions in humans. *Arch. Neurol.* 49: 137–143.

Damasio, A.R. and Tranel, D. (1993) Verbs and nouns are retrieved from separate neural systems. *Proc. Natl. Acad. Sci. USA,* 90: 4957–4960.

Damasio, A.R., Yamada, T., Damasio, H., Corbett, J., and McKee, J. (1980) Central achromatopsia: Behavioral, anatomic and physiologic aspects. *Neurology* 30: 1064–1071.

Damasio, A.R., Tranel, D., and Damasio, H. (1989) Amnesia caused by herpes simplex encephalitis, infarctions in basal forebrain, Alzheimer's disease, and anoxia. In: *Handbook of Neuropsychology*, Boller, F., and Grafman, J. (eds.). Vol. 3 (L. Squire, ed.). Elsevier, Amsterdam, pp. 149–166.

Damasio, A.R., Damasio, H., and Tranel, D. (1990a) Impairments of visual recognition as clues to the processes of memory. In: *Signal and Sense: Local and Global Order in Perceptual Maps*, Edelman, G., Gall, E., and Cowan, M. (eds.). Neuroscience Institute Monograph, Wiley-Liss, New York, pp. 451–473.

Damasio, A.R., Tranel, D., and Damasio, H. (1990b) Face agnosia and the neural substrates of memory. In: *Annual Review of Neuroscience*, Vol. 13, Annual Review, Palo Alto, CA, pp. 89–109.

Damasio, A.R., Damasio, H., Tranel, D., and Brandt, J.P. (1990c) The neural regionalization of knowledge access: Preliminary evidence. In: *Quantitative Biology*, Vol. 55, Cold Spring Harbor Laboratory Press, Cold Spring Harbor, NY, pp. 1039–1047.

Daugman J.G. (1988) Complete discrete 2-D Gabor transforms by neural networks for image analysis and compression. *IEEE Trans. Acoust. Speech Signal Proc.* 36: 1169–1179.

David, C., and Zucker, S.W. (1990) Potentials, valleys and dynamic global coverings. *Int. J. Comp. Vision* 5: 219–238.

Deacon, T. (1990) Rethinking mammalian brain evolution. *Am. Zool.* 30: 629–706.

Deiber, M.P., Bastuji, H., Fischer, M.D., and Maugiare, F. (1989) Changes of middle latency auditory evoked potentials during natural sleep in humans. *Neurology* 39: 806–813.

Dennett, D.C. (1992) *Consciousness Explained*. Little, Brown, and Co., New York.

Deppisch, J., Bauer, H.-U., Schillen, T.B., König, P., Pawelzik, K., and Geisel, T. (1992) Stochastic and oscillatory burst activities in a model of spiking neurons. In: *Artificial Neural Networks*, Vol. 23, Aleksander, J. and Taylor, J. (eds.). Elsevier, Amsterdam. pp. 921–924.

Deschênes, M., Madariaga-Domich, A., and Steriade, M. (1985) Dendrodendritic synapses in the cat reticularis thalami nucleus: A structural basis for thalamic spindle synchronization. *Brain Res.* 334: 165–168.

Desimone, R. (1992) Neural circuits for visual attention in the primate brain. In: *Neural Networks for Vision and Image Processing*, Carpenter, G. and Grossberg, S. (eds.). MIT Press, Cambridge, pp. 343–364.

Desimone, R. and Ungerleider, L.G. (1989) Neural mechanisms of visual processing in monkeys. In: *Handbook of Neuropsychology*, Vol. 2, Boller, F., and Grafman, J. (eds.). Elsevier, New York, pp. 267–299.

Desimone, R., Albright, T.D., Gross, C.G., and Bruce, C. (1984) Stimulus-selective properties of inferior temporal neurons in the macaque. *J. Neurosci.* 4: 2051–2062.

Desimone, R., Schein, S.J., Moran, J., and Ungerleider L.G. (1985) Contour, color and shape analysis beyond the striate cortex. *Vision Res.* 25: 441–452.

Desimone, R., Wessinger, M., Thomas, L., and Schneider, W. (1990) Attentional control of visual perception: cortical and subcortical mechanisms. *Cold Spring Harbor Symp. Quant. Biol.* 55: 963–971.

DeYoe, E.A. and Sisola, L.C. (1991) Distinct pathways link anatomical subdivisions of V4 with V2 and temporal cortex in the Macaque monkey. *Soc. Neurosci. Abstr.* 17: 1282.

Diamond, A. (1988) Differences between adult and infant cognition: Is the crucial variable presence or absence of language? In: *Thought Without Language*, Weiskrantz, L. (ed.). Oxford University Press, Oxford, pp. 337–370.

DiLollo, V., and Wilson, A.E. (1978) Iconic persistence and perceptual moment as determinants of temporal integration in vision. *Vision Res.* 18: 1607–1610.

Dobkins, K. R. and Albright, T.D. (1993) What happens if it changes color when it moves?: Psychophysical experiments on the nature of chromatic input to motion detectors. *Vision Res.* 33: 1019–1036.

Donoghue, J.P. and Sanes, J.N. (1991) Dynamic modulation of primate motor cortex output during movement. *Neurosci. Soc. Abstr.* 17: 407.5.

Doty, R.W. and Kimura, D.S. (1963) Oscillatory potentials in the visual system of cats and monkeys. *J. Physiol.* 168: 205–218.

Douglas, R.J. and Martin, K.A.C. (1990) Neocortex. In: *Synaptic Organization of the Brain*, Shepard, G.M. (ed.). Oxford University Press, New York, pp. 338-387.

Douglas, R.J., and Martin, K.A.C. (1992) Exploring cortical microcircuits. In: *Single Neuron Computation*, McKenna, T., Davis, J. and Zornetzer, S. (eds.). Academic Press, San Diego., pp. 381–412.

Douglas, R.J. and Martin, K.A.C. (1993) Integration in the neurons and microcircuits of the neocortex. In: *Exploring Brain Functions: Models in Neuroscience*, Poggio, T.A., and Glaser, D.A. (eds.). John Wiley, New York, pp. 43–56.

Douglas, R.J., Martin, K.A.C., and Whitteridge, D. (1988) Selective responses of visual cortical cells do not depend on shunting inhibition. *Nature* 332: 642–644.

Duffy, F.H. and Burchfield, J.L. (1975). Eye movement-related inhibition of primate visual neurons. *Brain Res.* 89: 121–132.

Duhamel, J.R., Colby, C.L., and Goldberg, M.E. (1992) The updating of the representation of visual space in parietal cortex by intended eye movements. *Science* 255: 90–92.

Dumenko, W.N. (1961) Veränderungen der elektrischen Rindenaktivitt bei Hunden bei der Bildung eines Stereotyps motorischer bedingter Reflexe. *Pawlow-Zeitschrift Höhere Nerventtigkeit* 11: 184–191.

Duncan, J. and Humphreys, G.W. (1989) Visual search and stimulus similarity. *Psychol. Rev.* 96: 433–458.

Eccles, J.C. (1984) The cerebral neocortex. A theory of its operation. In: *Cerebral Cortex, Vol. 2, Functional Properties of Cortical Cells*, Jones, E.G., and Peters, A. (eds.). Plenum Press, New York, pp. 1–36.

Eckhorn, R., Bauer, R., Jordan, W., Brosch, M., Kruse, W., Munk, M., and Reitboeck, H.J. (1988) Coherent oscillations: A mechanism of feature linking in the visual cortex? *Biol. Cybern.* 60: 121–130.

Eckhorn, R., Schanze, T., Brosch, M., Salem, W., and Bauer, R. (1992) Stimulus-specific synchronizations in cat visual cortex: Multiple microelectrode and correlation studies from several cortical areas. In: *Induced Rhythms in the Brain*, Basar, E., and Bullock, T.H. (eds.). Birkhauser, Boston, pp. 47–80.

Edelman, G.M. (1978) Group selection and phasic re-entrant signalling: A theory of higher brain functions. In: *The Mindful Brain*, Edelman, G.M. and Mountcastle, V.B. (eds.). MIT Press, Cambridge, MA.

Edelman, G.M. (1987) *Neural Darwinism: The Theory of Neuronal Group Selection*. Basic Books, New York.

Edelman, G.M. (1989) *The Remembered Present*. Basic Books, New York.

Edelman, G.M., and Mountcastle, V.B. (1978) *The Mindful Brain*. MIT Press, Cambridge, MA.

Edelman, S. (1991) Features of recognition. In: *Proceedings of the International Workshop on Visual Form, Capri, Italy*. Plenum Press, New York.

Edelman, S. and Bülthoff, H.H. (1992) Orientation dependence in the recognition of familiar and novel views of 3D objects. *Vision Res.* 32: 2385–2400.

Edwards, F.A., Konnerth, A., and Sakmann, B. (1990) Quantal analysis of inhibitory synaptic transmission in the dentate gyrus of rat hippocampal slices: A patch-clamp study. *J. Physiol.* 430: 213–249.

Efron, R. (1970a) The relationship between the duration of a stimulus and the duration of a perception. *Neuropsychologia* 8: 37–55.

Efron, R. (1970b) The minimum duration of a perception. *Neurophysiologia* 8: 57–63.

Efron, R. (1973) Conservation of temporal information by perceptual systems. *Percept. Psychophys.* 14: 518–530.

Elliot Smith, G. (1924) *The Evolution of Man*. Oxford University Press, Oxford.

Engel, A.K., König, P., Gray, C.M., and Singer, W. (1990) Stimulus-dependent neuronal oscillations in cat visual cortex: Inter-columnar interaction as determined by cross-correlation analysis. *Eur. J. Neurosci.* 2: 588–606.

Engel, A.K., Kreiter, A.K., König, P., and Singer, W. (1991a) Synchronization of oscillatory neuronal responses between striate and extrastriate visual cortical areas of the cat. *Proc. Natl. Acad. Sci. U.S.A* 88: 6048–6052.

Engel, A.K., König, P., Kreiter, A.K., and Singer, W. (1991b) Interhemispheric synchronization of oscillatory neuronal responses in cat visual cortex. *Science* 252: 1177–1179.

Engel, A.K., König, P., and Singer, W. (1991c) Direct physiological evidence for scene segmentation by temporal coding. *Proc. Natl. Acad. Sci. USA* 88: 9136–9140.

Engel, A.K., König, P., Kreiter, A.K., Schillen, T.B., and Singer, W. (1992) Temporal coding in the visual cortex: new vistas on integration in the nervous system. *Trends Neurosci.* 15: 218–226.

Eriksen, C.W. and Murphy, T.D. (1987) Movement of attentional focus across the visual field: A critical look at the evidence. *Percept. Psychophys.* 42: 299–305.

Esguerra, M. and Sur, M. (1990) Corticogeniculate feedback gates retinogeniculate transmission by activating NMDA receptors. *Soc. Neurosci. Abst.* 16: 159.

Eskandar, E.N., Richmond, B.J., and Optican, L.M. (1992) Role of inferior temporal neurons in visual memory: I. Temporal encoding of information about visual images, recalled images, and behavioral context. *J. Neurophysiol.* 68: 1277–1295.

Eslinger, P. and Damasio, A.R. (1986) Preserved motor learning in Alzheimer's disease. *J. Neurosci.* 6: 3006–3009.

Facon, E., Steriade, M., and Wertheim, N. (1958) Hypersomnie prolongée engendrée par des lésions bilatérales due systèm activateur médial le syndrome thrombotique de la biffurcation du tronc basilaire. *Rev. Neurol.* 98: 117–133.

Felleman, D.J. and Van Essen, D.C. (1991) Distributed hierarchical processing in the primate visual cortex. *Cerebral Cortex* 1: 1–47.

Ferster, D. and LeVay, S. (1978) The axonal arborizations of lateral geniculate neurons in the striate cortex of the cat. *J. Comp. Neurol.* 182: 923–944.

Ferster, D. and Lindström, S. (1985) Synaptic excitation of neurones in area 17 of the cat by intracortical axon collaterals of cortico-geniculate cells. *J. Physiol.* 367: 233–252.

Fetz, E., Toyama, K., and Smith, W. (1991) Synaptic interactions between cortical neurons. In: *Cerebral Cortex*, Vol. 9, Peters, A., and Jones, E.G. (eds.). Plenum, New York City, pp. 1–47.

Feynman, R.P. and Hibbs, A.R. (1965) *Quantum Mechanics and Path Integrals*. McGraw-Hill, New York.

Fields, H.L. and Basbaum, A.I. (1978) Brainstem control of spinal pain-transmission neurons. *Annu. Rev. Physiol.* 40: 217–248.

Fiez, J.A. and Petersen, S.E. (1993) PET as a part of an interdisciplinary approach to understanding processes involved in reading. *Psychol. Sci.*, 4: 287–292.

Fiorani, M., Jr., Rosa, M. G. P., Gattass, R., and Rocha-Miranda, C.E. (1992) Dynamic surrounds of receptive fields in primate striate cortex: A physiological basis for perceptual completion. *Proc. Natl. Acad. Sci. USA* 89: 8547–8551.

Fiorentini, A. and Berardi, N. (1981) Learning in grating waveform discrimination: Specificity for orientation and spatial frequency. *Vision Res.* 21: 1149–1158.

Fisher, R.A. (1925) *Statistical Methods for Research Workers*. Oliver and Boyd, Edinburgh.

Fitts, P.M. and Posner, M.I. (1967) *Human Performance*. Brooks/Cole, Belmont, CA.

Fodor, J.A. (1981) *Representations*. MIT Press, Cambridge, MA.

Fodor, J.A. (1983) *The Modularity of Mind*. MIT Press, Cambridge, MA.

Földiak, P. (1991) Learning invariance from transformation sequences. *Neural Comp.* 3: 194–200.

Fox, K., Sato, H. and Daw N.W. (1989) The location and function of NMDA receptors in cat and kitten visual cortex. *J. Neurosci.* 9: 2443–2454.

Fox, K., Sato, H., and Daw N.W. (1990) The effect of varying stimulus intensity on NMDA-receptor activity in cat visual cortex. *J. Neurophysiol.* 64: 1413–1428.

Freeman, W.J. (ed.) (1975) *Mass Action in the Nervous System.* Academic Press, New York.

Freeman, W. J., and van Dijk, B. W. (1987) Spatial patterns of visual cortical fast EEG during conditioned reflex in a rhesus monkey. *Brain Res.* 422: 267–276.

Freeman, W.J., and Skarda, C.A. (1985) Spatial EEG-patterns, non-linear dynamics and perception: The neo-Sherringtonian view. *Brain Res. Rev.* 10: 147–175.

Freund, T.F., Martin, K.A.C., Somogyi, P., and Whitteridge, D. (1985) Innervation of cat visual areas 17 and 18 by physiologically identified X- and Y-type thalamic afferents. II. Identification of postsynaptic targets by GABA immunocytochemistry and Golgi impregnation. *J. Comp. Neurol.* 242: 275–291.

Freund, T.F., Martin, K.A.C., Soltesz, I., Somogyi, P., and Whitteridge, D. (1989) Arborisation pattern and postsynaptic targets of physiologically identified thalamocortical afferents in striate cortex of the Macaque monkey. *J. Comp. Neurol.* 289: 315–336.

Friedman, D.P. (1983) Laminar patterns of termination of corticocortical afferents in the somatosensory system. *Brain Res.* 273: 147–151.

Fujita, I. and Tanaka, K. (1992) Columns for visual features of objects in monkey inferotemporal cortex. *Nature* 360: 343–346.

Fujita, I., Tanaka, K., Ito, M., and Cheng, K. (1992) Columns for visual features of objects in monkey inferotemporal cortex. *Nature* 360: 343–346.

Fukushima, K. (1980) Neocognitron: A self-organizing neural network model for a mechanism of pattern recognition unaffected by shift in position. *Biol.Cybern.* 36: 193–202.

Fukushima, K. (1987) Neural network model for selective attention in visual pattern recognition and associative recall. *Appl. Optics* 26: 4985–4992.

Fuster, J.M. (1990) Inferotemporal units in selective visual attention and short-term memory. *J. Neurophysiol.* 64: 681–697.

Fuster, J.M. and Jervey, J.P. (1981) Inferotemporal neurons distinguish and retain behaviorally relevant features of visual stimuli. *Science* 212: 952–955.

Fuster, J.M., Herz, A., and Creutzfeldt, O.D. (1965) Interval analysis of cell discharge in spontaneous and optically modulated activity in the visual system. *Arch. Ital. Biol.* 103: 159–177.

Gaal, G., Sanes, J.N., and Donoghue, J.P. (1992) Motor cortex oscillatory neural activity during voluntary movement in macaca fascicularis. *Soc. Neurosci. Abstr.* 18: 355.14.

Gabrieli, J.D.E., Cohen, N.J., and Corkin, S. (1988) The impaired learning of semantic knowledge following bilateral medial temporal lobe resection. *Brain Cogn.* 7: 157–177.

Galambos, R., and Makeig, S. (1988) In: *Dynamics of Sensory and Cognitive Processing by the Brain*, Basar, E., and Bullock, T. (eds.). Springer, Berlin, pp. 103–122

Galambos, R., Makeig, S., and Talmachoff, P.J. (1981) A 40-Hz auditory potential recorded from the human scalp. *Proc. Natl. Acad. Sci. USA* 78: 2643–2647.

Gallant, J.L., Braun, J., and Van Essen, D.C. (1993) Selectivity for polar, hyperbolic and cartesian gratings in macaque visual cortex. *Science* 259: 100–103.

Galletti, C. and Battaglini, P.P. (1989) Gaze-dependent visual neurons in area V3A of monkey prestriate cortex. *J. Neurosci.* 9: 1112–1125.

Gardner-Medwin, A.R., and Barlow, H.B. (1992) The effect of sparseness in distributed representations on the detectability of associations betweeen sensory events. *J. Physiol. (Lond.)* 452: 282P.

Gardner-Medwin, A.R. and Barlow, H.B. (1994) The factors determining the efficiency of associative learning. In preparation.

Gattass, R., Sousa, A.P.B., and Covey, E. (1985) Cortical visual areas of the macaque: Possible substrates for pattern recognition mechanisms. In: *Pattern Recognition Mechanisms*, Chagas C., Gattass, R., Gross, C. (eds.) Springer-Verlag, Berlin.

Georgopoulos, A.P. (1990) Neural coding of the direction of reaching and a comparison with saccadic eye movements. *Cold Spring Harbor Symp. Quant. Biol.*, LV, pp. 849–859.

Georgopolous, A.P., Kettner, R.E., and Schwartz, A.B. (1988) Primate motor cortex and free arm movements to visual targets in three-dimensional space. II. Coding of the direction of movement by a neuronal population. *J. Neurosci.* 8: 2928–2937.

Georgopolous, A.P., Lurito, J.T., Petrides, M., Schwartz, A.B., and Massey, J.T. (1989) Mental rotation of the neuronal population vector. *Science* 243: 234–236.

Gerstein, G.L., and Aertsen, A.M.H.J. (1985) Representation of cooperative firing activity among simultaneously recorded neurons. *J. Neurophysiol.* 54: 1513–1528.

Gerstein, G.L., and Perkel, D.H. (1972) Mutual temporal relationship among neuronal spike trains. Statistical techniques for display and analysis. *Biophys. J.* 12: 453–473.

Gerstein, G.L., Perkel, D.H., and Dayhoff, J.E. (1985) Cooperative firing activity in simultaneously recorded populations of neurons: Detection and measurement. *J. Neurosci.* 5: 881–889.

Ghose, G.M., and Freeman, R.D. (1990) Origins of oscillatory activity in the cortex. *Soc. Neurosci. Abstr.* 16: 523.4.

Ghose, G.M., and Freeman, R.D. (1992) Oscillatory discharge in the visual system: Does it have a functional role? *J. Neurophysiol.* 68: 1558–1574.

Giard, M.H., Perrin, F., Pernier, J., and Perronnet, F. (1988) Several attention-related waveforms in auditory areas: a topographic study. *Electroencephalogr. Clin. Neurophysiol.* 69: 371–384.

Gibson, J.J. (1966) *The Senses Considered as Perceptual Systems*. Houghton Mifflin, Boston.

Gibson, J.J. (1979) *The Ecological Approach to Visual Perception*. Houghton Mifflin, Boston.

Gilbert, C.D. (1983) Microcircuitry of the visual cortex. *Annu. Rev. Neurosci.* 6: 217–247.

Gilbert, C.D. and Wiesel, T.N. (1979) Morphology and intracortical projections of functionally identified neurons in cat visual cortex. *Nature* 280: 120–125.

Gilbert, C.D. and Wiesel, T.N. (1989) Columnar specificity of intrinsic horizontal and corticocortical connections in cat visual cortex. *J. Neurosci.* 9: 2432–2442.

Glass, L. (1969) Moirée effect from random dots. *Nature* 223: 578–580.

Glimcher, P.W. and Sparks, D.L. (1992) Movement selection in advance of action in the superior colliculus. *Nature* 355: 542–545.

Gochin, P.M., Miller, E.K., Gross, C.G., and Gerstein, G.L. (1991) Functional interactions among neurones in inferior temporal cortex of the awake macaque. *Exp. Brain Res.* 84: 505–516.

Goff, W.R., Allison, T., Shapiro, A., and Rosner, B.S. (1966) Cerebral somatosensory responses evoked during sleep in man. *Electroencephalogr. Clin. Neurophysiol.* 21: 1–9.

Goldberg, M.E., Colby, C.L., and Duhamel, J.-R. (1990) Representation of visuomotor space in the parietal lobe of the monkey. *Cold Spring Habor Symp. Quant. Biol.* LV: 729–739.

Goldman-Rakic, P.S. (1988a) Topography of cognition: Parallel distributed networks in primate association cortex. *Annu. Rev. Neurosci.* 11: 137–156.

Goldman-Rakic, P.S. (1988b) Changing concepts of cortical connectivity: parallel distributed cortical networks. In: *Neurobiology of Neocortex*, Rakic, P., and Singer, W. (eds.). John Wiley, New York, pp. 177–202.

Goldman-Rakic, P.S., Funahashi, S., and Bruce, C.J. (1990) Neocortical memory circuits. *Cold Spring Harbor Symp. Quant. Biol.* LV: 1025–1038.

Goodale, M.A. and Milner, A.D. (1992) Separate visual pathways for perception and action. *Trends Neurosci.* 15: 20–55.

Goodale, M.A., Milner, A.D., Jakobson, L.S., and Carey, D.P. (1991) A neurological dissociation between perceiving objects and grasping them. *Nature London* 349:154–156.

Gould, J.L. (1984) Natural history of honey bee learning. In: *The Biology of Learning*, Marler, P., and Terrace, H.S. (eds.). Springer-Verlag, Berlin, pp. 149–180.

Graff-Radford, N.R., Damasio, A.R., Hyman, B.T., Hart, M.N., Tranel, D., Damasio, H., Van Hoesen, G.W., and Rezai, K. (1990) Progressive aphasia in a patient with Pick's disease: a neuropsychological, radiologic, and anatomic study. *Neurology* 40: 620–626.

Gray, C.M. (1993) Rhythmic activity in neuronal systems: Insights into integrative functions. In: *Lectures at the Santa Fe Institute Summer School on Complex Systems*, Nadel, L., and Stein, D. (eds.). Addison-Wesley, Reading, MA, pp. 89–161.

Gray, C.M. and Singer, W. (1987) Stimulus-specific neuronal oscillations in the cat visual cortex: A cortical functional unit. *Soc. Neurosci. Abstr.* 13: 404.3.

Gray, C.M. and Singer, W. (1989) Stimulus-specific neuronal oscillations in orientation columns of cat visual cortex. *Proc. Natl. Acad. Sci. USA* 86: 1698–1702.

Gray, C.M. and Viana di Prisco, G. (1993) Properties of stimulus-dependent rhythmic activity of visual cortical neurons in the adult cat. *Soc. Neurosci. Abstr.* 19: 868.

Gray, C.M., König, P., Engel, A.K., and Singer, W. (1989) Oscillatory responses in cat visual cortex exhibit inter-columnar synchronization which reflects global stimulus properties. *Nature* 338: 334–337.

Gray, C.M, Engel, A.K., König, P., and Singer, W. (1990) Stimulus-dependent neuronal oscillations in cat visual cortex: receptive field properties and feature dependence. *Eur. J. Neurosci.* 2: 607–619.

Gray, C.M., Engel, A.K., König, P., and Singer, W. (1992a) Mechanisms underlying the generation of neuronal oscillations in cat visual cortex. In: *Induced Rhythms in the Brain*, Brain Dynamic Series, Basar, E. and Bullock, T.H. (eds.). Birkhauser, Boston, pp. 29–45.

Gray, C.M., Engel, A.K., König, P., and Singer, W. (1992b) Synchronization of oscillatory neuronal responses in cat striate cortex: Temporal properties. *Vis. Neurosci.* 8: 337–347.

Grenander, U. (1976–81) *Lectures in Pattern Theory I, II and III: Pattern Analysis, Pattern Synthesis and Regular Structures*. Springer-Verlag, Berlin.

Grieve, K.L., Murphy, P.C., and Sillito, A.M. (1991) Inhibitory and excitatory components to the subcortical and cortical influence of layer VI cells in the cat visual cortex. *Soc. Neurosci. Abst.* 17: 629.

Griffin, D.R. (1984) *Animal Thinking*. Harvard University Press, Cambridge, MA.

Grosof, D.H., Shapley, R.M., and Hawken, M.J. (1992) Macaque striate responses to anomalous contours? *Invest. Ophthalmol. Vis. Sci.* 33: 1257.

Gross, C.G. (1992) Representation of visual stimuli in inferior temporal cortex. *Phil. Trans. R. Soc. Lond. (Biol.)* 335: 3–10.

Gross, C.G., Rocha-Miranda, E.C., and Bender, D.B. (1972) Visual properties of neurons in inferotemporal cortex of the macaque. *J. Neurophysiol.* 35: 96–111.

Gross, C.G., Bender, D.B., and Gerstein, G.L. (1979) Activity of inferior temporal neurons in behaving monkeys. *Neuropsychologia* 17: 215–229.

Grossberg, S. (1980) How does the brain build a cognitive code? *Psychol. Rev.* 87: 1–51.

Grossberg, S. (ed.) (1987) *The Adaptive Brain I*. North-Holland, Amsterdam.

Grossberg, S. (ed.) (1988) *Neural Networks and Natural Intelligence*, MIT Press, Cambridge, MA.

Grossenbacher, P.G., Compton, P.E., Posner, M.I., and Tucker, D.M. (1991) Integrating isolated physical and semantic codes of words: a cognitive anatomical analysis. Poster presented at the 32nd Annual Meeting of the Psychonomics Society, San Francisco, CA.

Haberly, L.B. and Price, J.L. (1978) Association and commissural fiber systems of the olfactory cortex of the rat. *J. Comp. Neurol.* 178: 711–740.

Haenny, P.E., Maunsell, J.H.R., and Schiller, P.H. (1988) State dependent activity in monkey visual cortex. II. Retinal and extraretinal factors in V4. *Exp. Brain Res.* 69: 245–259.

Hansel, D., and Sompolinsky, H. (1992) Synchronization and computation in a chaotic neural network. *Phys. Rev. Lett.* 68: 718–721.

Hardy, S.G.P. and Lynch, J.C. (1992) The spatial distribution of pulvinar neurons that project to two subregions of the inferior parietal lobule in the macaque. *Cerebral Cort.* 2: 217–230.

Harris, C.S. (1980) Insight or out of sight? Two examples of perceptual plasticity in the human adult. In: *Visual Coding and Adaptability*, Harris, C.S. (ed.). Erlbaum, Hillsdale, NJ, pp. 95–149.

Harris, W.A. (1987) Neurogenetics. In: *Encyclopedia of Neuroscience*, Adelman, G. (ed.). Birkhäuser, Basel, pp. 791–793.

Harting, J.K., Updyke, B.V., and Lieshout, D.P.V. (1992) Corticotectal projections in the cat: Anterograde transport studies of twenty-five cortical areas. J. Comp. Neurol. 324: 379–414.

Hasselmo, M.E., Rolls, E.T., Baylis, G.C., and Nalwa, V. (1989) Object-centered encoding by by face-selective neurons in the cortex of in the superior temporal sulcus of the monkey. *Exp. Brain Res.* 75: 417–429.

Hassenstein, B. and Reichardt, W. (1956) Systemtheoretische Analyse der Zeit, Reihenfolgen, und Vorzeichenauswertung bei der Bewegungsperzepion des Rüsselkäfers. *Chlorophanus. Z. Naturforsch.* 11b: 513–524.

Hata, Y., Tsumoto, T., Sato, H., Hagihara, K., and Tamura, H. (1988) Inhibition contributes to orientation selectivity in visual cortex of cat. *Nature* 335: 815–817.

Hebb, D.O. (1949) *The Organization of Behavior*. John Wiley, New York.

Henderson, J.M. (1992) Visual attention and eye movement control during reading and picture viewing. In: *Eye Movements and Visual Cognition: Scene Perception and Reading*, Rayner, K. (ed.). Springer-Verlag, New York.

Henderson, J.M. (1993) Visual attention and the perception-action interface. In: *Vancouver Studies in Cognitive Science, Vol. V, Problems in perception*, Oxford University Press, Oxford (in press).

Henderson, J.M., Pollatsek, A., and Rayner, K. (1989) Covert visual attention and extrafoveal information use during object identification. *Percep. Psychophys.* 45: 196–208..

Hentschel, H.G.E., and Barlow, H.B. (1991) Minimum entropy coding with Hopfield networks. *Network* 2: 135–148.

Herrick, C.J. (1926) *Brains of Rats and Men*. University of Chicago Press, Chicago.

Hillyard, S. A. and Picton, T.W. (1987) Electrophysiology of cognition. In: *Handbook of Physiology*, Vol. 5, Plum, F. (ed.). American Physiological Society, Baltimore, pp. 519–584.

Hinton, G.E. (1981a) A parallel computation that assigns canonical object-based frames of reference. In: *Proceedings of the Seventh International Joint Conference on Artificial Intelligence*, 2, Vancouver B.C., Canada, pp. 683–685.

Hinton, G.E. (1981b) Shape representation in parallel systems. In: *Proceedings of the Seventh International Joint Conference on Artificial Intelligence*, 2, Vancouver B.C., Canada, pp. 1088-1096.

Hinton, G.E. and Lang, K.J. (1985) Shape recognition and illusory conjunctions. In: *Proceedings of the Ninth International Joint Conference on Artificial Intelligence*. Los Angeles, pp. 252–259.

Hirsch, J.A. and Gilbert, C.D. (1991) Synaptic physiology of horizontal connections in the cat's visual cortex. *J. Neurosci.* 11: 1800–1809.

Hobson, J.A. (1988) *The Dreaming Brain*, Basic Books, New York.

Hodgkin, A.L. and Huxley, A.F. (1952) A quantitative description of membrane current and its application to conduction and excitation in nerve. *J. Physiol.* 117: 500–544.

Hohmann, C.F., Kwiterovich, K.K., Oster-Granite, M.L., and Coyle, J.T., (1991). Newborn basal forebrain lesions disrupt cortical cytodifferentiation as visualized by rapid Golgi staining. *Cerebral Cortex*, 1: 143–157.

Holland, J.H. (1975) *Adaption in Natural and Artificial Systems*, Univ. Michigan Press, Ann Arbor.

Hong, T.H. and Rosenfeld, A. (1984) Compact region extraction using weighted pixel linking in a pyramid. *IEEE Trans. Pattern Anal. Mach. Int.* 6: 222–229.

Hoover, J.E. and Strick, P.L. (1993) Multiple output channels in the basal ganglia. *Science* 259: 819–821.

Hopfield, J.J. (1984) Neurons with graded response have collective computational properties like those of two-state neurons. *Proc. Natl. Acad. Sci. USA* 81: 3088–3092.

Horn, B.K.P. (1986) *Robot Vision*. MIT Press, Cambridge, MA.

Houk, J.C., Keifer, J., and Barto, A.G. (1993) Distributed motor commands in the limb premotor network. *Trends Neurosci.* 16: 27–33.

Hubel, D.H. and Wiesel, T.N. (1965a) Binocular interaction in striate cortex of kittens reared with artificial squint. *J. Neurophysiol.* 28: 1041–1059.

Hubel, D.H. and Wiesel, T.N. (1965b) Receptive fields and functional architecture in two non-striate visual areas (18 and 19) of the cat. *J. Neurophysiol.* 18: 229–289.

Hubel, D.H. and Wiesel, T.N. (1970) The period of susceptibility to the physiological effects of unilateral eye closure in kittens. *J. Physiol. (Lond.)* 206: 419–436.

Hübener, M., Schwarz, C., and Bolz, J. (1990) Morphological types of projection neurons in layer 5 of cat visual cortex. *J. Comp. Neurol.* 301: 655–674.

Huber, P.J. (1985) Projection pursuit. *Ann. Statistics* 13: 435–475.

Humphrey, N.K. (1976) The social function of intellect. In: *Growing Points in Ethology*, Bateson, P.P. , and Hinde, R.A. (eds.). Cambridge University Press, Cambridge.

Hurlbert, A. and Poggio, T. (1986) Do computers need attention? *Nature* 321: 651–652.

Innocenti, G.M. (1981) Growth and reshaping of axons in the establishment of visual callosal connections. *Science* 212: 824–827.

Innocenti, G.M. and Frost, D.O. (1979) Effects of visual experience on the maturation of the efferent system to the corpus callosum. *Nature* 280: 231–234.

Intrator, N. (1992) Feature extraction using an unsupervised neural network. *Neural Comp.* 4: 98–107.

Intrator, N. and Cooper, L.N. (1992) Objective function formulation of the BCM theory of visual cortical plasticity: Statistical connections, stability conditions. *Neural Networks* 5: 3–17.

Jackson, S. and Houghton, G. (1992) Basal ganglia function in the control of visuospatial attention: A neural-network model. Institute of Cognitive and Decision Science Technical Report. University of Oregon, Eugene.

Jagadeesh, B., Ferster, D., and Gray, C.M. (1991) Visually evoked oscillations of membrane potential in neurons of cat area 17. *Soc. Neurosci. Abstr.* 17: 73.2.

Jagadeesh, B., Gray, C.M., and Ferster, D. (1992) Visually evoked oscillations of membrane potential in neurons of cat striate cortex studied with in vivo whole cell patch recording. *Science* 257: 552–554.

James, W. (1890) *The Principles of Psychology.* Henry Holt, London.

Jeannerod, M., and Decety, J. (1990) The accuracy of visuo-motor transformation: An investigation into the mechanisms of visual recognition of objects. In: *Vision and Action: The Control of Grasping*, Goodale, M.A. (ed.). Ablex, Norwood, NJ.

Jerison, H.J. (1991) *Brain Size and the Evolution of Mind.* American Museum of Natural History, New York.

Johnson-Laird, P. N. (1988) *The Computer and the Mind.* Harvard University Press, Cambridge, MA.

Jones, E.G. (1984) Laminar distribution of cortical efferent cells. In: *Cerebral Cortex: Cellular Components of the Cerebral Cortex*, Peters, A. and Jones, E.G. (eds.). Plenum Press, New York, pp. 521–552.

Jones, E.G. (1985) *The Thalamus.* Plenum Press, New York.

Jones, E.G., Coulter, J.D., Burton, H., and Porter, R. (1977) Cells of origin and terminal distribution of corticostriatal fiibers arising in the sensory-motor cortex of monkeys. *J. Comp. Neurol.* 173: 53–80.

Jordan, M.I., and Jacobs, R.A. (1993) Hierarchical mixtures of experts and the EM algorithm. *MIT Comp. Cog. Sci. Technical Report* 9301, Cambridge, MA.

Jouvet, M. and Michel, F. (1959) Corrélations électromyographiques du sommeil chez le chat décortiqué et mésendcéphalique chronique. *C.R. Soc. Biol. Paris* 153: 422–425.

Julesz, B. (1991) Early vision and focal attention. *Rev. of Modern Phys.* 63: 735–772.

Kaas, J.H. (1989) Why does the brain have so many visual areas? *J. Cogn. Neurosci.* 1: 121–135.

Kapoula, Z. (1984). Aiming precision and characteristics of saccades. In: *Theoretical and Applied Aspects of Eye Movement Research*, Gale, A.G. and Johnson, F. (eds.). North-Holland, Amsterdam.

Karni, A. and Sagi, D. (1990) Human texture discrimination learning—evidence for low-level neuronal plasticity in adults. *Perception* 19: 335.

Karni, A. and Sagi, D. (1991) Where practice makes perfect in texture discrimination: evidence for primary visual cortex plasticity. *Proc. Natl. Acad. Sci. USA* 88: 4966–4970.

Karten, H.J. and Shimizu, T. (1989) The origins of neocortex: Connections and lamination as distinct events in evolution. *J. Cog. Neurosci.* 1: 291–301.

Kasper, E., Larkman, A., Blakemore, C., and Judge, S. (1991) Physiology and morphology of identified projection neurons in rat visual cortex studied *in vitro. Soc. Neurosci. Abstr.* 17: 114.

Katz, B. (1969) *The Release of Neural Transmitter Substances*. Charles C Thomas, Springfield, IL.

Katz, L.C., Burkhalter, A., and Dreyer, W.J. (1984) Fluorescent latex microspheres as a retrograde neuronal marker for *in vivo* and *in vitro* studies of visual cortex. *Nature* 310: 498–500.

Keele, S.W. (1981) Behavioral analysis of movement. In *Handbook of Physiology: Section 1: The Nervous System, Vol. II, Motor Control, Part 2*, Brooks, V.B. (ed.) American Physiological Society, Bethesda. pp.1391–1414.

Keele, S.W. (1986) Motor control. In: *Handbook of Human Perception and Performance*, Boff, J.K., Kaufman, L., and Thomas, J.P. (eds.). John Wiley, New York, Vol. II, 30: 1–60.

Keele, S.W., Pokorny, R.A., Corcos, D.M., and Ivry, R. (1985) Do perception and motor production share common timing mechanisms: A correlational analysis. *Acta Psychol.* 60: 173–191.

Kihlstrom, J. F. (1987) The cognitive unconscious. *Science* 237: 1445–1452.

Knierim, J.J. and Van Essen, D.C. (1992) Neuronal responses to static texture patterns in area V1 of the alert macaque monkey. *J. Neurophysiol.* 67: 961–980.

Kobatake, E., Tanaka, K., and Tamori, Y. (1993) Learning new shapes changes the stimulus selectivity of cells in the inferotemporal cortex of the adult monkey. *Invest. Ophthalm. Vis. Sci. Suppl.* 34: 814.

Koch, C. (1993) Computational approaches to cognition: The bottom-up view. *Curr. Opinion Neurobiol.* 3: 203–208.

Koch, C. (1987) The action of the corticofugal pathway on sensory thalamic nuclei: a hypothesis. *Neuroscience* 23: 399–406.

Koch, C. and Poggio, T. (1987) Biophysics of computational systems: neurons, synapses, and membranes. In: *Synaptic Function*, Edelman, G.M., Gall, W.E., and Cowan, W.M. (eds.). John Wiley, New York.

Koch, C. and Poggio, T. (1992) Multiplying with synapses and neurons. In: *Single Neuron Computation*, McKenna, T., Davis, J.L., and Zornetzer, S.F. (eds.). Academic Press, Cambridge, MA.

Koch, C. and Schuster, H.G. (1992) A simple network showing burst synchronization without frequency locking. *Neural Comp.* 4: 211–223.

Koch, C. and Ullman, S. (1985) Shifts in selective visual attention: Towards the underlying neural circuitry. *Human Neurobiol.* 4: 219–227.

Koenderink, J.J. and van Doorn, A.J. (1990) Receptive field families. *Biol. Cybern.* 63: 291–297.

Kofka, K. (1935) *Principles of Gestalt Psychology*. Harcourt, Brace and World, New York.

Köhler, W. (1969) *The Task of Gestalt Psychology*. Princeton University Press, Princeton.

König, P. (1994) A method for the quantification of synchrony and oscillatory properties of neuronal activity. *J. Neurosci. Meth.*, in press.

König, P. and Schillen, T.B. (1991) Stimulus-dependent assembly formation of oscillatory responses: I. synchronization. *Neural Comp.* 3: 155–166.

König, P., Engel, A.K., Löwel, S., and Singer, W. (1990) Squint affects occurrence and synchronization of oscillatory responses in cat visual cortex. *Soc. Neurosci. Abstr.* 16: 523.2.

König, P., Engel, A.K., and Singer, W. (1992a) Gamma-oscillations as a vehicle for synchronization. *Soc. Neurosci. Abstr.* 18: 11.9.

König, P., Janosch, B., and Schillen, T.B. (1992b) Stimulus-dependent assembly formation of oscillatory responses. III. Learning. *Neural Comp.* 4: 666–681.

König, P., Engel, A.K., Löwel, S., and Singer, W. (1993) Squint affects synchronization of oscillatory responses in cat visual cortex. *Eur. J. Neurosci.*, 5: 501–508.

Kosslyn, S.M. (1980) *Image and Mind*. Harvard University Press, Cambridge, MA.

Kosslyn, S.M. (1988) Aspects of a cognitive neuroscience of mental imagery. *Science* 240: 1621–1626.

Kosslyn, S.M. and Koenig, O. (1992) *Wet Mind, The New Cognitive Neuroscience*. Free Press, New York.

Kosslyn, S.M., Alpert, N.M., Thompson, W.L., Maljkovic, V., Weise, S.B., Chabris, C.F., Hamilton, S.E., Rauch, S.L., and Buonanno, F.S. (1993) Visual mental imagery activates topographically organized visual cortex: PET investigations. *J. Cog. Neurosci.* 5: 263–287.

Krebs, J.R., Kacelnik, A., and Taylor, P. (1978) Test of optimal sampling by foraging great tits. *Nature* 275: 27–31.

Kreiter, A.K. and Singer, W. (1992) Oscillatory neuronal responses in the visual cortex of the awake macaque monkey. *Eur. J. Neurosci.* 4: 369–375.

Kreiter, A.K., Engel, A.K., and Singer, W. (1992) Stimulus-dependent synchronization of oscillatory neuronal activity in the superior temporal sulcus of the macaque monkey. *Eur. Neurosci. Assoc. Abst.*, 15: 1076.

Krieg, W.J.S. (1966) *Functional Neuroanatomy*. Brain Books, Pantagraph Printing, Bloomington, IL.

Kristofferson, A.B. (1984) Quantal and deterministic timing in human duration discrimination. *Ann. N.Y. Acad. Sci.* 423: 3–15.

Kröse, B.J.A. and Julesz, B. (1989) The control and speed of shifts of attention. *Vision Res.* 29: 1607–1619.

Krüger, J. and Aiple, F. (1988) Multimicroelectrode investigation of monkey striate cortex: Spike train correlations in the infragranular layers. *J. Neurophysiol.* 60:798–828.

Kuperstein, M., Eichenbaum, H., and VanDeMark, T. (1986) Neural group properties in the rat hippocampus during the theta rhythm. *Exp. Brain Res.* 61: 438–442.

LaBerge, D. and Buchsbaum, M.S. (1990) Positron emission tomographic measurements of pulvinar activity during an attention task. *J. Neurosci.* 10: 613–319.

Lakoff, G. (1987) *Women, Fire and Dangerous Things*. University of Chicago Press, Chicago.

Lal, R. and Friedlander, M.J. (1989) Gating of the retinal transmission by afferent eye position and movement signals. *Science* 243: 93–96.

Lande, R. (1979) Quantitative genetic analysis of multivariate evolution, applied to brain-body size allometry. *Evolution* 33: 400–416.

Larkman, A. and Mason, A. (1990) Correlations between morphology and electrophysiology of pyramidal neurons in slices of rat visual cortex. I Establishment of cell classes. *J. Neurosci.* 10: 1407–1414.

Le Bihan, D., Turner, R., Jezzard, P., Cuenod, C.A., and Zeffiro, T. (1992) Activation of human visual cortex by mental representation of visual patterns. In: *Book of Abstracts of Work in Progress*, 11th Annual Meeting of the Society for Magnetic Resonance in Medicine, p. 311.

LeCun, Y., Boser, B., Denker, J.S., Henderson, D., Howard, R.E., Hubbard, W., and Jackel L.D. (1990) Backpropagation applied to handwritten zip code recognition. *Neural Comp.* 1: 541–551.

Lee, C., Rohrer, W.H., and Sparks, D.L. (1988) Population coding of saccadic eye movements by neurons in the superior colliculus. *Nature* 332: 357–360.

Lee, T.S. (1992) A computational framework for understanding the primary visual cortex. *Harvard Robotics Lab Report*, Cambridge, MA.

Lee, T.S., Mumford, D., and Yuille, A. (1992) Texture segmentation by minimizing vector-valued energy functionals. In: *Computer Vision—ECCV '92, Springer Lecture Notes in Computer Science* 588: 165–173.

LeVay, S. and Gilbert, C.D. (1976) Laminar patterns of geniculocortical projection in the cat. *Brain Res.* 113: 1–19.

Li, L., Miller, E.K., and Desimone, R. (1993) The representation of stimulus familiarity in anterior inferior temporal cortex of macaques. *J. Neurophysiol.*, 69: 1918–1929.

Liberman, A. (1982) On finding that speech Is special. *Amer. Psychol.* 37: 148–167.

Linsker, R. (1986) From basic network principles to neural architecture (series). *Proc. Natl. Acad. Sci. USA*, 83: 7508–7512, 8390–8394, 8779–8783.

Linsker, R. (1990) Perceptual neural organization: Some approaches based on network models and information theory. *Annu. Rev. Neurosci.* 13: 257–281.

Liu, Z., Gaska, J.P., Jacobson, L.D., and Pollen, D.A. (1992) Functional interactions between members of adjacent cell pairs in the monkey's visual cortex. *Soc. Neurosci. Abstr.* 18: 11.

Livingstone, M.S. (1991) Visually evoked oscillations in monkey striate cortex. *Soc. Neurosci. Abstr.* 17: 73.3.

Livingstone, M. S. and Hubel, D. H. (1981) Effects of sleep and arousal on the processing of visual information in the cat. *Science* 291: 554–561.

Ljunberg, T., Apicella, P., and Schultz, W. (1992) Responses to monkey dopamine neurons during learning of behavioral reactions. *J. Neurophysiol.* 67: 145–163.

Llinas, R.R. (1987) 'Mindness' as a functional state of the brain. In: *Mindwaves*, Blakemore, C. and Greenfield, S. (eds.). Basil Blackwell, Oxford.

Llinás, R.R. (1988a) The intrinsic electrophysiological properties of mammalian neurons: Insights into central nervous system function. *Science* 242: 1654–1664.

Llinás, R.R. (1988b) "Mindness" as a functional state of the brain. In: *Mindwaves*, Blakemore, C., and Greenfield, S A. (eds.). Basil Blackwell, Oxford, pp. 339–358.

Llinás, R.R. (1990a) Intrinsic electrical properties of mammalian neurons and CNS function. In: *Fidia Research Foundation Neuroscience Award Lectures*. Raven Press, New York, pp. 175–194.

Llinás, R.R. (1990b) Intrinsic electrical properties of nerve cells and their role in network oscillation. *Cold Spring Habor Symp. Quant. Biol.*, LV, 933–938.

Llinás, R.R. and Paré, D. (1991) Of dreaming and wakefulness. *Neuroscience* 44: 521–535.

Llinás, R.R. and Ribary, U. (1992) Rostrocaudal scan in human brain: a global characteristic of the 40-Hz response during sensory input. In: *Induced Rhythms in the Brain*, Chapter 7, Basar, E., and Bullock, T. (eds.). Birkhäuser, Boston, pp. 147–154.

Llinás, R. and Ribary, U. (1993) Coherent 40-Hz oscillation characterizes dream state in humans. *Proc. Natl. Acad. Sci. USA* 90: 2078–2081.

Llinás, R.R., Grace, A.A., and Yarom, Y. (1991) *In vitro* neurons in mammalian cortical layer 4 exhibit intrinsic activity in the 10 to 50 Hz frequency range. *Proc. Natl. Acad. Sci. USA* 88: 897–901.

Logothetis, N.K. and Charles, E.R. (1990) V4 responses to gratings defined by random dot motion. *Invest. Ophthalmol. Vis. Sci. Suppl.* 31: 90.

Logothetis, N. K., and Schall, J. D. (1989) Neuronal correlates of subjective visual perception. *Science* 245: 761–763.

Lorente de Nó, R. (1932) Studies on the structure of the cerebral cortex. *J. F. Psychol. Neurol.* 45: 381–438.

Lorente de Nó, R. (1949) Cerebral cortex: Architecture, intracortical connections, motor projections. In: *Physiology of the Nervous System*, 3rd ed., Fulton, J.F. (ed.). Oxford University Press, Oxford, pp. 288–315.

Lowe, D.G. (1987) The viewpoint consistency constraint. *Int. J. Computer Vision* 1: 57–72.

Löwel, S. and Singer, W. (1992) Selection of intrinsic horizontal connections in the visual cortex by correlated neuronal activity. *Science* 255: 209–212.

Luck, S., Chelazzi, L., Hillyard, S., and Desimone, R. (1993) Effects of spatial attention in area of V4 of the macaque. *Soc. Neurosci. Abstr.*, 19: 27.

Luhmann, H.J., Martinez-Millan, L., and Singer, W. (1986) Development of horizontal intrinsic connections in cat striate cortex. *Exp. Brain Res.* 63: 443–448.

Luhmann, H.J., Singer, W., and Martinez-Millan, L. (1990) Horizontal interactions in cat striate cortex: I. Anatomical substrate and postnatal development. *Eur. J. Neurosci.* 2: 344–357.

Lund, J.S. (1988a) Excitatory and inhibitory circuitry and laminar mapping strategies in primary visual cortex of the monkey. Excitatory and inhibitory circuitry and laminar mapping strategies in primary visual cortex of the monkey. In: *Signal and Sense: Local and Global Order in Perceptual Maps*, Edelman, G.M., Gall, W.E., and Cowan, W.M. (eds.). John Wiley, New York.

Lund, J.S. (1988b) Anatomical organization of macaque monkey striate visual cortex. *Ann. Rev. Neurosci.* 11: 253–288.

Lund, J.S. and Boothe, R.G. (1975) Interlaminar connections and pyramidal neuron organization in the visual cortex, area 17, of the macaque monkey *Macaca mulatta*. *J. Comp. Neurol.* 159: 305–334.

Luria, A.R. (1962) *Higher Cortical Functions in Man.* transl. by Basic Books, New York, 1966.

MacKay D.M. (1955) The epistemological problem for automata. In: *Automata Studies*, Shannon, C.E., and McCarthy, J. (eds.). Princeton University Press, Princeton, pp. 235–250.

Mackintosh, N.J. (1974) *The Psychology of Animal Learning.* Academic Press, New York.

Magleby, K. L. (1987) Short-term changes in synaptic efficacy. In: *Synaptic Function*, Edelman, G. M., Gall, W. E., and Cowan, W. M. (eds.). John Wiley, New York, pp. 21–56.

Maloney, R.K., Mitchison, G.J., and Barlow, H.B. (1987) The limit to the detection of glass patterns in the presence of noise. *J. Opt. Soc. Am. A* 4: 2336–2341.

Manabe, T. Renner, P., and Nicoll, R.A. (1992) Postsynaptic contribution to long-term potentiation revealed by the analysis of miniature synaptic currents. *Nature* 355: 50–55.

Mangun, G.R., Hillyard, S.A., and Luck, S. (1993) Electrocortical substrates of visual selective attention. In: *Attention and Performance XIV*, Meyer, D. and Kornblum, S. (eds.)., MIT Press, Cambridge, MA, pp. 219–244.

Marin-Padilla, M. (1978) Dual origin of the mammalian neocortex and evolution of the cortical plate. *Anat. Embryol.* 152: 109–126.

Marr, D. (1970) A theory for cerebral neocortex. *Proc. Roy. Soc. London B* 176: 161–234.

Marr, D. (1982) *Vision.* W. H. Freeman, San Francisco, CA.

Marr, D. and Poggio, T. (1977) From understanding computation to understanding neural circuitry. *Neurosci. Res. Prog. Bull.* 15: 470–488.

Martin, K.A.C. (1984) Neuronal circuits in cat striate cortex. *Cerebral Cortex* 2: 241–284.

Martin, K.A.C. (1988) From single cells to simple circuits in the cerebral cortex. *Quart. J. Exp. Physiol.* 73: 637–702.

Maruyama, M., Girosi, F., and Poggio, T. (1991) A connection between HBF and MLP. A.I. Memo 1291, Artificial Intelligence Laboratory, Massachusetts Institute of Technology.

Mason, A., Nicoll, A., and Statford, K. (1991) Synaptic transmission between individual pyramidal neurons of the rat visual cortex *in vitro. J. Neurosci.* 11: 72–84.

Mastronarde, D.N. (1983) Correlated firing of cat retinal ganglion cells. I. Spontaneously active inputs to X- and Y-cells to single quantal response. *J. Neurophysiol.* 49: 325–345.

Matsubara, J., Cynader, M., Swindale, N.V., and Stryker, M.P. (1985) Intrinsic projections within visual cortex: Evidence for orientation-specific local connections. *Proc. Natl. Acad. Sci. USA* 82: 935–939.

Maunsell, J.H.R., and Gibson, J.R. (1992) Visual response latencies in striate cortex of the macaque monkey. *J. Neurophysiol.* 68: 1332–1334.

Maunsell, J.H., and Newsome, W.T. (1987) Visual processing in monkey extrastriate cortex. *Annu. Rev. Neurosci.* 10: 363–401.

Maunsell, J.H.R., and Van Essen, D.C. (1983) Functional properties of neurons in middle temporal visual area (MT) of macaque monkey. I. Selectivity for stimulus direction, velocity and orientation. *J. Neurophysiol.* 49: 1127–1147.

Maunsell, J.H., Sclar, G., Nealey, T.A., and DePriest, D.D. (1991) Extraretinal representations in area V4 in the macaque monkey. *Vis. Neurosci.* 7: 561–573.

McCarthy, R.A., and Warrington, E.K. (1988) Evidence for modality-specific meaning systems in the brain. *Nature* 334: 428–430.

McConkie, G.W. (1979) On the role and control of eye movements in reading. In: *Processing of Visible Language*, Kolers, A., Wrolstad, M.E., and Bouma, H. (eds.). Plenum Press, New York.

McConkie, G.W. (1990) Where vision and cognition meet. Paper presented at the HFSP Workshop on Object and Scene Perception. Leuven, Belgium.

McConkie, G.W. and Zola, D. (1987) Visual attention during eye fixations in reading. In: *Attention and Performance XII*. Erlbaum, London.

McConkie, G.W., Kerr, P.W., Reddix, M.D., and Zola, D. (1988) Eye movement control during reading. I. The location of initial eye fixations on words. *Vision Res.* 28: 1107–1118.

McCormick, D.A. (1990) Membrane properties and neurotransmitter actions. In: *The Synaptic Organization of the Brain*, 3rd ed., Shepard, G.M. (ed.). Oxford University Press, New York.

McCormick, D.A., Connors, B.W., Lighthall, J.W., and Prince, D.A. (1985) Comparative electrophysiology of pyramidal and sparsely spiny stellate neurons of the neocortex. *J. Neuroophysiol.* 54: 782–806.

McGuire, B.A., Hornung, J.P., Gilbert, C.D., and Wiesel, T.N. (1984) Patterns of synaptic input to layer 4 of cat striate cortex. *J. Neurosci.* 4: 3021–3033.

Mel, B.W. (1988) MURPHY: A robot that learns by doing. In: *Neural Information Processing Systems*, Anderson, D.Z. (ed.). American Institute of Physics, University of Colorado, Denver.

Mel, B.W. (1990) The Sigma-Pi column: A model of associative learning in cerebral neocortex. Computation and Neural Systems Memo 6, California Institute of Technology, Pasadena, CA.

Mel, B.W. (1992) NMDA-based pattern discrimination in a modeled cortical neuron. *Neural Comp.* 4: 502–517.

Melzack, R. and Wall, P.D. (1965) Pain mechanisms: A new theory. *Science* 150: 971–979.

Mendel, M.I. and Goldstein, R. (1971) Early components of the averaged electroencephalographic response to constant-level clicks during all-night sleep. *J. Speech Hear. Res.* 14: 829–840.

Mendel, M.I. and Kuperman, G.L. (1974) Early components of the averaged electroencephalographic response to constant level clicks during rapid eye movement sleep. *Audiology* 13: 23–32.

Michalski, A., Gerstein, G.L., Czarkowska, J., and Tarnecki, R. (1983) Interactions between cat striate cortex neurons. *Exp. Brain Res.* 51: 97–107.

Mignard, M. and Malpeli, J.G. (1991) Paths of information flow through visual cortex. *Science* 251: 1249–1251.

Mikami, A. and Kubota, K. (1980) Inferotemporal neuron activities and color discrimination with delay. *Brain Res.* 182: 65–78.

Miller, E.K. and Desimone, R. (1994) Parallel neuronal mechanisms of short-term memory. *Science* 263: 520–522.

Miller, E.K., Gochin, P.M., and Gross, C.G. (1991a) Habituation-like decrease in the responses of neurons in inferior temporal cortex of the macaque. *Vis. Neurosci.* 7: 357–362.

Miller, E.K., Li, L., and Desimone, R. (1991b) A neural mechanism for working and recognition memory in inferior temporal cortex. *Science* 254: 1377–1379.

Miller, E.K., Li, L., and Desimone, R. (1993) Activity of neurons in anterior inferior temporal cortex during a short-term memory task. *J. Neurosci.* 13: 1460–1478.

Miller, K.D. and Stryker, M.P. (1990) Ocular dominance column formation: Mechanisms and models. In: *Connectionist Modeling and Brain Function: The Developing Interface*, Hanson, S.J. and Olson, C.R. (eds.). MIT Press, Cambridge, MA.

Miller, K.D., Chapman, B., and Stryker, M.P. (1989) Visual responses in adult cat visual cortex depend on N-methyl-D-aspartate receptors. *Proc. Natl. Acad. Sci. USA.* 86: 5183–5187.

Milner, P.M. (1974) A model for visual shape recognition. *Psychol. Rev.* 81: 521–535.

Milner, B., Corkin, S., and Teuber, H.-L. (1968) Further analyses of the hippocampal amnesic syndrome: Fourteen year follow-up study of HM. *Neuropsychologia* 6: 215–234.

Minsky, M. (1985) *The Society of Minds.* Simon and Schuster, New York.

Minsky, M. and Papert, S. (1969) *Perceptrons: An Introduction to Computational Geometry.* MIT Press, Cambridge, MA.

Miyashita, Y. (1988) Neuronal correlate of visual associative long-term memory in the primate temporal cortex. *Nature* 335: 817–820.

Miyashita, Y., and Chang, H.S. (1988) Neuronal correlate of pictorial short-term memory in the primate temporal cortex. *Nature* 331: 68–70.

Moller, A.R., and Burgess, J. (1986) Neural generators of the brain stem auditory-evoked potentials (BAEPs) in the rhesus monkey. *Electroencephalogr. Clin. Neurophysiol.* 65: 361–372.

Montague, P.R., Dayan, P., Nowlan, S.J., Pouget, A., and Sejnowski, T.J. (1993) Using aperiodic reinforcement for directed self-organization during development. In: *Advances in Neural Information Processing Systems 5*, Hanson, S. J., Cowan, J. D., and Giles, C. L. (eds.). Morgan Kaufmann, San Mateo, CA, pp. 969–976.

Montague, P.R., Dayan, P., and Sejnowski, T.J. (1994) Foraging in an uncertain environment using predictive Hebbian learning. In: *Advances in Neural Information Processing 6*, Cowan, J.D., Tesauro, G., and Alspector, J. (eds.). Morgan Kaufmann, San Mateo, CA.

Montaron, M.P., Bouyer, J.J., Rougeul, A., and Buser, P. (1982) Ventral mesencephalic tegmentum (VMT) controls electrocortical beta rhythms and associated attentive behavior in cat. *Behav. Brain Res.* 6: 129–145.

Moran, J., and Desimone, R. (1985) Selective attention gates visual processing in the extrastriate cortex. *Science* 229: 782–784.

Motter, B. and Poggio, G. (1984) Binocular fixation in the Rhesus monkey: Spatial and temporal characteristics. *Exp. Brain Res.* 54: 304–314.

Mountcastle, V.B. and Hennemann, E. (1949) Pattern of tactile representation in thalamus of cat. *J. Neurophysiol.* 12: 85–100.

Mountcastle, V.B. and Hennemann, E. (1952) The representation of tactile sensibility in the thalamus of the monkey. *J. Comp. Neurol.* 97: 409–440.

Mountcastle, V.B., Andersen, R.A., and Motter, B.C. (1981) The influence of attentive fixation upon the excitability of the light-sensitive neurons of the posterior parietal cortex. *J. Neurosci.* 1: 1218–1235.

Movshon, J.A. and Van Sluyters, R.C. (1981) Visual neural development. *Annu. Rev. Psychol.* 32: 477–522.

Mozer, M.C. and Behrmann, M. (1990) On the interaction of selective attention and lexical knowledge: A connectionist account of neglect dyslexia. *J. Cog. Neurosci.* 2: 96–123.

Mumford, D. (1991) On the computational architecture of the neocortex. I. The role of the thalamo-cortical loop. *Biol. Cybern.* 65: 135–145.

Mumford, D. (1992) On the computational architecture of the neocortex. II. The role of cortico-cortical loops. *Biol. Cybern.* 66: 241–251.

Mumford, D. (1993) Pattern theory: A unifying perspective. In: *Proceedings of the 1st European Congress of Mathematics.* Birkhauser, Boston, in press.

Munemori, J., Hara, K., Kimura, M., and Sato, R. (1984) Statistical features of impulse trains in cat's lateral geniculate neurons. *Biol. Cybern.* 50: 167–172.

Munk, M.H.J., Nowak, L.G., Chouvet, G., Nelson, J.I., and Bullier, J. (1992) The structural basis of cortical synchronization. *Eur. J. Neurosci. Suppl.* 5: 21.

Murthy, V.N. and Fetz, E.E. (1992) Coherent 25- to 35-Hz oscillations in the sensorimotor cortex of awake behaving monkeys. *Proc. Natl. Acad. Sci. USA* 89: 5670–5674.

Mussa-Ivaldi, F.A., Giszter, S.F., and Bizzi, E. (1990) Motor-space coding in the central nervous system. *Cold Spring Habor Symp. Quant. Biol.,* LV: 827–835.

Nakayama, K. and Mackeben, M. (1989) Sustained and transient components of focal visual attention. *Vision Res.* 29: 1631–1647.

Nakayama, K. and Paradiso, M.A. (1991) Brightness perception and filling-in. *Vision Res.* 31: 1221–1236.

Nakayama, K. and Shimojo, S. (1992) Experiencing and perceiving visual surfaces. *Science* 257: 1357–1362.

Nakamura, K., Mikami, A., and Kubota, K. (1991) Unique oscillatory activity related to visual processing in the temporal pole of monkeys. *Neurosci. Res.* 12: 293–299.

Nakamura, K., Mikami, A., and Kubota, K. (1992) Oscillatory neuronal acti 'ty related to visual short-term memory in monkey temporal pole. *NeuroReport* 3: 117–120.

Nault, B., Michaud, Y., Morin, C., Casanova, C., and Molotchnikoff, S. (1990) Responsiveness of cells in area 17 after local interception of the descending path from area 18. *Soc. Neurosci. Abst.* 16: 1219.

Nazir, T. and O'Regan, J.K. (1990) Some results on translation invariance in the human visual system. *Spatial Vis.* 5: 81–100.

Nelson, D.L., Schreiber, T.A., and Mcevoy, C.L. (1992a) Processing implicit and explicit representations. *Psychol. Rev.* 99: 322–348.

Nelson, J.I., Nowak, L.G., Chouvet, G., Munk, M.H.J., and Bullier, J. (1992b) Synchronization between cortical neurons depends on activity in remote areas. *Soc. Neurosci. Abstr.* 11.8.

Nelson, J.I., Salin, P.A., Munk, M.H.J., Arzi, M., and Bullier, J. (1992c) Spatial and temporal coherence in cortico-cortical connections: A cross-correlation study in areas 17 and 18 in the cat. *Vis. Neurosci.* 9: 21–38.

Nelson, S.B. (1991) Temporal interactions in the cat visual system. I. Orientation-selective suppression in the visual cortex. *J. Neurosci.* 11: 344–356.

Nelson S.B., and Sur, M. (1992) NMDA receptors in sensory information processing. *Curr. Opinion Neurobiol.* 2: 484–488.

Neuenschwander, S., and Varela, F.J. (1990) Sensory-triggered oscillatory activity in the avian optic tectum. *Soc. Neurosci. Abstr.* 16: 47.6.

Neuenschwander, S., and Varela, F.J. (1993) Visually-triggered neuronal oscillations in birds: an autocorrelation study of tectal activity. *Eur. J. Neurosci.* 5: 870–881.

Newsome, W.T., Britten, K.H., and Movshon, J.A. (1989) Neuronal correlates of a perceptual decision. *Nature* 341: 52–54.

Newsome, W.T., Britten, K.H., Salzmann, C.D., and Movshon, J.A. (1990) Neuronal mechanisms of motion perception. *Cold Spring Habor Symp. Quant. Biol.*, LV: 697–705.

Niebur, E., Koch, C., and Rosin, C. (1993) An oscillation-based model for the neuronal basis of attention. *Vision Res.* 33: 2789–2802.

Nishihara, H.K. and Poggio, T. (1984) Stereo vision for robotics. In: *Robotics Research: The First International Symposium*, Brady, M. (ed.). MIT Press, Cambridge, MA.

Nitzberg, M., Mumford, D., and Shiota, T. (1993) Filtering, segmentation and depth. *Springer Lecture Notes in Computer Science*, 662: 1–11.

Nowlan, S. J. and Sejnowski, T.J. (1993) Filter selection model for generating visual motion signals. In: *Advance in Neural Information Processing Systems 5*, Hanson, S. J., Cowan, J. D., and Giles, C. L. (eds.). Morgan Kaufmann, San Mateo, CA, pp. 369–376.

Nunez, A., Amzica, F., and Steriade, M. (1992) Voltage-dependent fast (20–40 Hz) oscillations in long-axoned neocortical neurons. *Neuroscience* 51: 7–10.

Ogren, M.P., and Hendrickson, A.E. (1979) The structural organization of the inferior and lateral subdivisions of the *Macaca* monkey pulvinar. *J. Comp. Neurol.* 188: 147–178.

O'Kusky, J., and Colonnier, M. (1982) A laminar analysis of the number of neurons, glia, and synapses in the visual cortex (area 17) of adult macaque monkeys. *J. Comp. Neurol.* 210: 178–290.

Olshausen, B.A., Anderson, C.H., and Van Essen, D.C. (1992) A neural model of visual attention and invariant pattern recognition. *CNS Memo 18*. California Institute of Technology, Pasadena, CA.

Olshausen, B.A., Anderson, C.H., and Van Essen, D.C. (1993) A neurobiological model of visual attention and invariant pattern recognition based on dynamic routing of information flow. *J. Neurosci.* 13: 4700–4719.

Orban, G.A. (1984) *Neuronal Operations in the Visual Cortex*. Springer, Berlin.

O'Regan, J.K. (1990) Eye movements and reading. In: *Eye Movements and Their Role in Visual and Cognitive Processes*, Kowler, E. (ed.). Elsevier, Amsterdam.

O'Regan, J.K. (1992) Solving the "real" mysteries of visual perception: the world as an outside memory. *Cana. J. Psychol.* 46: 461–488.

Osaka, N., and Oda, K. (1991) Effective visual size necessary for vertical reading during Japanese text processing. *Bull. Psychon. Soc.* 29: 345–347.

Palm, G. (1982) *Neural Assemblies*. Springer Verlag, Berlin.

Palm, G. (1990) Cell assemblies as a guideline for brain research. *Concepts Neurosci.* 1: 133–137.

Pantev, C., Makeig, S., Hoke, M., Galambos, R., Hampson, S., and Gallen, C. (1991) Human auditory evoked gamma-band magnetic fields. *Proc. Natl. Acad. Sci. USA* 88: 8996–9000.

Pardo, J.V., Pardo, P.J., Janer, K.W., and Raichle, M.E. (1990) The anterior cingulate cortex mediates processing selection in the Stroop attentional conflict paradigm. *Proc. Natl. Acad. Sci. USA*. 87: 256–259.

Parker, A. and Hawken, M. (1985) Capabilities of monkey cortical cells in spatial resolution tasks. *J. Opt. Soc. Am. A* 2: 1101–1114.

Pece A.E.C. (1992) Redundancy reduction of a Gabor representation: A possible computational role for feedback from primary visual cortex to lateral geniculate nucleus. *Artificial Neural Networks* 2: 865–868.

Pei, X., Volgushev, M., and Creutzfeldt, O. (1992) A comparison of directional sensitivity with the excitatory and inhibitory field structure in cat striate cortical simple cells. *Perception* 21, (suppl. 2): 26.

Pellionisz, A. and Llinás, R.R. (1982) Space-time representation in the brain. The cerebellum as a predictive space-time metric tensor. *Neuroscience* 7: 2949–2970.

Penfield, W. and Rasmussen, T. (1950) *The Cerebral Cortex of Man*. Macmillan, New York.

Perez-Borja, C., Tyce, F.A., McDonald, C., and Uihlein, A. (1961) Depth electrographic studies of a focal fast response to sensory stimulation in the human. *Electroencephalogr. Clin. Neurophysiol.* 13: 695–702.

Perrett, D.I., and Harries, M.H. (1988) Characteristic views and the visual inspection of simple faceted and smooth objects: Tetrahedra and potatoes. *Perception* 17: 703–720.

Perrett, D.I. and Oram, S. (1992) The neurophysiology of shape processing. *Image Vis. Comp.*, submitted.

Perrett, D.I., Rolls, E.T., and Caan, W. (1982) Visual neurones responsive to faces in the monkey temporal cortex. *Exp. Brain Res.* 47: 329–342.

Perrett, D.I., Smith, P.A.J., Potter, D.D., Mistlin, A.J., Head, A.S., Milner, A.D., and Jeeves, M.A. (1985) Visual cells in the temporal cortex sensitive to face view and gaze direction. *Proc. Roy. Soc. London B* 223: 293–317.

Perrett, D.I., Mistlin, A.J., and Chitty, A.J. (1989) Visual neurones responsive to faces. *Trends Neurosci.* 10: 358–364.

Peters, A. (1987) Number of neurons and synapses in primary visual cortex. *Cerebral Cortex* 6: 267–294.

Petersen, S.E., Robinson, D.L., and Keys, W. (1985) Pulvinar nuclei of the behaving rhesus monkey: Visual responses and their modulation. *J. Neurophysiol.* 54: 867–886.

Petersen, S.E., Fox, P.T., Posner, M.I., Mintun, M., and Raichle, M.E. (1989) Positron emission tomographic studies of the processing of single words. *J. Cog. Neurosci.* 1: 153–170.

Petersen, S.E., Fox, P.T., Snyder, A.Z., and Raichle, M.E. (1990) Activation of extrastriate and frontal cortical areas by visual words and word like stimuli. *Science* 249: 1041–1044.

Peterson, C. (1990) Parallel distributed approaches to combinatorial optimization: Benchmark studies on traveling salesman problem. *Neural Comp.* 2: 261–269.

Peterson, M.A. and Gibson, B. (1991). The initial identification of figure-ground relationship: Contributions from shape recognition processes. Bell. Psychonomic Soc. 29: 199–202.

Phillips, C.G., Zeki, S., and Barlow, H.B. (1984) Localisation of function in the cerebral cortex. *Brain* 107: 327–361.

Phillips, W.A. and Singer, W. (1974) Function and interaction of on- and off-transients in vision. I. Psychophysics. *Exp. Brain Res.* 19: 493–506.

Picton, T.W. and Hillyard, S.A. (1974) Human AEPs. II. Effect of attention. *Electroencephalogr. Clin. Neurophysiol.* 36: 191–199.

Pinault, D. and Deschênes, M. (1992a) Voltage-dependent 40-Hz oscillations in rat reticular thalamic neurons *in vivo*. *Neuroscience* 51: 245–258.

Pinault, D. and Deschênes (1992b) Control of 40-Hz firing of reticular thalamic cells by neurotransmitters. *Neuroscience* 51: 259–268.

Poggio, T. (1984) Routing thoughts. MIT A.I. Lab Working Paper 258.

Poggio, T. (1990) A theory of how the brain might work. *Cold Spring Harbor Symp. Quant. Biol.*, 55: 899–910.

Poggio, T. (1991) 3d object recognition and prototypes: One 2d view may be sufficient. Technical Report 9107–02, I.R.S.T., Italy.

Poggio, T., and Edelman, S. (1990) A network that learns to recognize three-dimensional objects. *Nature* 343: 263–266.

Poggio, T., and Girosi, F. (1989) A theory of networks for approximation and learning. A.I. Memo 1140, Artificial Intelligence Laboratory, Massachusetts Institute of Technology, Cambridge, MA.

Poggio, T. and Girosi, F. (1990a) Regularization algorithms for learning that are equivalent to multilayer networks. *Science* 247: 978–982.

Poggio, T. and Girosi, F. (1990b) Extension of a theory of networks for approximation and learning: Dimensionality reduction and clustering. A.I. Memo 1167, Artificial Intelligence Laboratory, Massachusetts Institute of Technology, Cambridge, MA.

Poggio, T. and Girosi, F. (1990c) Networks for approximation and learning. *Proc. IEEE* 78: 1481–1497.

Poggio, T. and Girosi, F. (1990d) Theory of networks for learning. *Science* 247: 987–979.

Poggio, T. and Glaser, D. (eds.) (1993) *Exploring Brain Functions: Models in Neuroscience*. John Wiley, New York.

Poggio, T. and Torre, V. (1978) A theory of synaptic interactions. In: *Theoretical Approaches in Neurobiology*, Reichardt, W.E., and Poggio, T. (eds.). MIT Press, Cambridge, MA, pp. 28–38.

Poggio, T. and Vetter, T. (1992) Recognition and structure from one 2D model view: Observations on prototypes, object classes and symmetries. A.I. Memo 1347, Artificial Intelligence Laboratory, Massachusetts Institute of Technology, Cambridge, MA.

Poggio, T., Torre, V., and Koch, C. (1985) Computational vision and regularization theory. *Nature* 317: 314–319.

Poggio, T., Fahle, M., and Edelman, S. (1992) Fast perceptual learning in visual hyperacuity. *Science* 256: 1018–1021.

Pollatsek, A., Bolozky, S., Well, A.D., and Rayner, K. (1981) Asymmetries in perceptual span for Israeli readers. *Brain Lang.* 14: 174–180.

Pollen, D.A., Lee, J.R., and Taylor, J.H. (1971) How does the striate cortex begin the reconstruction of the visual world? *Science* 173: 74–77.

Posner, M.I. (1978) *Chronometric Explorations of Mind.* Erlbaum, Hillsdale, NJ.

Posner, M.I. and Cohen, Y. (1984) Components of visual orienting. In: *Attention and Performance X: Control of Language Processes*, Bouma, H. and Bouwhuis, D.G. (eds.) Erlbaum, Hillsdale, NJ.

Posner, M.I. and Driver, J. (1992) The neurobiology of selective attention. *Curr. Opin. Neurobiol.* 2: 165–169.

Posner, M.I. and Petersen, S.E. (1990) The attention system of the human brain. *Annu. Rev. Neurosci.* 13: 25–42.

Posner, M.I. and Rothbart, M. K. (1992) Attention and conscious experience. In: *The Neuropsychology of Consciousness*, Milner, A.D., and Rugg, M.D. (eds.). Academic Press, London, pp. 91–112.

Postma, E.O., van den Herik, H.J., and Hudson, P.T.W. (1992) The gating lattice: A neural substrate for dynamic gating. In: *Computation and Neural Systems'92*, Eeckman, F.H., and Bower, J.M. (ed.). Kluwer Academic Publishers, Boston pp. 221–226.

Pouget, A., Fisher, S.A., and Sejnowski, T.J. (1993) Egocentric spatial representation in early vision. *J. Cog. Neurosci.* 5: 150–161.

Powell, T.P.S. (1981) Certain aspects of the intrinsic organization of the cerebral cortex. In: *Brain Mechanisms and Perceptual Awareness*, Pompeiano, O. and Ajmone Marsan, C. (eds.). Raven Press, New York, pp. 1–19.

Powers, W.T. (1973) *Behavior: The Control of Perception.* Aldine, Chicago.

Price, D.J. and Blakemore, C. (1985a) The postnatal development of the association projection from visual cortical area 17 to area 18 in the cat. *J. Neurosci.* 5: 2443–2452.

Price, D.J. and Blakemore, C. (1985b) Regressive events in the postnatal development of association projections in the visual cortex. *Nature* 316: 721–724.

Purcell, D.G. and Stewart, A.L. (1988) The face-detection effect: Configuration enhances detection. *Percep. Psychophys.* 43: 355–366.

Purves, D., Riddle, D.R., and LaMantia, A-S. (1992) Iterated patterns of brain circuitry (or how the cortex gets its spots). *Trends Neurosci.* 15: 362–368.

Quillian, M.R. (1967) Word concepts: A theory and simulation of some basic semantic capabilities. *Behav. Sci.* 12: 410–430.

Raastad, M., Storm, J.F., and Andersen, P. (1992) Putative single quantum and single fiber excitatory postsynaptic currents show similar amplitude range and variability in rat hippocampal slices. *Eur. J. Neurosci.* 4: 113–117

Raether, A., Gray, C.M., and Singer, W. (1989) Intercolumnar interactions of oscillatoary neuronal responses in the visual cortex of alert cats. *Eur. Neurosci. Assoc.* 12: 72.5.

Raichle, M.E. (1987) Circulatory and metabolic correlates of brain function in normal humans. In: *Handbook of Physiology*, Vol. 5, Plum, F. (ed.). American Physiological Society, Baltimore, MD, pp. 643–674.

Ramachandran, V.S. (1985) Apparent motion of subjective surfaces. *Perception* 14: 127–134.

Ramachandran, V.S. (1986) Capture of stereopsis and apparent motion by illusory contours. *Percep. Psychophys.* 39:361–373.

Ramachandran, V.S. (1988) Perception of depth from shading. *Sci. Am.* 269: 76–83.

Ramachandran, V.S. and Anstis, S.M. (1983) Perceptual organization in moving patterns. *Nature* 304: 529–531.

Ramachandran, V.S. and Anstis, S.M. (1986) Perception of apparent motion. *Sci. Am.* 254: 102–109.

Ramachandran, V.S., Cobb, S., and Valenti, C. (1993) Plasticity of binocular correspondence, stereopsis, and egocentric localization in human vision. Unpublished.

Ratliff, F. (1965) *Mach Bands: Quantitative Studies on Neural Networks in the Retina.* Holden-Day, Oakland, CA.

Rauschecker, J.P. (1991) Mechanisms of visual plasticity: Hebb synapses, NMDA receptors, and beyond. *Physiol. Rev.* 71: 587–615.

Rayner, K., Well, A.D., and Pollatsek, A. (1980) Asymmetry of the effective visual field in reading. *Percep. Psychophys.* 27: 537–544.

Rayner, K., Slowiaczek, M.L., Clifton, C., and Bertera, J.H. (1983). Latency of sequential eye movements: Implications for reading. *J. Exp. Psychol.: Hum. Percep. Perform.* 9: 912–922.

Real, L.A. (1991) Animal choice behavior and the evolution of cognitive architecture. *Science* 253: 980–986.

Real, L.A., Ellner, S., and Hardy, L.D. (1990) Short-term energy maximization and risk aversion in bumblebees: A reply to Possingham. *Ecology* 71: 1625–1628.

Rechtschaffen, A., Hauri, P., and Zeitlin, M. (1966) Auditory awakening thresholds in REM and NREM sleep stages. *Percept. Motor. Skills* 22: 927–942.

Reichardt, W.E. and Poggio, T. (eds.) (1981) *Theoretical Approaches in Neurobiology.* MIT Press, Cambridge, MA.

Reid, R.C., Soodak, R.E., and Shapley, R.M. (1991) Directional selectivity and spatiotemporal structure of receptive fields of simple cells in cat striate cortex. *J. Neurophysiol.* 66: 505–529.

Rescorla, R.A., and Wagner, A.R. (1972) A theory of Pavlovian conditioning: The effectiveness of reinforcement and non-reinforcement. In: *Classical Conditioning II: Current Research and Theory*, Black, A.H., and Prokasy, W.F. (eds.). Appleton-Century-Crofts, New York, pp. 64–69.

Ribary, U., Ioannides, A.A., Singh, K.D., Hasson, R., Bolton, J.P.R., Lado, F., Mogilner, A., and Llinas, R. (1991) Magnetic field tomography of coherent thalamocortical 40-Hz oscillations in humans. *Proc. Natl. Acad. Sci. USA.* 88: 11037–11041.

Riches, I.P., Wilson, F.A., and Brown, M.W. (1991) The effects of visual stimulation and memory on neurons of the hippocampal formation and the neighboring parahippocampal gyrus and inferior temporal cortex of the primate. *J. Neurosci.* 11: 1763–1779.

Richmond, B.J., and Optican, L.M. (1990) Temporal encoding of two-dimensional patterns by single units in primate primary visual cortex. II. Information transmission. *J. Neurophysiol.* 64: 370–380.

Richmond, B.J., and Optican, L.M. (1992) The structure and interpretation of neuronal codes in the visual system. In: *Neural Networks for Perception*, Wechsler, H. (ed.). Academic Press, New York, pp. 104–119.

Robinson, D.L., and Petersen, S.E., (1992) The pulvinar and visual salience. *Trends in Neurosci.* 15: 127–132.

Robinson, D.L., Bowman, E.M., and Kertzman, C. (1991) Covert orienting of attention in Macaque: II. A signal in parietal cortex to disengage attention. *Soc. Neurosci. Abstr.* 17: 442.

Rockel, A.J., Hirons, R.W., and Powell, T.P.S. (1980) The basic uniformity in structure of the neocortex. *Brain* 103: 221–244.

Rockland, K.S. (1992a) Laminar distribution of neurons projecting from area V1 to V2 in macaque and squirrel monkeys. *Cerebral Cortex* 2: 38–47.

Rockland, K.S. (1992b) Configuration, in serial reconstruction, of individual axons projecting from area V2 to V4 in the macaque monkey. *Cerebral Cortex* 2: 353–374.

Rockland, K.S. (1993) The organization of feedback connections from area V2 to V1. *Cerebral Cortex*. 10.

Rockland, K.S. and Lund, J.S. (1982) Widespread periodic intrinsic connections in the tree shrew visual cortex. *Science* 215: 1532–1534.

Rockland, K.S. and Lund, J.S. (1983) Intrinsic laminar lattice connections in primate visual cortex. *J. Comp. Neurol.* 216: 303–318.

Rockland, K.S. and Pandya, D.N. (1979) Laminar origins and terminations of cortical connections of the occipital lobe in the rhesus monkey. *Brain Res.* 179: 3–20.

Rockland, K.S., and Virga, A. (1989) Terminal arbors of individual "feedback" axons projecting from area V2 to V1 in the macaque monkey: A study using immunohistochemistry of anterogradely transported Phaseolus vulgaris leucoagglutinin. *J. Comp. Neurol.* 285: 54–72.

Rockland, K.S., and Virga, A. (1990) Organization of individual cortical axons projecting from area V1 (area 17) to V2 (area 18) in the macaque monkey. *Vis. Neurosci.* 4: 11–28.

Rockland, K.S., Saleem, K.S., and Tanaka, K. (1992) Widespread feedback connections from areas V4 and TEO. *Soc. Neurosci. Abst.* 18: 390.

Roe, A.W., and Ts'o, D.Y. (1992) Functional connectivity between V1 and V1 in the primate. *Soc. Neurosci. Abstr.* 18: 11.4

Roelfsema, P.R., Engel, A.K., König, P., Sireteanu, R., and Singer, W. (1993) Squint-induced amblyopia is associated with reduced synchronization of cortical responses. *Soc. Neurosci. Abstr.* 19: 359.1.

Rolls, E.T. (1991) Neural organization of higher visual functions. *Curr. Opin. Neurobiol.* 1: 274–278.

Rolls, E.T., Tovee, M.J., and Lee, B. (1991) Temporal response properties of neurons in the macaque inferior temporal cortex. *Eur. J. Neurosci.* 4: 84.

Rosch, E. (1978) Principles of categorization. In: *Cognition and Categorization*, Rosch, E. and Lloyd, B. (eds.). Erlbaum, Hillsdale, NJ.

Rosenblith, W.A. (ed.) (1961) *Sensory Communication*. MIT Press, Cambridge, MA.

Rougeul, A., Bouyer, J.J., Dedet, L., and Debray, O. (1979) Fast somato-parietal rhythms during combined focal attention and immobility in baboon and squirrel monkey. *Electroencephalogr. Clin. Neurophysiol.* 46: 310–319.

Sacks, O. (1991) Neurological dreams. *Med. Doctor* 35: 29–32.

Sakai, K., and Miyashita, Y. (1991) Neural organization for the long-term memory of paired associates. *Nature* 354: 152–155.

Salzman, C.D., Murasugi, C.M., Britten, K.H., and Newsome, W.T. (1992) Microstimulation in visual area MT: Effects on direction discrimination performance. *J.Neurosci.* 12: 1231–1255.

Sandon, P.A. (1990) Simulating visual attention. *J. Cog. Neurosci.* 2: 213–231.

Sanes, J. N., and Donoghue, J. P. (1993) Oscillations in local field potentials of the primate motor cortex. *Proc. Natl. Acad. Sci. USA* 90: 4470–4474.

Sato, T. (1988) Effects of attention and stimulus interaction on visual responses of inferior temporal neurons in macaque. *J. Neurophysiol.* 60: 344–364.

Schacter, D.L., Cooper, L.A., and Delaney, S.M. (1990) Implicit memory for unfamiliar objects depends on access to structural descriptions. *J. Exp. Psychol.* 119: 5–24.

Schacter, D.L., Cooper, L.A., Tharan, M., and Rubens, A.B. (1991) Preserved priming of novel objects in patients with memory disorders. *J. Cog. Neurosci.* 3: 117–130.

Schall, J.D. (1991) Neural basis of saccadic eye movements in primates. In: *Eye Movements*, Carpenter, R.H.S. (ed.). CRC Press, Boca Raton, FL.

Schein, S.J. and Desimone, R. (1990) Spectral properties of V4 neurons in the macaque. *J. Neurosci.* 10: 3369–3389.

Schieber, M.H. (1990) How might the motor cortex individuate movements? *Trends Neurosci.* 13: 440–445.

Schieber, M.H. (1992) Widely distributed neuron activity in primary motor cortex hand area during individuated finger movements. *Soc. Neurosci. Abstr.* 18: 504.

Schillen, T.B. and König, P. (1990) Coherency detection by coupled oscillatory responses—synchronization connections in neural oscillator layers. In: *Parallel Processing in Neural Systems and Computers*, Eckmiller, R. (ed.). Elsevier, Amsterdam, pp. 139–142.

Schillen, T.B., and König, P. (1991) Stimulus-dependent assembly formation of oscillatory responses. II. Desynchronization. Neural Comp. 3: 167–178.

Schillen, T.B. and König, P. (1993) Temporal structure can solve the binding problem for multiple feature domains. In: *Computation and Neural Systems*, Eeckman, F.H., and Bower, J.M. (eds.) Kluwer, Norwell, MA, pp. 509–513.

Schuster, H.G. and Wagner, P. (1990a) A model for neuronal oscillations in the visual cortex. 1. Mean-field theory and derivation of the phase equations. *Biol. Cybern.* 64: 77–82.

Schuster, H.G. and Wagner, P. (1990b) A model for neuronal oscillations in the visual cortex. 2. Phase description of the feature dependent synchronization. *Biol. Cybern.* 64: 83–85.

Schwarz, C. and Bolz, J. (1991) Functional specificity of the long-range horizontal connections in cat visual cortex: A cross-correlation study. *J. Neurosci.* 11: 2995–3007.

Sejnowski, T.J. (1986) Open questions about computation in cerebral cortex. In: *Parallel Distributed Processing*, Vol. 2, McClelland, J.L., and Rumelhart, D.E. (eds.). MIT Press, Cambridge, MA, pp. 372–389.

Sejnowski, T.J. and Churchland, P.S. (1992) Computation in the age of neuroscience. In: *A New Era in Computation*, Metropolis, N. and Rota, G.-C. (eds.). MIT Press, Cambridge, MA, pp. 167–190.

Sejnowski, T.J. and Tesauro, G. (1989) The Hebb rule for synaptic plasticity: Algorithms and implementations. In: *Neural Models of Plasticity: Experimental and Theoretical Approaches*, Byrne, J.H., and Berry, W.O. (eds.). Academic Press, San Diego, pp. 94–103.

Semenza, C. and Zettin, M. (1989) Evidence from aphasia for the role of proper names as pure referring expressions. *Nature* 342: 678–679.

Sem-Jacobsen, C.W., Petersen, M.C., Dodge, H.W., Lazarte, J.A., and Holman, C.B. (1956) Electroencephalographic rhythms for the depths of the parietal, occipital and temporal lobes in man. *Electroencephalogr. Clin. Neurophysiol.* 8: 263–278.

Shannon, C.E. and Weaver, W. (eds.). (1949) *The Mathematical Theory of Communication*. University of Illinois Press, Urbana.

Shatz, C.J. (1990) Impulse activity and the patterning of connections during CNS development. *Neuron* 5: 745–756.

Sheer, D.E. (1984) Focused arousal, 40 Hz EEG, and dysfunction. In: *Self-Regulation of the Brain and Behavior*, Elbert, T., Rockstroh, B., Lutzenberger, W., and Birbaumer, N. (eds.). Springer, Berlin, pp. 66–84.

Sheer, D.E. (1989) Sensory and cognitive 40-Hz event-related potentials. In: *Brain Dynamics*, Basar, E. and Bullock, T.H. (eds.). Springer, Berlin, pp. 338–374.

Sherman, S.M. and Koch, C. (1986) The control of retinogeniculate transmission in the mammalian lateral geniculate nucleus. *Exp. Brain Res.* 63: 1–20.

Sherman, S.M. and Koch, C. (1990) Thalamus. In: *Synaptic Oragnization of the Brain*, 3rd ed. Shepherd, G. (ed.). Oxford University Press, Oxford.

Sherman, S.M., Scharfman, H.E., Lu, S.M., Guide, W., and Adams, P.R. (1990) N-methyl-D-aspartate (NMDA) and non-NMDA receptors participate in EPSPs of cat lateral geniculate neurons recorded in thalamic slices. *Soc. Neurosci. Abst.* 16: 159.

Shimizu, H., Yamaguchi, Y., Tsuda, I., and Yano, M. (1986) Pattern recognition based on holonic information dynamics: towards synergetic computers. In: *Complex Systems-Operational Approaches*, Haken, H. (ed.). Springer-Verlag, Berlin, pp. 225–240.

Sillar, K.T. (1991) Spinal pattern generation and sensory gating mechanisms. *Curr. Opin. Neurobiol.* 1: 583–589.

Silva, L.S., Amitai, Y., and Connors, B.W. (1991) Intrinsic oscillations of neocortex generated by layer 5 pyramidal neurons. *Science* 251: 432–435.

Silver, R.A., Traynelis, S.F., and Cull-Candy, S.G. (1992) Rapid time-course miniature and evoked excitatory currents at cerebellar synapses *in situ*. *Nature* 355: 163–166.

Singer, W. (1977) Control of thalamic transmission by cortico-fugal and ascending reticular pathways in the visual system. *Physiol. Rev.* 57: 386–420.

Singer, W. (1979) Central-core control of visual cortex functions. In: *The Neurosciences, Fourth Study Program*, Schmitt, F.O., and Worden, F.G. (eds.). MIT Press, Cambridge, MA, pp. 1093–1110.

Singer, W. (1985) Activity-dependent self-organization of the mammalian visual cortex. In: *Models of the Visual Cortex*, Rose, D. and Dobson, V.G. (eds.). John Wiley, Chichester, pp. 123–136.

Singer, W. (1990) Search for coherence: A basic principle of cortical self-organization. *Concepts Neurosci.* 1: 1–26.

Singer, W. (1993) Synchronization of cortical activity and its putative role in information processing and learning. *Annu. Rev. Physiol.* 55: 349–374.

Singer, W., Gray, C., Engel, A., Konig, P., Artola, A., and Brocher, S. (1990) Formation of cortical cell assemblies. *Cold Spring Harbor Symp. Quant. Biol.* 55: 939–952.

Singer, W., Artola, A., Engel, A.K., König, P., Kreiter, A.K., Löwel, S., and Schillen, T.B. In: *Exploring Brain Functions*, Poggio, T.A. and Glaser, D.A. (eds.). J. Wiley, New York, pp. 179–194.

Sloper, J.J. and Powell, T.P.S. (1979) An experimental electron microscopic study of afferent connections to the primate motor and somatic sensory cortices. *Phil. Trans. Roy. Soc. London. B* 285: 199–226.

Somogyi, P. and Cowey, A. (1981) Combined Golgi and electron microscopic study on the synapses formed by double bouquet cells in the visual cortex of the cat and monkey. *J. Comp. Neurol.* 195: 547–566.

Sparks, D.L. and Jay, M.F. (1986) The functional organization of the primate superior colliculus: A motor perspective. *Prog. Brain Res.* 64: 235–241.

Sparks, D.L., Lee, C., and Rohrer, W.H. (1990) Population coding of the direction, amplitude and velocity of saccadic eye movements by neurons in the superior colliculus. *Cold Spring Harbor Symp. Quant. Biol.*, LV: 805–811.

Sperling, G. (1960) The information available in brief visual presentations. *Psych. Monographs* 74: 498.

Sperling, G., and Dosher, B.A. (1986) Strategy and optimization in human information processing. In: *Handbook of Perception and Human Performance, I*, Boff, K.R., Kaufmann, L., and Thomas, J.P. (eds.) John Wiley, New York, pp. 1–65.

Sperry, R.W. (1952) Neurology and the mind-brain problem. *Am. Scientist* 40: 291–312.

Sporns, O., Gally, J.K., Reeke, G., and Edelman, G. (1989) Reentrant signaling among simulated neuronal groups leads to coherency in their oscillatory activity. *Proc. Natl. Acad. Sci. USA.* 86: 7265–7269.

Sporns, O., Tononi, G., and Edelman, G.M. (1991) Modeling perceptual grouping and figure-ground segregation by means of active reentrant connections. *Proc. Natl. Acad. Sci. USA* 88: 129–133.

Spydell, J.D., Ford, M.R., and Sheer, D.E. (1979) Task dependent cerebral lateralization of the 40 Hertz EEG rhythm. *Psychophysiology* 16: 347–350.

Squire, L.R. (1992) Declarative and nondeclarative memory: Multiple brain systems supporting learning and memory. *J. Cog. Neurosci.* 4: 232–243.

Squire, L.R. and Zola-Morgan, S. (1991) The medial temporal lobe memory system. *Science* 253: 1380–1386.

Squire, L.R., Ojemann, J.G., Miezin, F.M., Petersen, S.E., Videen, T.O., and Raichle, M.E. (1992) Activation of the hippocampus in normal humans: A functional anatomical study of memory. *Proc. Natl. Acad. Sci. USA* 89: 1837–1841.

Steinmetz, M.A., Connor, C.E., and MacLeod, K.M. (1992) Focal spatial attention suppresses responses of visual neurons in monkey posterior parietal cortex. *Soc. Neurosci. Abstr.* 18: 148.

Stent, G.S. (1973) A physiological mechanism for Hebb's postulate of learning. *Proc. Natl. Acad. Sci. USA* 70: 997–1001.

Steriade, M. (1991) Alertness, quiet sleep, dreaming. In: *Cerebral Cortex*, Vol. 9, Peters, A., and Jones, E.G. (eds.). Plenum Press, New York, pp. 279–357.

Steriade, M. and McCarley, R.W. (eds.) (1990) *Brainstem Control of Wakefulness and Sleep.* Plenum Press, New York.

Steriade, M., Parent, A., and Hada, J. (1984) Thalamic projections of nucleus reticularis thalami of cat: A study using retrograde transport of horseradish peroxidase and double fluorescent tracers. *J. Comp. Neurol.* 229: 531–547.

Steriade, M., Jones, E.G., and Llinas, R. (1990) *Thalamic Oscillations and Signalling.* John Wiley, New York.

Steriade, M., Curró Dossi, R., Paré, D., and Oakson, G. (1991) Fast oscillations (20–40 Hz) in thalamocortical systems and their potentiation by mesopontine cholinergic nuclei in the cat. *Proc. Natl. Acad. Sci. USA* 88: 4396–4400.

Steriade, M., Curro-Dossi, R., and Contreras, D. (1993a) Properties of intralaminar thalamocortical cells discharging rhythmic (40 Hz) spike-bursts at 1000 Hz. *Neuroscience* 56: 1–9.

Steriade, M., McCormick, D., and Sejnowski, T.J. (1993b) The sleeping and aroused brain: Thalamocortical oscillations in neurons and networks. *Science* 262: 679–685.

Steriade, M., Nuñez, A., and Amzica, F. (1993c) *J. Neuroscience* 13: 3252–3265.

Stevens, C.F. (1987) Specific consequences of general brain properties: In: *Synaptic Function*, chapter 24, Edelman, J.M., Gall, W.E., and Cowan, W.M. (eds.). John Wiley, New York, pp. 699–709.

Stevens, C.F. (1989) How cortical interconnectedness varies with network size. *Neural Comp.* 1: 473–479.

Stevens, C.F. (1993) Theories on the brain. *Nature* 361: 500–501.

Stone, J., Dreher, B., and Leventhal A. (1979) Hierarchical and parallel mechanisms in the organization of individual cortex. *Brain Res. Rev.* 1: 345–394.

Störig, P. and Cowey, A. (1991) Wavelength sensitivity in blindsight. *Nature* 342: 916–918

Stratford, K., Mason, A., Larkman, A., Major, G., and Jack, J. (1989) The modeling of pyramidal neurones in the visual cortex. In: *The Computing Neuron*, Durbin, R., Miall, C., and Mitchison, G. (eds.). Addison-Wesley, Reading, MA.

Stringa, L. (1992a) Eyes detection for face recognition. Technical Report 9203–07, I.R.S.T., Italy.

Stringa, L. (1992b) Automatic face recognition using directional derivatives. Technical Report 9205–04, I.R.S.T., Italy.

Sutton, R.S. and Barto, A.G. (1981) Toward a modern theory of adaptive networks: Expectation and prediction. *Psychol. Rev.* 88: 135–170.

Sutton, R.S. and Barto, A.G. (1987) Toward a modern theory of adaptive networks: Expectation and prediction. *Proc. Ninth Ann. Conf. Cognitive Science Society*, Seattle, WA.

Swain, M.J. and Ballard, D.H. (1990) Indexing via color histograms. In: *Proceedings of the International Conference on Computer Vision*, Osaka, Japan, pp. 390–393.

Swindale, N.V. (1990) Is the cerebral cortex modular? *Trends Neurosci.* 13: 487–492.

Talbot, J.D., Marrett, A., Evans, A.C., Meyer, E., Bushnell, M.C., and Duncan, G.H. (1991) Multiple representation of pain in human cerebral cortex. *Science* 251: 1355–1357.

Tanaka, K. (1983) Cross-correlation analysis of geniculostriate neuronal relationships in cats. *J. Neurophysiol.* 9: 1303–1318.

Tanaka, K., Saito, H., Fukada, Y., and Moriya, M. (1991) Coding visual images of objects in the inferotemporal cortex of the macaque monkey. *J. Neurophys.* 66: 170–189.

Thatcher, A.R. (1983) How many people have ever lived on earth? In: *44th Session of the International Statistical Institute*, Vol. 2, Madrid, pp. 841–843.

Thompson, A.M., Girdlestone, D., and West, D.C. (1988) Voltage-dependent currents prolong single-axon potentials in layer III pyramidal neurons in rat neocortical slices. *J. Neurophysiol.* 60:1895–1907.

Thorpe, S.J., Celebrinin, S., Trotter, Y., and Imbert, M. (1991) Dynamics of stereo processing in area V1 of the awake primate. *Eur. J. Neurosci.* 4: 83.

Tiitinen, H., Sinkkonen, J., Reinikainen, K., Alho, K., Lavikainen, J., and Ntnen, R. (1993) Selective attention enhances the auditory 40-Hz transient response in humans. *Nature* 364: 59–60.

Tolhurst, D.J. and Barlow, H.B. (1994) The kurtotic statistical distribution of pixel values at edges. In preparation.

Tolman, E.C. (1948) Cognitive maps in rats and men. *Psychol. Rev.* 55: 189–208.

Tononi, G., Sporns, O., and Edelman, G. (1992) Reentry and the problem of integrating multiple cortical areas: Simulation of dynamic integration in the visual system. *Cerebral Cortex* 2: 310–335.

Torre, V. and Poggio, T. (1978) A synaptic mechanism possibly underlying directional selectivity to motion. *Proc. Roy. Soc. London B* 202: 409–416.

Tovee, J.M., and Rolls, E.T. (1992) Oscillatory activity is not evident in the primate temporal visual cortex with static stimuli. *NeuroReport* 3: 369–372.

Toyama, K., Kimura, M., and Tanaka, K. (1981a) Cross-correlation analysis of interneuronal connectivity in cat visual cortex. *J. Neurophysiol.* 46: 191–201.

Toyama, K., Kimura, M., and Tanaka, K. (1981b) Organization of cat visual cortex as investigated by cross-correlation techniques. *J. Neurophysiol.* 46: 202–214.

Treisman, A. (1988) Features and objects: The fourteenth Bartlett memorial lecture. *Quart. J. Exp. Psychol. [A]* 40: 201–237.

Trojanowski, J.Q. and Jacobson, S. (1976) Areal and laminar distribution of some pulvinar cortical efferents in rhesus monkey. *J. Comp. Neurol.* 169: 371–392.

Trotter, Y., Celibrini, S., Stricanne, B., Thorpe, S., and Imbert, M. (1992) Modulation of neural stereoscopic processing in primate area V1 by the viewing distance. *Science* 257: 1279–1281.

Ts'o, D. and Gilbert, C. (1988) The organization of chromatic and spatial interactions in the primate striate cortex. *J. Neurosci.* 8: 1712–1727.

Ts'o, D., Gilbert, C., and Wiesel, T.N. (1986) Relationship between horizontal interactions and functional architecture in cat striate cortex as revealed by cross-correlation analysis. *J. Neurosci.* 6: 1160–1170.

Tsotsos, J.K. (1987) In: *Encyclopedia of Artificial Intelligence*, Shapiro, S. (ed.). John Wiley, New York, pp. 389–409.

Tsotsos, J.K. (1991) Localizing stimuli in a sensory field using an inhibitory attention beam. Technical Report, RBCV-TR-91-37, Dept. of Computer Science, University of Toronto.

Tsumoto, T. (1990) Excitatory amino acid transmitters and their receptors in neural circuits of the cerebral neocortex. *Neurosci. Res.* 9: 79–102.

Tsumoto, T., Creutzfeldt, O.D., and Legendy, C.R. (1978) Functional organization of the corticofugal system from visual cortex to lateral geniculate nucleus in the cat. *Exp. Brain Res.* 32: 345–364.

Tulving, E. and Schacter, D. L. (1990) Priming and human memory systems. *Science* 247: 301–306.

Turk, M. and Pentland, A. (1991) Eigenfaces for recognition. *J. Cog. Neurosci.* 3: 71–86.

Ullman, S. (1979) *The Interpretation of Visual Motion*, MIT Press, Cambridge, MA.

Ullman, S. (1991) Sequence seeking and counter streams: a model for information processing in the cortex. *MIT AI Memo 1331*, MIT, Cambridge, MA.

Ullman, S. and Richards, W. (1984) *Image Understanding*. Ablex, Norwood, NJ.

Ungerleider, L.G. and Mishkin, M. (1982) Two cortical visual systems. In: *Analysis of Visual Behavior*, Ingle, J., Goodale, M.A., and Mansfield, R.J.W. (eds.). MIT Press, Cambridge, MA, pp. 549–586.

Ungerleider, L.G., Galkin, T.W., and Mishkin, M. (1983) Visuotopic organization of projections from striate cortex to inferior and lateral pulvinar in rhesus monkey. *J. Comp. Neurol.* 217: 137–157.

Vaadia, E., Ahissar, E., Bergman, H., and Lavner, Y. (1991) Correlated activity of neurons: A neural code for higher brain functions? In: *Neuronal Cooperativity*, Springer Series in Synergistics, Krüger, J. (ed.). Springer-Verlag, Berlin, pp. 249–279.

Van Essen, D.C. (1985) Functional organization of primate visual cortex. In: *Cerebral Cortex*, Vol. 3, Peters, A. and Jones, E.J. (eds.). Plenum Press, New York.

Van Essen, D.C. and Anderson, C.H. (1990) Information processing strategies and pathways in the primate retina and visual cortex. In: *An Introduction to Neural and Electronic Networks*, Zornetzer, S.F., Davis, J.L., and Lau, C. (eds.). Academic Press, New York, pp. 43–72.

Van Essen, D.C. and Maunsell, J.H.R. (1980) Two-dimensional maps of the visual cortex. *J. Comp. Neurol.* 191: 255–281.

Van Essen, D. and Maunsell, J.H.R. (1983) Hierarchical organization and functional streams in the visual cortex. *Trends Neurosci.* 6: 370–375.

Van Essen, D.C., Newsome, W.T., Maunsell, J.H.R., and Bixby, J.L. (1986) The projections from striate cortex (V1) to areas V2 and V3 in the Macaque monkey: Asymmetries, areal boundaries, and patchy connections. *J. Comp. Neurol.* 244: 451–480.

Van Essen, D.C., Olshausen, B., Anderson, C.H., and Gallant, J.L. (1991) Pattern recognition, attention, and information bottlenecks in the primate visual system. *Proc. SPIE Conf. on Visual Information Processing: From Neurons to Chips* 1473: 17–28.

Van Hoesen, G.W. (1982) The primate parahippocampal gyrus: New insights regarding its cortical connections. *Trends Neurosci.* 5: 345–350.

Van Hoesen, G.W. (1993) The modern concept of association cortex. *Curr. Opin. Neurobiol.* 3: 150–154.

Van Hoesen, G.W. and Damasio, A.R. (1987) Neural correlates of cognitive impairment in Alzheimer's disease. In: *Handbook of Physiology: Higher Functions of the Nervous System*, Mountcastle, V., and Plum, F. (eds.). American Physiological Society, Bethesda, MD, pp. 871–898.

Varela, F.J. and Singer, W. (1987) Neuronal dynamics in the visual corticothalamic pathway revealed through binocular rivalry. *Exp. Brain Res.* 66: 10–20.

Velasco, F., Velasco, M., Cepeda, C., and Munoz, H. (1980) Wakefulness sleep modulation of cortical and subcortical somatic evoked potentials in man. *Electroencepahlogr. Clin. Neurophysiol.* 48: 64–72.

Verghese, P. and Pelli, D.G. (1992) The information capacity of visual attention. *Vision Res.* 32: 983–995.

Vetter, T., Poggio, T., and Bülthoff, H.B.B. (1992) 3d object recognition: symmetry and virtual views. Artificial Intelligence Laboratory Memo 1409, Massachusetts Institute of Technology, Cambridge, MA.

Vogels, R. and Orban, G.A. (1990) Effects of task related stimulus attributes on infero-temporal neurons studied in the discriminating monkey. *Soc. Neurosci. Abstr.* 16: 621.

Volgushev, M., Pei, X., Vidyasagar, T.R., and Creutzfeldt, O.D. (1992) Orientation-selective inhibition in cat visual cortex: An analysis of postsynaptic potentials. *Perception* 21, (Suppl. 2): 26.

Volterra, V. (1959) *Theory of Functionals and of Integral and Integro-Differential Equations*. Dover, New York.

von der Heydt, R. and Peterhans, E. (1989) Cortical contour mechanisms and geometrical illusions. In: *Neural Mechanisms of Visual Perception*, Lam, D., and Gilbert, C. (eds.). Gulf Publ. Co., Houston.

von der Heydt, R., Peterhans, E., and Baumgartner, G. (1984) Illusory contours and cortical neuron responses. *Science* 224: 1260–1262, von der Malsburg, C. (1981) The correlation theory of brain function. Internal Report 81-2, Max-Planck-Institute for Biophysical Chemistry, Göttingen, Germany.

von der Malsburg, C. (1985) Nervous structures with dynamical links. *Ber. Bunsenges. Phys. Chem.* 89: 703–710.

von der Malsburg, C. and Bienenstock, E. (1986) Statistical coding and short-term synaptic plasticity: A scheme for knowledge representation in the brain. In: *Disordered Systems and Biological Organization (NATO ASI Series, Vol. F20)*, Bienenstock, E. et al.(eds.). Springer-Verlag, Berlin, pp. 247–272.

von der Malsburg, C. and Schneider, W. (1986) A neural cocktail-party processor. *Biol. Cybern.* 54: 29–40.

von Noorden, G.K. (1990) *Binocular Vision and Ocular Motility, Theory and Management of Strabismus.* C.V. Mosby, St. Louis.

Wang, Z. and McCormick, D.A. (1993) Control of firing mode of corticotectal and corticopontine layer V burst-generating neurons by norepinephrine, acetylcholine and 1S,3R-ACPD. *J. Neurosci.* 13: 2199–2216.

Warrington, E.T. and Shallice, T. (1984) Category specific semantic impairments. *Brain* 107: 829–854.

Watanabe, S. (1960) Information-theoretical aspects of inductive and deductive Inference. *IBM J. Res. Dev.* 4: 208–231.

Weatherburn, C.E. (1961). *A First Course in Mathematical Statistics.* Cambridge University Press, Cambridge.

Weiskrantz, L. (1986) *Blindsight.* Oxford University Press, Oxford.

Weyand, T.G., and Malpeli, J.G. (1989) Responses of neurons in primary visual cortex are influenced by eye position. *Soc. Neurosci. Abstr.* 15: 1016.

White, E.L. (1978) Identified neurons in mouse Sm1 cortex which are postsynaptic to thalamocortical axon terminals: A combined Golgi-electronmicroscopic and degeneration study. *J. Comp. Neurol.* 181: 627–662.

White E.L. (1989) *Cortical Circuits.* Birkhauser, Boston.

White, E.L. and Hersch, S.M. (1981) Thalamocortical synapses of pyramidal cells which project Sm1 to Ms1 cortex in the mouse. *J. Comp. Neurol.* 198: 167–181.

White, E.L. and Hersch, S.M. (1982) A quantitative study of thalamocortical and other synapses involving the apical dendrites of corticothalamic projection cells in mouse Sm1 cortex. *J. Neurocytol.* 11: 137–157.

White, E.L. and Keller, A. (1987) Intrinsic circuitry involving the local axonal collaterals of corticothalamic projection cells in mouse SmI cortex. *J. Comp. Neurol.* 262: 13–26.

Whitehead, S.D. and Ballard, D.H. (1990) Active perception and reinforcement learning. *Neural Comp.* 2: 409–419.

Whitehead, S.D. and Ballard, D. (1991) Connectionist designs on planning. In: *Neural Information Processing Systems 3*, Lippmann, R.P., Moody, J.E., and Touretsky, D. (eds.). Morgan Kaufmann, San Mateo, CA, pp. 357–370.

Williams, H.L., Hammack, J.T., Daly, R.L., Dement, W.C., and Lubin, A. (1964) Responses to auditory stimulation, sleep loss and the EEG stages of sleep. *Electroencephalogr. Clin. Neurophysiol.* 16: 269–279.

Wilson, M.A. and Bower, J.M. (1991) A computer simulation of oscillatory behavior in primary visual cortex. *Neural Comp.* 3: 498–509.

Wilson, M.A. and Bower, J.M. (1992) Cortical oscillations and temporal interactions in a computer simulation of piriform cortex. *J. Neurophysiol.* 67: 981–995.

Wilson, J.R., Friedlander, M.J., and Sherman, S.M. (1984) Ultrastructural morphology of identified X- and Y-cells in the cat's lateral geniculate nucleus. *Proc. R. Soc. B* 221: 411–436.

Wise, R.A. (1982) Neuroleptics and operant behavior: the anhedonia hypothesis. *Behav. Brain Sci.* 5: 39–87.

Wong-Riley, M. (1978) Reciprocal connections between striate and prestriate cortex in squirrel monkey as demonstrated by combined peroxidase histochemistry and autoradiography. *Brain Res.* 147: 159–164.

Wurtz, R.H. and Mohler, C.W. (1976) Enhancement of visual response in monkey striate cortex and frontal eye fields. *J. Neurophysiol.* 39: 766–772.

Wurtz, R.H., Yamasaki, D.S., Duffy, C.J., and Roy, J.-P. (1990) Functional specialization for visual motion processing in primate cerebral cortex. *Cold Spring Habor Symp. Quant. Biol.*, LV: 717–727.

Yamada, T., Kameyama, S., Fuchigami, Z., Nakazumi, Y., Dickins, Q.S., and Kimura, J. (1988) Changes of short-latency somatosensory evoked potential in sleep. *Electroencephalogr.. Clin. Neurophysiol.* 70: 126–136.

Yarbus, A.L. (1967) *Eye Movements and Vision.* Plenum Press, New York.

Yen, C.T., Conley, M., Hendry, S.H.C., and Jones, E.G. (1985) The morphology of physiologically identified GABAergic neurons in the somatic sensory part of the thalamic reticular nucleus in the cat. *J. Neurosci.* 5: 2254–2268.

Young, M.P. (1992) Objective analysis of the topological organization of the primate cortical visual system. *Nature* 358: 152–155.

Young, M.P. (1993) The organization of neural systems in the primate cerebral cortex. *Proc. R. Soc. Lond. B.* 252: 13–18.

Young, M.P. and Yamane, S. (1992) Sparse population coding of faces in the inferotemporal cortex. *Science* 256: 1327–1331.

Young, M.P., Tanaka, K., and Yamane, S. (1992) On oscillating neuronal responses in the visual cortex of the monkey. *J. Neurophysiol.* 67: 1464–1474.

Yuille, A.L. and Ullman, S. (1990) Computational theories of low-level vision. In: *Visual Cognition and Action*, Osherson, D.N., Kosslyn, S.M., and Hollerbach, J.M. (eds.). MIT Press, Cambridge, MA.

Zeki, S.M. (1973) Colour coding in the rhesus monkey prestriate cortex. *Brain Res.* 53: 422–427.

Zeki, S. and Shipp, S. (1988) The functional logic of cortical connections. *Nature* 335: 311–317.

Zeki, S., Watson, J.D.G., Lueck, C.J., Friston, K.J., Kennard, C., and Frackowiak, R.S.J. (1991) A direct demonstration of functional specialization in human visual cortex. *J. Neurosci.* 11: 641–649.

Zijang, J.H. and Nakayama, K. (1992) Surfaces vs. features in visual search. *Nature* 359: 231–233.

Zipser, D. and Andersen, R.A. (1988) A back-propagation programmed network that simulates response properties of a subset of posterior parietal neurons. *Nature* 331: 679–684.

Zola-Morgan, S., Squire, L.R., and Amaral, D.G. (1986) Human amnesia and the medial temporal region: Enduring memory impairment following a bilateral lesion limited to field CA1 of the hippocampus. *J. Neurosci.* 6: 2950–2967.

Contributors

Charles H. Anderson
Department of Anatomy and
Neurobiology
Washington University Medical
School
St. Louis, Missouri

Horace Barlow
NEC Research Institute
Princeton, New Jersey and
Kenneth Craik Laboratory
Cambridge, England

Leonardo Chelazzi
Laboratory of Neuropsychology
National Institute of Mental Health
Bethesda, Maryland

Patricia S. Churchland
Department of Philosophy
University of California, San Diego
La Jolla, California

Francis Crick
The Salk Institute
La Jolla, California

Antonio R. Damasio
Department of Neurology
University of Iowa College of
Medicine
Iowa City, Iowa and
The Salk Institute
La Jolla, California

Hanna Damasio
Department of Neurology
University of Iowa College of
Medicine
Iowa City, Iowa and
The Salk Institute
La Jolla, California

Joel L. Davis
Office of Naval Research
Arlington, Virginia

Robert Desimone
Laboratory of Neuropsychology
National Institute of Mental Health
Bethesda, Maryland

Anya Hurlbert
Physiological Sciences
Medical School
University of Newcastle-upon-Tyne
England

Christof Koch
Division of Biology
California Institute of Technology
Pasadena, California

Rodolfo R. Llinás
Department of Physiology and
Biophysics
New York University Medical
Center
New York, New York

Earl K. Miller
Laboratory of Neuropsychology
National Institute of Mental Health
Bethesda, Maryland

David Mumford
Department of Mathematics
Harvard University
Cambridge, Massachusetts

Bruno A. Olshausen
Department of Anatomy and
Neurobiology
Washington University Medical
School
St. Louis, Missouri

Tomaso A. Poggio
Artificial Intelligence Laboratory
Massachusetts Institute of
Technology
Cambridge, Massachusetts

Michael I. Posner
Department of Psychology
University of Oregon
Eugene, Oregon

Vilayanur S. Ramachandran
Department of Psychology
University of California,
San Diego
La Jolla, California

Urs Ribary
Department of Physiology and
Biophysics
New York University Medical
Center
New York, New York

Mary K. Rothbart
Department of Psychology
University of Oregon
Eugene, Oregon

Terrence J. Sejnowski
Howard Hughes Medical Institute
The Salk Institute
La Jolla, California

Wolf Singer
Max Planck Institute for Brain
Research
Frankfurt, Germany

Charles F. Stevens
The Salk Institute
La Jolla, California

Shimon Ullman
Department of Applied
Mathematics
The Weizmann Institute of Science
Rehovot, Israel

David C. Van Essen
Department of Anatomy and
Neurobiology
Washington University Medical
School
St. Louis, Missouri

Index